She sits, inclining forward as to speak,
Her lips half-open, and her finger up,
As though she said, "Beware!"

(*Page* 341.)

ECLECTIC EDUCATIONAL SERIES.

McGUFFEY'S®

SIXTH

ECLECTIC READER.

REVISED EDITION.

McGuffey Editions and Colophon are Trademarks of
VAN NOSTRAND REINHOLD COMPANY INC.
New York Cincinnati Toronto London Melbourne

M'G REV. 6TH ED.

IN the SIXTH READER, the general plan of the revision of McGUFFEY'S SERIES has been carefully carried out to completion.

That plan has been to retain, throughout, those characteristic features of McGUFFEY'S READERS, which have made the series so popular, and caused their widespread use throughout the schools of the country. At the same time, the books have been enlarged; old pieces have been exchanged for new wherever the advantage was manifest; and several new features have been incorporated, which it is thought will add largely to the value of the series.

In the revision of the SIXTH READER, the introductory matter has been retained with but little change, and it will be found very valuable for elocutionary drill. In the preparation of this portion of the work, free use was made of the writings of standard authors upon Elocution, such as Walker, McCulloch, Sheridan Knowles, Ewing, Pinnock, Scott, Bell, Graham, Mylins, Wood, Rush, and many others.

In making up the Selections for Reading, great care and deliberation have been exercised. The best pieces of the old book are retained in the REVISED SIXTH, and to these have been added a long list of selections from the best English and American literature. Upwards of one hundred leading authors are represented (see "Alphabetical List of Authors," page ix), and thus a wide range of specimens of the best style has been secured. Close scrutiny revealed the fact that many popular selections, common to several series of Readers, had been largely *adapted*, but in McGUFFEY'S REVISED READERS, wherever it was possible to do so, the selections have been compared, and made to conform strictly with the originals as they appear in the latest editions authorized by the several writers.

The character of the selections, aside from their elocutionary value, has also been duly considered. It will be found, upon examination, that they present the same instructive merit and healthful moral tone which gave the preceding edition its high reputation.

Two new features of the REVISED SIXTH deserve especial attention—the explanatory notes, and the biographical notices of authors. The first, in the absence of a large number of books of reference, are absolutely necessary, in some cases, for the intelligent reading of the piece; and it is believed that in all cases they will add largely to the interest and usefulness of the lessons.

The biographical notices, if properly used, are hardly of less value than the lessons themselves. They have been carefully prepared, and are intended not only to add to the interest of the pieces, but to supply information usually obtained only by the separate study of English and American literature.

The illustrations of the REVISED SIXTH READER are presented as specimens of fine art. They are the work of the best artists and engravers that could be secured for the purpose in this country. The names of these gentlemen may be found on page ten.

The publishers would here repeat their acknowledgments to the numerous friends and critics who have kindly assisted in the work of revision, and would mention particularly President EDWIN C. HEWETT, of the State Normal University, Normal, Illinois, and the HON. THOMAS W. HARVEY, of Painesville, Ohio, who have had the revision of the SIXTH READER under their direct advice.

Especial acknowledgment is due to Messrs. Houghton, Osgood & Co., for their permission to make liberal selections from their copyright editions of many of the foremost American authors whose works they publish.

January, 1880.

INTRODUCTION.

SELECTIONS FOR READING.

ALPHABETICAL LIST OF AUTHORS.

LIST OF ILLUSTRATIONS.

INTRODUCTION.

THE subject of Elocution, so far as it is deemed applicable to a work of this kind, will be considered under the following heads, viz:

1. ARTICULATION.	4. READING VERSE.
2. INFLECTION.	5. THE VOICE.
3. ACCENT AND EMPHASIS.	6. GESTURE.

I. ARTICULATION.

Articulation is the utterance of the elementary sounds of a language, and of their combinations.

As words consist of one or more elementary sounds, the first object of the student should be to acquire the power of uttering those sounds with *distinctness*, *smoothness*, and *force*. This result can be secured only by careful practice, which must be persevered in until the learner has acquired a perfect control of his organs of speech.

ELEMENTARY SOUNDS.

An **Elementary Sound** is a simple, distinct sound made by the organs of speech.

The Elementary Sounds of the English language are divided into *Vocals*, *Subvocals*, and *Aspirates*.

VOCALS.

Vocals are sounds which consist of pure tone only. They are the most prominent elements of all words, and it is proper that they should first receive attention. A vocal may be represented by one letter, as in the word *hat*, or by two or more letters, as in *heat*, *beauty*. A *diphthong* is a union of two vocals, commencing with one and ending with the other. It is usually represented by two letters, as in the words *oil*, *boy*, *out*, *now*.

Each of these can be uttered with great force, so as to give a distinct expression of its sound, although the voice be suddenly suspended, the moment the sound is produced. This is done by putting the lips, teeth, tongue, and palate in their proper position, and then expelling each sound from the throat in the same manner that the syllable "ah!" is uttered in endeavoring to deter a child from something it is about to do; thus, a'—a'—a'—.

Let the pupil be required to utter every one of the elements in the Table with all possible suddenness and percussive force, until he is able to do it with ease and accuracy. This must not be considered as accomplished until he can give each sound with entire clearness, and with all the suddenness of the "crack" of a rifle. Care must be taken that the *vocal alone* be heard; there must be no consonantal sound, and no vocal sound other than the one intended.

At first, the elementary sounds may be repeated by the class in concert; then separately.

TABLE OF VOCALS.

Long Sounds.

ā, as in hāte.	ẽ, as in ẽrr.
â, " hâre.	ī, " pīne.
ȧ, " pȧss.	ō, " nō.
ä, " fär.	o͞o, " co͞ol.
a̤, " fa̤ll.	ū, " tūbe.
ē, " ēve.	û, " bûrn.

Short Sounds.

ă, as in măt.	ŏ, as in hŏt.
ĕ, " mĕt.	o͝o, " bo͝ok.
ĭ, " ĭt.	ŭ, " ŭs.

Diphthongs.

oi, oy, as in oil, boy.	ou, ow, as in out, now.

REMARK I.—In this table, the short sounds are nearly or quite the same, in *quantity*, as the long sounds. The difference consists chiefly in *quality*. Let the pupil determine this fact by experiment.

REMARK II.—The vocals are often represented by other letters or combinations of letters than those used in the table: for instance, ā is represented by *ai* in *hail*, by *ea* in *steak*, etc.

REMARK III.—As a general rule, the long vocals and the diphthongs should be articulated with a full, clear utterance; but the short vocals have a sharp, distinct, and almost explosive utterance. Weakness of speech follows a failure to observe the first point, while drawling results from carelessness with respect to the second.

SUBVOCALS AND ASPIRATES.

Subvocals are those sounds in which the vocalized breath is more or less obstructed.

Aspirates consist of breath only, modified by the vocal organs.

Words ending with subvocal sounds should be selected for practice on the subvocals; words beginning or ending with aspirate sounds may be used for practice on the aspirates. Pronounce these words forcibly and distinctly, several times in succession; then drop the other sounds, and repeat the subvocals and aspirates alone. Let the class repeat the words and elements, at first, in concert; then separately.

TABLE OF SUBVOCALS AND ASPIRATES.

Subvocals.	*Aspirates.*
b, as in bābe.	p, as in răp.
d, " băd.	t, " ăt.
g̅, " năg̅.	k, " bo͝ok.
j, " jŭdġe.	ch, " rĭch.
v, " mo̤ve.	f, " līfe.
th, " wĭth.	th, " smĭth.
z, " bŭzz.	s, " hĭss.
z, " ăzure (ăzh-).	sh, " rŭsh.
w, " wīne.	wh, " wha̤t.

REMARK.—These eighteen sounds make nine pairs of *cognates*. In articulating the aspirates, the vocal organs are put in the position required in the articulation of the corresponding subvocals; but the breath is expelled with some force, without the utterance of any vocal sound. The pupil should first verify this by experiment, and then practice on these cognates.

The following subvocals and aspirate have no cognates:

SUBVOCALS.

l, as in mĭll.	ng, as in sĭng.
m, " rĭm.	r, " rṳle.
n, " rŭn.	y, " yĕt.

ASPIRATE.

h, as in hat.

SUBSTITUTES.

Substitutes are characters used to represent sounds ordinarily represented by other characters.

TABLE OF SUBSTITUTES.

ạ	for ŏ,	as in	whạt.	y̆	for ĭ,	as in	hy̆mn.
ê	" â,	"	thêre.	ç	" s,	"	çīte.
e̱	" ā,	"	fre̱ight.	c	" k,	"	căp.
ï	" ē,	"	pōlïçe.	çh	" sh,	"	maçhïne.
ĩ	" ẽ,	"	sĩr.	ch	" k,	"	chôrd.
ȯ	" ŭ,	"	sȯn.	ġ	" j,	"	cāġe.
o̤	" o͞o,	"	to̤.	n̲	" ng,	"	rĭn̲k.
ọ	" o͝o,	"	wọuld.	ş	" z,	"	rōşe.
ô	" a̤,	"	côrn.	s	" sh	"	sugar.
õ	" û,	"	wõrm.	x̱	" gz,	"	ex̱ămine.
ụ	" o͝o,	"	pụll.	gh,	" f,	"	läugh.
ṳ	" o͞o,	"	rṳde.	ph	" f,	"	sy̆lph.
ȳ	" ī,	"	mȳ.	qu	" k	"	pïque.

qu for kw, as in quĭck.

FAULTS TO BE REMEDIED.

The most common faults of articulation are dropping an unaccented vowel, sounding incorrectly an unaccented vowel, suppressing final consonants, omitting or mispronouncing syllables, and blending words.

1. Dropping an unaccented vocal.

EXAMPLES.

CORRECT.	INCORRECT.	CORRECT.	INCORRECT.
Gran′a-ry	gran'ry	a-ban′don	a-ban-d'n.
im-mor′tal	im-mor-t'l.	reg′u-lar	reg'lar.
in-clem′ent	in-clem'nt.	par-tic′u-lar	par-tic'lar.
des′ti-ny	des-t'ny.	cal-cu-la′tion	cạl-cl'a-sh'n.
un-cer′tain	un-cer-t'n.	oc-ca′sion	oc-ca-sh'n.
em′i-nent	em'nent.	ef′fi-gy	ef'gy.
ag′o-ny	ag'ny.	man′i-fold	man'fold.
rev′er-ent	rev'rent.	cul′ti-vate	cult'vate.

2. Sounding incorrectly an unaccented vowel.

EXAMPLES.

CORRECT.	INCORRECT.	CORRECT.	INCORRECT.
Lam-en-ta′-tion	lam-*u*n-ta-tion	ter′ri-ble	ter-r*u*b-ble.
e-ter′nal	e-ter-n*u*l.	fel′o-ny	fel-*e*r-ny.
ob′sti-nate	ob-st*u*n-it.	fel′low-ship	fel-l*e*r-ship.
e-vent′	*uv*-ent.	cal′cu-late	cal-k*e*r-late.
ef′fort	*u*f-fort.	reg′u-lar	reg-g*y*-l*u*r.

EXERCISES.

The vocals most likely to be dropped or incorrectly sounded are italicized.

He *a*ttended d*i*vine serv*i*ce reg*u*l*a*rly.
This is my p*a*rticular r*e*quest.
She *i*s un*i*vers*a*lly *e*steemed.
George *i*s sens*i*ble *o*f his fault.
This calc*u*lation *i*s inc*o*rrect.
What a terr*i*ble c*a*lam*i*ty.
His eye through vast *i*mmens*i*ty c*a*n pierce.
*O*bserve these nice d*e*pend*e*ncies.
He *i*s a form*i*d*a*ble adv*e*rsary.
He *i*s gen*e*rous to his friends.
A temp*e*st des*o*lated the land.
He pr*e*ferred death to serv*i*tude.
God *i*s th*e* auth*o*r *o*f all things vis*i*ble *a*nd invis*i*ble.

3. Suppressing the final subvocals or aspirates.

EXAMPLES.

John an' James are frien's o' my father.
Gi' me some bread.
The want o' men is occasioned by the want o' money.
We seldom fine' men o' principle to ac' thus.
Beas' an' creepin' things were foun' there.

EXERCISES.

He learn*ed* to write.
The mas*ts* of the ship were cas*t* down.
He entered the lis*ts* at the head of his troo*ps.*
He is the merries*t* fellow in existence.
I regard not the worl*d*'s opinion.
He has three assistan*ts*.
The dep*ths* of the sea.
She trus*ts* too much to servan*ts*.
His attemp*ts* were fruitless.
He chance*d* to see a bee hoverin*g* over a flower.

4. Omitting or mispronouncing whole syllables.

EXAMPLES.

Lit′er-a-ry is	*improperly*	*pronounced*	lit-rer-ry.
co-tem′po-ra-ry	"	"	co-tem-po-ry.
het-er-o-ge′ne-ous	"	"	het-ro-ge-nous.
in-quis-i-to′ri-al	"	"	in-quis-i-to-ral.
mis′er-a-ble	"	"	mis-rer-ble.
ac-com′pa-ni-ment	"	"	ac-comp-ner-ment.

EXERCISES.

He devoted his attention chiefly to lit*er*ary pursuits.
He is a mis*era*ble creature.
His faults were owing to the degen*era*cy of the times.
The manuscript was undeciph*era*ble.
His spirit was unconqu*er*able.
Great industry was nec*essa*ry for the performance of the task.

5. Blending the end of one word with the beginning of the next.

EXAMPLES.

I court thy gif *s*no more.
The grove *s*were God *s*fir *s*temples.

My hear *t*was a mirror, that show' *d*every treasure.
It reflecte *d*each beautiful blosso *m*of pleasure.

Han *d'*me the slate.

This worl *d*is all a fleeting show,
  For man' *s*illusion given.

EXERCISES.

The magistrate*s* ought to arrest the rogue*s* *s*peedily.
The whirlwind*s* *s*weep the plain.
Link*ed* to thy side, through every chan*ce* I go.
But ha*d* he seen a*n* actor i*n* our day*s* enacting Shakespeare.
Wha*t* awful sound*s* assail my ears?
We caug*ht* a glimp*se* of her.
Old age ha*s* on their templ*es* sh*ed* her silver frost.

Our eagle shall rise mid the whirlwin*ds* of war,
  And dart through the dun clou*d* of battle hi*s* eye.
Then honor shall wea*ve* of the laurel a crown,
  That beauty shall bin*d* on the brow of the brave.

II. INFLECTION.

Inflection is a bending or sliding of the voice **either** *upward* or *downward.*

The **upward** or **rising inflection** is an upward slide of the voice, and is marked by the acute accent, thus, (´); as,

Did you call´? Is he sick´?

The **downward** or **falling inflection** is a downward slide of the voice, and is marked by the grave accent, thus, (`); as,

Where is London`? Where have you been`?

Sometimes both the rising and falling inflections are given to the same sound. Such sounds are designated by the *circumflex*, thus, (⌣) or thus, (∧). The former is called the *rising circumflex*; the latter, the *falling circumflex*; as,

But nôbody can bear the death of Clŏdius.

When several successive syllables are uttered without either the upward or downward slide, they are said to be uttered in a *monotone*, which is marked thus, (—); as,

R̄ōl̄l̄ ōn̄, t̄h̄ōū d̄ēēp̄ ān̄d̄ d̄ār̄k̄ b̄l̄ūē Ōc̄ēān̄—r̄ōl̄l̄!

EXAMPLES.

Does he read correctly′ or incorrectly‵?

In reading this sentence, the voice should slide somewhat as represented in the following diagram:

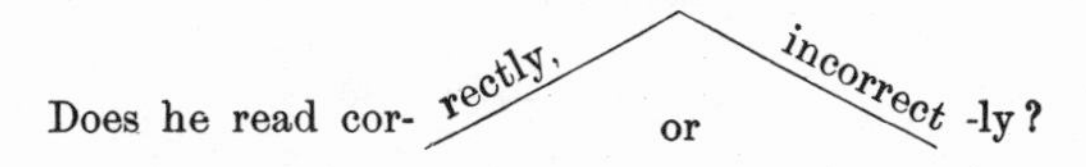

If you said vĭnegar, I said sûgar.

To be read thus:

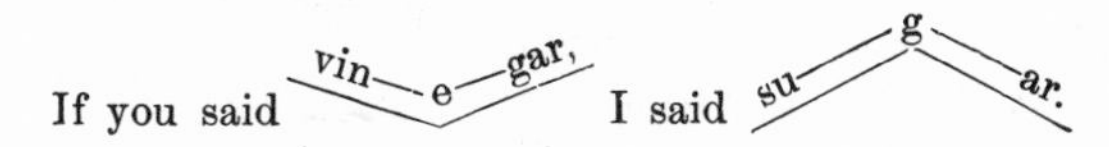

If you said yĕs, I said nô.

To be read thus:

If you said y e s, I said n o.

What′! did he say no′?

To be read thus:

What! did he say no?

He didˋ; he said noˋ,
To be read thus:

He did; he said no.

Did he do it voluntarily′, or involuntarilyˋ?
To be read thus:

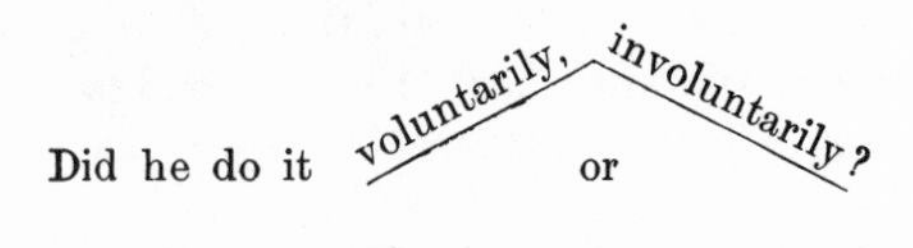

He did it voluntarilyˋ, not involuntarily′.
To be read thus:

He did it voluntarily, not involuntarily.

EXERCISES.

Do they act prudently′, or imprudentlyˋ?
Are they at home′, or abroadˋ?
Did you say Europe′, or Asiaˋ?
Is he rich′, or poorˋ?
He said painˋ, not pain′.
Are you engaged′, or at leisureˋ?
Shall I say plain′, or painˋ?
He went homeˋ, not abroad′.
Does he say able′, or tableˋ?
He said hazyˋ, not lazy′?
Must I say flat′, or flatˋ?
You should say flatˋ, not flat′.
My father′, must I stay′?
Oh! but he paûsed upon the brink.
It shall go hard with me, but I shall ûse the weapon.

Heard ye those loud contending waves,
That shook Cecropia's pillar'd state′?
Saw ye the mighty from their graves
Look up′, and tremble at your fate′?

First′ Fear\\, his hand, its skill to try′,
Amid the chords bewildered laid\\;
And back recoiled\\, he knew not why′
E'en at the sound himself had made.

Where be your gibes\\ now? your gambols\\? your songs\\? your flashes of merriment, that were wont to set the table on a roar\\?

Thus saith the high and lofty One that inhabiteth eternity, whose name is Holy; "I dwell in the high and holy place."

FALLING INFLECTION.

RULE I.—Sentences, and parts of sentences which make complete sense in themselves, require the *falling inflection.*

EXAMPLES.

1. By virtue we secure happiness\\.
2. For thou hast said in thine heart, I will ascend into heaven\\: I will exalt my throne above the stars of God\\: I will sit, also, upon the mount of the congregation, in the sides of the north\\.
3. The wind and the rain are over\\; calm is the noon of the day\\: the clouds are divided in heaven\\; over the green hills flies the inconstant sun\\; red through the stormy vale comes down the stream\\.
4. This proposition was, however, rejected,\\ and not merely rejected, but rejected with insult\\.

Exception.—Emphasis sometimes reverses this rule, and requires the *rising inflection*, apparently for the purpose of calling attention to the idea of an unusual manner of expressing it.

EXAMPLES.

1. I should not like to ride in that car′.
2. Look out! A man was drowned there yesterday′.
3. Presumptuous man! the gods\\ take care of Cato′.

Rule II.—The language of emphasis generally requires the *falling inflection.*

EXAMPLES.

1. Charge\`, Chester, charge\`; on\`, Stanley, on\`.

2. Were I an American, as I am an Englishman, while a single\` foreign troop\` remained\` in my country, I would never\` lay down my arms\`—never\`, never\`, never.\`

3. Does any one suppose that the payment of twenty shillings, would have ruined Mr. Hampden's fortune? No\`. But the payment of half\` twenty shillings, on the principle\` it was demanded, would have made him a slave\`.

4. I insist\` upon this point\`: I urge\` you to it; I press\` it, demand\` it.

5. All that I have\`, all that I am\`, and all that I hope\` in this life, I am now ready\`, here\`, to stake\` upon it.

Rule III.—Interrogative sentences and members of sentences, which *can not* be answered by *yes* or *no*, generally require the *falling inflection.*

EXAMPLES.

1. How many books did he purchase\`?
2. Why reason ye these things in your hearts\`?
3. What see\` you, that you frown so heavily to-day\`?
4. Ah! what is that flame which now bursts on his eye\`?
5. Whence this pleasing hope\`, this fond desire\`,
   This longing after immortality\`?

Exception.—When questions usually requiring the falling inflection are *emphatic* or *repeated,* they take the *rising inflection.*

EXAMPLES.

1. Where did you say he had gone′?
2. To whom did you say the blame was to be imputed′?
3. What is\` he? A knave. What′ is he? A knave, I say.

RISING INFLECTION.

RULE IV.—The *rising inflection* is generally used where the sense is dependent or incomplete.

REMARK.—This inflection is generally very slight, requiring an acute and educated ear to discern it, and it is difficult to teach pupils to distinguish it, though they constantly use it. Care should be taken not to exaggerate it.

EXAMPLES.

1. Nature being exhausted′, he quietly resigned himself to his fate.

2 A chieftain to the Highlands bound′,
Cries′, "Boatman, do not tarry!"

3. As he spoke without fear of consequences′, so his actions were marked with the most unbending resolution.

4. Speaking in the open air′, at the top of the voice′, is an admirable exercise.

5. If then, his Providence′, out of our evil, seek to bring forth good′, our labor must be to prevent that end.

6. He′, born for the universe′, narrowed his mind,
And to party gave up what was meant for mankind.

REMARK.—The names of persons or things addressed, when not used emphatically, are included in this rule.

7. Brother′, give me thy hand; and, gentle Warwick′,
Let me embrace thee in my weary arms.

8. O Lancaster′, I fear thy overthrow.

9. Ye crags′ and peaks′, I'm with you once again.

Exception 1.—*Relative emphasis* often reverses this and the first rule, because emphasis is here expressed in part by changing the usual inflections.

EXAMPLES.

1. If you care not for your *property*\, you surely value your life′.

2. If you will not labor for your *own*\ advancement, you should regard that of your children′.

3. It is your place to *obey*\`, not to command′.

4. Though by that course he should not destroy his *reputation*\`, he will lose all self-respect′.

Exception 2.—The names of persons addressed in a formal speech, or when used emphatically, have the *falling inflection.*

EXAMPLES.

1. Romans, countrymen, and lovers\`, hear me for my cause, etc.

2. Gentlemen of the jury\`, I solicit your attention, etc.

3. O Hubert\`, Hubert\`, save me from these men.

RULE V.—Negative sentences and parts of sentences, usually require the *rising inflection.*

EXAMPLES.

1. It is not by starts of application that eminence can be attained′.

2. It was not an eclipse that caused the darkness at the crucifixion of our Lord′; for the sun and moon were not relatively in a position\` to produce an eclipse′.

3. They are not fighting′: do not disturb′ them: this man is not expiring with agony′: that man is not dead′: they are only pausing′.

4. My Lord, we could not have had such designs′.

5. You are not left alone to climb the steep ascent′: God is with you, who never suffers the spirit that rests on him to fail.

Exception 1.—Emphasis may reverse this rule.

EXAMPLE.

We repeat it, we do *not*\` desire to produce discord; we do *not*\` wish to kindle the flames of a civil war.

Exception 2.—General propositions and commands usually have the *falling inflection.*

EXAMPLES.

God is not the author of sin\\. Thou shalt not kill.

RULE VI.—Interrogative sentences, and members of sentences which can be answered by *yes* or *no*, generally require the *rising inflection.*

EXAMPLES.

1. Are fleets and armies necessary to a work of love and reconciliation′?

2. Does the gentleman suppose it is in his power′, to exhibit in Carolina a name so bright′ as to produce envy′ in my bosom?

3. If it be admitted, that strict integrity is not the shortest way to success, is it not the surest′, the happiest′, the best′?

4. Is there not rain enough in the sweet heavens,
To wash this crimson hand as white as snow′?

Exception.—Emphasis may reverse this rule.

EXAMPLES.

1. *Can*\\ you be so blind to your interest? *Will*\\ you rush headlong to destruction?

2. I ask again, *is*\\ there no hope of reconciliation? *Must*\\ we abandon all our fond anticipations?

3. Will you deny′ it? Will you *deny*\\ it?

4. Am I Dromio′? Am I your man′? Am I *myself*\\?

RULE VII.—Interrogative exclamations, and words repeated as a kind of echo to the thought, require the *rising inflection.*

EXAMPLES.

1. Where grows′, where grows it not′?

2. What′! Might Rome have been taken′? Rome taken when I was consul′?

3. Banished from Rome′! Tried and convicted traitor′!

4. *Prince Henry.* What's the matter‵?

Falstaff. What's the matter′? Here be four of us have taken a thousand pounds this morning.

Prince H. Where is‵ it, Jack, where is‵ it?

Fal. Where is′ it? Taken from us, it is.

5. Ha′! laughest thou, Lochiel, my vision to scorn?

6. And this man is called a statesman. A statesman′? Why, he never invented a decent humbug.

7. I can not say, sir, which of these motives influence the advocates of the bill before us; a bill′, in which such cruelties are proposed as are yet unknown among the most savage nations.

RISING AND FALLING INFLECTIONS.

RULE VIII.—Words and members of a sentence expressing *antithesis* or *contrast*, require *opposite inflections.*

EXAMPLES.

1. By honor′ and dishonor‵; by evil′ report and good‵ report; as deceivers′ and yet true‵.

2. What they know by reading′, I know by experience‵.

3. I could honor thy courage′, but I detest thy crimes‵.

4. It is easier to forgive the weak‵, who have injured us′, than the powerful′ whom we‵ have injured.

5. Homer was the greater genius′, Virgil the better artist‵.

6. The style of Dryden is capricious and varied′; that of Pope is cautious and uniform‵. Dryden obeys the emotions of his own mind′; Pope constrains his mind to his own rules of composition.‵ Dryden is sometimes vehement and rapid′; Pope is always smooth, uniform, and gentle‵. Dryden's page is a natural field, rising into inequalities, varied by exuberant vegetation′; Pope's is a velvet lawn, shaven by the scythe and leveled by the roller‵.

7. If the flights of Dryden are higher′, Pope continues longer on the wing\. If the blaze of Dryden's fire is brighter′, the heat of Pope's is more regular and constant\. Dryden often surpasses′ expectation, and Pope never falls below\ it.

REMARK 1.—Words and members connected by *or* used *disjunctively*, generally express contrast or antithesis, and *always* receive *opposite inflections.*

EXAMPLES.

1. Shall we advance′, or retreat\?
2. Do you seek wealth′, or power\?
3. Is the great chain upheld by God′, or thee\?
4. Shall we return to our allegiance while we may do so with safety and honor′, or shall we wait until the ax of the executioner is at our throats\?
5. Shall we crown′ the author of these public calamities with garlands′, or shall we wrest\ from him his ill-deserved authority\?

REMARK 2.—When the antithesis is between affirmation and negation, the latter *usually* has the *rising inflection*, according to Rule V.

EXAMPLES.

1. You were paid to fight\ against Philip, not to rail′ at him.
2. I said rationally\, not irrationally′.
3. I did not say rationally′, but irrationally\.
4. I said an elder\ soldier, not a better′.
5. Let us retract while we can\, not when we must′.

REMARK 3.—The more emphatic member *generally* receives the *falling inflection.*

EXAMPLES.

1. A countenance more in sorrow\, than anger′.
2. A countenance less in anger′, than sorrow\.
3. You should show your courage by deeds\, rather than by words′.
4. If we can not remove\ pain, we may alleviate′ it.

OF SERIES.

A **series** is a number of particulars immediately following one another in the same grammatical construction.

A *commencing series* is one which *commences* a sentence or clause.

EXAMPLE.

Faith, hope, love, joy, are the fruits of the spirit.

A *concluding series* is one which *concludes* a sentence or a clause.

EXAMPLE.

The fruits of the spirit are *faith, hope, love, and joy.*

RULE IX.—All the members of a commencing series, when not emphatic, usually require the *rising inflection.*

EXAMPLES.

1. War′, famine′, pestilence′, storm′, and fire′ besiege mankind.
2. The knowledge′, the power′, the wisdom′, the goodness′ of God, must all be unbounded.
3. To advise the ignorant′, to relieve the needy′, and to comfort the afflicted′ are the duties that fall in our way almost every day of our lives.
4. No state chicanery′, no narrow system of vicious politics′, no idle contest for ministerial victories′, sank him to the vulgar level of the great.
5. For solidity of reasoning′, force of sagacity′, and wisdom of conclusion′, no nation or body of men can compare with the Congress at Philadelphia.
6. The wise and the foolish′, the virtuous and the evil′, the learned and the ignorant′, the temperate and the profligate′, must often be blended together.

7. Absalom's beauty′, Jonathan's love′, David's valor′, Solomon's wisdom′, the patience of Job′, the prudence of Augustus′, and the eloquence of Cicero′ are found in perfection in the Creator.

REMARK.—Some elocutionists prefer to give the *falling inflection* to the *last* member of a commencing series.

Exception.—In a commencing series, forming a climax, the *last* term usually requires the *falling inflection.*

EXAMPLES.

1. Days′, months′, years′, and ages‵, shall circle away,
 And still the vast waters above thee shall roll.

2. Property′, character′, reputation′, everything‵, was sacrificed.

3. Toils′, sufferings′, wounds′, and death‵ was the price of our liberty.

RULE X.—All the members of a concluding series, when not at all emphatic, usually require the *falling inflection.*

EXAMPLES.

1. It is our duty to pity‵, to support‵, to defend‵, and to relieve‵ the oppressed.

2. At the sacred call of country, they sacrifice property‵, ease‵, health‵, applause‵, and even life‵.

3. I protest against this measure as cruel‵, oppressive‵, tyrannous‵, and vindictive‵.

4. God was manifest in the flesh‵, justified in the Spirit‵, seen of angels‵, preached unto the Gentiles‵, believed on in the world‵, received up into glory‵.

5. Charity vaunteth not itself‵, is not puffed up‵, doth not behave itself unseemly‵, seeketh not her own‵, is not easily provoked‵, thinketh no evil‵; beareth‵ all things, believeth‵ all things, hopeth‵ all things, endureth‵ all things.

REMARK.—Some authors give the following rule for the reading of a concluding series: "All the particulars of a concluding series, except the last but one, require the *falling inflection.*"

Exception 1.—When the particulars enumerated in a concluding series are not at all emphatic, all except the last requre the *rising inflection.*

EXAMPLE.

He was esteemed for his kindness′, his intelligence′, his self-denial′, and his active benevolence‵.

Exception 2.—When all the terms of a concluding series are strongly emphatic, they all receive the *falling inflection.*

EXAMPLES.

1. They saw not one man‵, not one woman‵, not one child‵, not one four-footed beast‵.
2. His hopes‵, his happiness‵, his life‵, hung upon the words that fell from those lips.
3. They fought‵, they bled‵, they died‵, for freedom.

PARENTHESIS.

RULE XI.—A *parenthesis* should be read more rapidly and in a lower key than the rest of the sentence, and should terminate with the same inflection that next precedes it. If, however, it is complicated, or emphatic, or disconnected from the main subject, the inflections must be governed by the same rules as in the other cases.

REMARK.—A smooth and expressive reading of a parenthesis is difficult of acquisition, and can be secured only by careful and persistent training.

EXAMPLES.

1. God is my witness′ (whom I serve with my spirit, in the gospel of his Son′), that, without ceasing, I make mention of you always in my prayers; making request′ (if, by any means, now at length, I might have a prosperous journey by the will of God′), to come unto you.

2. When he had entered the room three paces, he stood still; and laying his left hand upon his breast′ (a slender, white staff with which he journeyed being in his right′), he introduced himself with a little story of his convent.

3. If you, Æschines, in particular, were persuaded′ (and it was no particular affection for me, that prompted you to give up the hopes, the appliances, the honors, which attended the course I then advised; but the superior force of truth, and your utter inability to point any course more eligible‵) if this was the case, I say, is it not highly cruel and unjust to arraign these measures now, when you could not then propose a better?

4. As the hour of conflict drew near′ (and this was a conflict to be dreaded even by him‵), he began to waver, and to abate much of his boasting.

CIRCUMFLEX.

Rule XII.—The *circumflex* is used to express *irony, sarcasm, hypothesis,* or *contrast.*

Note.—For the reason that the circumflex always suggests a double or doubtful meaning, it is appropriate for the purposes expressed in the rule. It is, also, frequently used in sportive language; jokes and puns are commonly given with this inflection.

EXAMPLES.

1. Man never ĭs, but always to bê, blest.

2. Thêy follow an adventurer whom they fĕar; wĕ serve a monarch whom we lôve. They boast, they come but to improve our state, enlarge our thoughts, and free us from the yoke of error. Yes, thĕy will give enlightened freedom to our minds, who are themsĕlves the slaves of passion, avarice, and pride. They offer us their protection: yes, sŭch protection as vŭltures give to lămbs, cŏvering and devôuring them.

MONOTONE.

RULE XIII.—The use of the *monotone* is confined chiefly to grave and solemn subjects. When carefully and properly employed, it gives great dignity to delivery.

EXAMPLES.

1. The unbeliever! one who can gaze upon the sun, and moon, and stars, and upon the unfading and imperishable sky, spread out so magnificently above him, and say, "All this is the work of chance!"

2. God walketh upon the ocean. Brilliantly
The glassy waters mirror back his smiles;
The surging billows, and the gamboling storms
Come crouching to his feet.

3. I hail thee, as in gorgeous robes,
Blooming thou leav'st the chambers of the east,
Crowned with a gemmed tiara thick embossed
With studs of living light.

4. High on a throne of royal state, which far
Outshone the wealth of Ormus and of Ind,
Or where the gorgeous east, with richest hand,
Showers on her kings barbaric pearls and gold,
Satan exalted sat.

5. His broad expanded wings
Lay calm and motionless upon the air,
As if he floated there without their aid,
By the sole act of his unlorded will.

6. In dim eclipse, disastrous twilight sheds
On half the nations, and with fear of **change**
Perplexes monarchs.

III. ACCENT AND EMPHASIS.

ACCENT.

That syllable in a word which is uttered more forcibly than the others, is said to be *accented*, and is marked thus, (′); as the italicized syllables in the following words:

*morn′*ing.	*pos′*si-ble.
*ty′*rant.	re-*cum′*bent.
pro-*cure′*.	ex-*or′*bi-tant.
de-*bate′*.	com-pre-*hen′*sive.

Common usage *alone* determines upon what syllable the accent should be placed, and to the lexicographer it belongs, to ascertain and record its decision on this point.

In some few cases, we can trace the *reasons* for common usage in this respect. In words which are used as different parts of speech, or which have different meanings, the distinction is sometimes denoted by changing the accent.

EXAMPLES.

sub′ject.	sub-ject′.
pres′ent.	pre-sent′.
ab′sent.	ab-sent′.
cem′ent.	ce-ment.′
con′jure.	con-jure′.

There is another case, in which we discover the reason for changing the accent, and that is, when it is requred by emphasis, as in the following:

EXAMPLES.

1. His a*bil′*ity or *in′*ability to perform the act materially varies the case.

2. This cor*rup′*tion must put on *in′*corruption.

SECONDARY ACCENT.

In words of more than two syllables, there is often a second accent given, but more slight than the principal one, and this is called the *secondary* accent; as, *em*″ig*ra*′tion, *rep*″ar*tee*′, where the principal accent is marked (′), and the secondary, (″); so, also, this accent is obvious, in *nav*″ig*a*′tion, *com*″pre*hen*′sion, *plau*″si*bil*′ity, etc. The whole subject, however, properly belongs to dictionaries and spelling books.

EMPHASIS.

Emphasis consists in uttering a word or phrase in such a manner as to give it force and energy, and to draw the attention of the hearer particularly to the idea expressed.

This is most frequently accomplished by an increased stress of voice laid upon the word or phrase. Sometimes, though more rarely, the same object is effected by an unusual lowering of the voice, even to a whisper, and not unfrequently by a pause before the emphatic word.

The inflections are often made subsidiary to this object. To give emphasis to a word, the *inflection* is changed or increased in force or extent. When the rising inflection is ordinarily used, the word, when emphatic, frequently takes the falling inflection; and sometimes, also, the falling inflection is changed into the rising inflection, for the same purpose.

Emphatic words are often denoted by being written in *italics*, in SMALL CAPITALS, or in CAPITALS.

Much care is necessary to train the pupil to give clear and expressive emphasis, and at the same time to avoid an unpleasant "jerky" movement of the voice.

ABSOLUTE EMPHASIS.

Where the emphasis is independent of any contrast or comparison with other words or ideas, it is called *absolute emphasis.*

EXAMPLES.

1. We *praise* thee, O God; we acknowledge thee to be the *Lord.*
2. *Roll* on, thou deep and dark blue Ocean—*roll!*
3. *Arm*, warriors, *arm!*
4. You know that you are Brutus, that speak this,
   Or, *by the gods*, this speech were else your last.
5. *Hamlet.* Saw, *who?*
   Horatio. The *king*, your *father.*
   Hamlet. The *king*, my *father?*
6. *Strike*—till the last armed foe expires;
   Strike—for your altars and your fires;
   Strike—for the green graves of your sires;
   God, and your *native land!*

RELATIVE EMPHASIS.

Where there is antithesis, either expressed or implied, the emphasis is called *relative.*

EXAMPLES.

1. We can do nothing *against* the truth, but *for* the truth.
2. But I am describing *your* condition, rather than my *own.*
3. I fear not *death*, and shall I then fear *thee?*
4. **Hunting *men*, and not *beasts*, shall be his game.**

5. He is the propitiation for our sins; and not for *ours* only, but for the sins of the *whole world.*

6. It may *moderate* and *restrain*, but it was not designed to *banish* gladness from the heart of man.

In the following examples, there are *two sets* of antitheses in the same sentence.

7. To *err* is *human*, to *forgive, divine.*

8. *John* was *punished; William*, *rewarded.*

9. *Without* were *fightings*, *within* were *fears.*

10. *Business* sweetens *pleasure*, as *labor* sweetens *rest.*

11. Justice appropriates *rewards* to *merit*, and *punishments* to *crime.*

12. On the *one* side, all was *alacrity* and *courage;* on the *other*, all was *timidity* and *indecision.*

13. The *wise* man is happy when he gains his *own* approbation; the *fool*, when he gains the applause of *others.*

14. His care was to *polish* the country by *art*, as he had *protected* it by *arms.*

In the following examples, the relative emphasis is applied to *three sets* of antithetic words.

15. The difference between a madman and a fool is, that the *former* reasons *justly* from *false* data; and the *latter*, *erroneously* from *just* data.

16. *He* raised a *mortal to* the *skies*,
She drew an *angel down.*

Sometimes the antithesis is *implied*, as in the following instances.

17. The spirit of the *white* man's heaven,
Forbids not *thee* to weep.

18. I shall enter on no encomiums upon *Massachusetts.*

EMPHASIS AND ACCENT.

When words, which are the same in part of their formation, are contrasted, the emphasis is expressed by accenting the syllables in which they differ. See Accent, page 33.

EXAMPLES.

1. What is the difference between *prob*ability and *pos*sibility?
2. Learn to *un*learn what you have learned amiss.
3. John attends *reg*ularly, William, *ir*regularly.
4. There is a great difference between *giv*ing and *for*giving.
5. The conduct of Antoninus was characterized by justice and humanity; that of Nero, by *in*justice and *in*humanity.
6. The conduct of the former is deserving of *ap*probation, while that of the latter merits the severest *rep*robation.

EMPHASIS AND INFLECTION.

Emphasis sometimes changes the inflection from the rising to the falling, or from the falling to the rising. For instances of the former change, see Rule II, and Exception 1 to Rule IV. In the first three following examples, the inflection is changed from the rising to the falling inflection; in the last three, it is changed from the falling to the rising, by the influence of emphasis.

EXAMPLES.

1. If we have no regard for religion in *youth*\`, we ought to have respect for it in *age.*
2. If we have no regard for our *own*\` character, we ought to regard the character of *others.*

3. If content can not *remove*\ the disquietudes of life, it will, at least, *alleviate* them.

4. The sweetest melody and the most perfect harmony fall powerless upon the ear of one who is *deaf′*.

5. It is useless to expatiate upon the beauties of nature to one who is *blind′*.

6. And they that have believing masters, let them not despise them, because they are *brethren′;* but rather let them do them service.

EMPHATIC PHRASE.

When it is desired to give to a phrase great force of expression, each word, and even the parts of a compound word, are independently emphasized.

EXAMPLES.

1. *Cassius.* Must I endure all this?
Brutus. All this!—*Ay,—more.* Fret, till your *proud—heart —break.*

2. What! weep you when you but behold
Our Cæsar's *vesture* wounded? Look ye here,
Here is *himself, marred,* as you see, by *traitors.*

3. There was a time, my fellow-citizens, when the Lacedæmonians were sovereign masters, both by sea and by land; while this state had not one ship—no, NOT—ONE—WALL.

4. Shall I, the conqueror of Spain and Gaul; and not only of the Alpine nations, but of the Alps themselves; shall I compare myself with this HALF—YEAR—CAPTAIN?

5. You call me *misbeliever—cutthroat—dog.*
Hath a *dog—money?* Is it *possible—*
A *cur* can lend *three—thousand—ducats?*

EMPHATIC PAUSE.

A short pause is often made before or after, and sometimes both before and after, an emphatic word or phrase, —thus very much increasing the emphatic expression of the thought.

EXAMPLES.

1. May one be pardoned, and *retain*—the offense?
   In the corrupted currents of *this* world,
   Offense's gilded hand may shove by—*justice;*
   And oft 't is seen, the wicked prize itself
   Buys out the law: but 't is not so—*above:*
   There—is no shuffling: *there*—the action lies
   In its true nature.

2. He woke to hear his sentries shriek,
   "To arms! they come! the Greek! the Greek!
   He woke—to die—midst flame and smoke."

3. *This*—is no *flattery: These*—are counselors
   That *feelingly* persuade me what I am.

4. And *this*—our life, exempt from public haunt,
   Finds *tongues*—in *trees, books*—in the running *brooks,*
   Sermons—in *stones,* and—*good* in *everything.*

5. Heaven gave this Lyre, and thus decreed,
   Be thou a *bruised*—but not a *broken*—reed.

IV. INSTRUCTIONS FOR READING VERSE.

INFLECTIONS.

In reading verse, the inflections should be *nearly* the same as in reading prose; the chief difference is, that in **poetry, the *monotone* and *rising inflection* are more fre-**

quently used than in prose. The greatest difficulty in reading this species of composition, consists in giving it that measured flow which distinguishes it from prose, without falling into a *chanting* pronunciation.

If, at any time, the reader is in doubt as to the proper inflection, let him reduce the passage to earnest conversation, and pronounce it in the most familiar and prosaic manner, and thus he will generally use the proper inflection.

EXERCISES IN INFLECTION.

1. Meanwhile the south wind rose, and with black wings
Wide hovering′, all the clouds together drove
From under heaven‵: the hills to their supply′,
Vapor and exhalation dusk and moist
Sent up amain‵: and now, the thickened sky
Like a dark ceiling stood‵: down rushed the rain
Impetuous′, and continued till the earth
No more was seen‵: the floating vessel swam
Uplifted′, and, secure with beakèd prow′,
Rode tilting o'er the waves‵.

2. My friend′, adown life's valley′, hand in hand′,
With grateful change of grave and merry speech
Or song′, our hearts unlocking each to each′,
We'll journey onward to the silent land‵;
And when stern death shall loose that loving band,
Taking in his cold hand, a hand of ours′,
The one shall strew the other's grave with flowers′,
Nor shall his heart a moment be unmanned‵.
My friend and brother′! if thou goest first′,
Wilt thou no more revisit me below′?
Yea, when my heart seems happy causelessly′,
And swells′, not dreaming why′, my soul shall know
That thou′, unseen′, art bending over me‵.

3. Here rests his head upon the lap of earth′,
A youth, to fortune and to fame unknown‵;
Fair Science frowned not on his humble birth′,
And Melancholy marked him for her own‵.

4. Large was his bounty′, and his soul sincere‵,
   Heaven did a recompense as largely send‵;
   He gave to misery (all he had) a tear‵,
   He gained from heaven′ ('t was all he wished′) a friend‵.

5. No further seek his merits to disclose′,
   Or draw his frailties from their dread abode′;
   (There they alike′ in trembling hope repose′,)
   The bosom of his Father, and his God‵.

ACCENT AND EMPHASIS.

In reading verse, every syllable must have the same accent, and every word the same emphasis as in prose; and whenever the *melody* or *music* of the verse would lead to an *incorrect* accent or emphasis, this must be disregarded.

If a poet has made his verse deficient in melody, this must not be remedied by the reader, at the expense of sense or the established rules of accent and quantity. Take the following:

EXAMPLE.

O'er shields, and helms, and helmèd heads he rode,
Of thrones, and mighty Seraphim pros*trate.*

According to the *metrical* accent, the last word must be pronounced "pros-*trate*′." But according to the authorized pronunciation it is "*pros*′trate." Which shall yield, the poet or established usage? Certainly not the latter.

Some writers advise a compromise of the matter, and that the word should be pronounced without accenting either syllable. Sometimes this may be done, but where it is not practiced, the prosaic reading should be preserved.

In the following examples, the words and syllables which are *improperly accented* or *emphasized* in the poetry, are marked in *italics.* According to the principle stated above, the reader should avoid giving them that pronunciation which the correct reading of the *poetry* would require, but should read them as prose, except where he can throw off all accent and thus compromise the conflict between the poetic reading and the correct reading. That is, he must read the poetry *wrong*, in order to read the language *right.*

EXAMPLES.

1. Ask *of* thy mother earth why oaks are made
   Tal*ler* and stronger than the weeds they shade.

2. Their praise is still, "the style is excel*lent*,"
   The sense they humbly take upon content.

3. False elo*quence*, like *the* prismatic glass,
   Its fairy colors spreads on every place.

4. To do aught good, nev*er* will be our task,
   But ever *to* do ill is our sole delight.

5. Of all the causes which combine to blind
   Man's erring judgment, *and* mislead the mind,
   What *the* weak head with strongest bias rules,
   Is pride, the never-failing vice of fools.

6. Eye Nature's walks, shoot folly *as* it flies,
   And catch the manners living *as* they rise.

7. To whom then, first incensed, Ad*am* replied,
   "Is this thy love, is this the recompense
   Of mine to thee, ungrateful Eve?"

8. We may, with more successful hope, resolve
   To wage, by force or guile, successful war,
   Irreconcilable to our grand foe,
   Who now tri*umphs*, and in excess of joy
   Sole reigning holds the tyranny of Heaven.

9. Which, when Beëlzebub perceived (than whom,
   Satan except, none higher sat), with grave
   As*pect*, he rose, and in his rising seemed
   A pillar of state.

10. Thee, Sion, and the flowery brooks beneath,
That wash thy hallowed feet, and warbling flow,
Nightly I visit: nor sometimes forget
Those other two *equaled* with me in fate.

NOTE.—Although it would be necessary, in these examples, to violate the laws of accent or emphasis, to give perfect rhythm, yet a careful and well-trained reader will be able to observe these laws and still give the rhythm in such a manner that the defect will scarcely be noticed.

POETIC PAUSES.

In order to make the measure of poetry perceptible to the ear, there should generally be a *slight pause* at the end of each line, even where the sense does not require it.

There is, also, in almost every line of poetry, a pause at or near its middle, which is called the *cæsura.*

This should, however, never be so placed as to injure the sense of the passage. It is indeed reckoned a great beauty, where it naturally coincides with the pause required by the sense. The cæsura, though *generally* placed near the middle, may be placed at other intervals.

There are sometimes, also, two additional pauses in each line, called demi-cæsuras.

The cæsura is marked (||), and the demi-cæsura thus, (|), in the examples given.

There should be a marked accent upon the long syllable next preceding the cæsura, and a slighter one upon that next before each of the demi-cæsuras. When made too prominent, these pauses lead to a singsong style, which should be carefully avoided.

In the following examples, the cæsura is marked in each line; the demi-cæsura is not marked in every case.

EXAMPLES.

1. Nature | to all things || fixed | the limits fit,
And wisely | curbed || proud man's | pretending wit.

2. Then from his closing eyes || thy form shall part,
And the last pang || shall tear thee from his heart.

3. Warms in the sun, || refreshes in the breeze,
Glows in the stars, || and blossoms in the trees.

4. There is a land || of every land the pride,
Beloved by Heaven || o'er all the world beside,
Where brighter suns || dispense serener light,
And milder moons || imparadise the night;
Oh, thou shalt find, || howe'er thy footsteps roam,
That land—thy country, || and that spot—thy home.

5. In slumbers | of midnight || the sailor | boy lay;
His hammock | swung loose || at the sport | of the wind;
But, watch-worn | and weary, || his cares | flew away,
And visions | of happiness || danced | o'er his mind.

6. She said, | and struck; || deep entered | in her side
The piercing steel, || with reeking purple dyed:
Clogged | in the wound || the cruel | weapon stands,
The spouting blood || came streaming o'er her hands.
Her sad attendants || saw the deadly stroke,
And with loud cries || the sounding palace shook.

SIMILE.

Simile is the likening of anything to another object of a different class; it is a poetical or imaginative comparison.

A simile, in poetry, should usually be read in a lower key and more rapidly than other parts of the passage—somewhat as a parenthesis is read.

EXAMPLES.

1. Part curb their fiery steeds, or shun the goal
With rapid wheels, or fronted brigades form.
As when, to warn proud cities, war appears,
Waged in the troubled sky, and armies rush
To battle in the clouds.
Others with vast Typhoëan rage more fell,
Rend up both rocks and hills, and ride the air
In whirlwind. Hell scarce holds the wild uproar.
As when Alcides felt the envenomed robe, and tore,
Through pain, up by the roots, Thessalian pines,
And Lichas from the top of Œta threw
Into the Euboic sea.

2. Each at the head,
Leveled his deadly aim; their fatal hands
No second stroke intend; and such a frown
Each cast at th' other, *as when two black clouds,*
With heaven's artillery fraught, came rolling on
Over the Caspian, there stand front to front,
Hovering a space, till winds the signal blow
To join the dark encounter, in mid-air:
So frowned the mighty combatants.

3. Then pleased and thankful from the porch they go,
And, but the landlord, none had cause of woe:
His cup was vanished; for, in secret guise,
The younger guest purloined the glittering prize.
As one who spies a serpent in his way,
Glistening and basking in the summer ray,
Disordered, stops to shun the danger near,
Then walks with faintness on, and looks with fear,—
So seemed the sire, when, far upon the road,
The shining spoil his wily partner showed.

V. THE VOICE.

PITCH AND COMPASS.

The **natural pitch** of the voice is its keynote, or governing note. It is that on which the voice usually dwells, and to which it most frequently returns when wearied. It is also the pitch used in conversation, and the one which a reader or speaker naturally adopts — when he reads or speaks — most easily and agreeably.

The **compass** of the voice is its range above and below this pitch. To avoid monotony in reading or speaking, the voice should rise above or fall below this keynote, but always with reference to the sense or character of that which is read or spoken. The proper natural pitch is that above and below which there is most room for variation.

To strengthen the voice and increase its compass, select a short sentence, repeat it several times in succession in as low a key as the voice can sound naturally; then rise one note higher, and practice on that key, then another, and so on, until the highest pitch of the voice has been reached. Next, reverse the process, until the lowest pitch has been reached.

EXAMPLES IN PITCH.

High Pitch.

NOTE.—Be careful to distinguish *pitch* from *power* in the following exercises. Speaking in the open air, at the very top of the voice, is an exercise admirably adapted to strengthen the voice and give it compass, and should be frequently practiced.

1. Charge\! Chester\, charge\! On\! Stanley, on\!

2. A horse\! a horse\! my kingdom\ for a horse\!

3. Jump far out\, boy\, into the wave\!
Jump\, or I **fire**\!

4. Run\! run\! run for your lives!

5. Fire\! fire\! fire\! Ring the bell\!

6. Gentlemen may cry peace′! peace′! but there is no peace!

7. Rouse\, ye Romans! rouse\, ye slaves\!
Have ye brave sons′? Look in the next fierce brawl
To see them die\. Have ye fair daughters′? Look
To see them live, torn from your arms\, distained\,
Dishonored\, and if ye dare call for justice′,
Be answered by the lash\!

Medium Pitch.

NOTE.—This is the pitch in which we converse. To strengthen it, we should read or speak in it as loud as possible, without rising to a higher key. To do this requires long-continued practice.

1. Under a spreading chestnut tree,
The village smithy stands\;
The smith, a mighty man is he,
With large and sinewy hands\;
And the muscles of his brawny arms
Are strong as iron bands.

2. There is something in the thunder's voice that makes me tremble like a child. I have tried to conquer\ this unmanly weakness\. I have called pride\ to my aid\; I have sought for moral courage in the lessons of philosophy\, but it avails me nothing\. At the first moaning of the distant cloud, my heart shrinks and dies within me.

3. He taught the scholars the Rule of Three′,
Reading, and writing, and history\, too\;
He took the little ones on his knee′,
For a kind old heart in his breast had he′,
And the wants of the littlest child he knew\.
"Learn while you 're young\," he often said′,
"There is much to enjoy down here below′;
Life for the living′, and rest for the dead\,"
Said the jolly old pedagogue\, long ago\.

Low Pitch.

1. O, proper stuff!
This is the very painting of your fear:
This is the air-drawn dagger which, you said,
Led you to Duncan. O, these flaws and starts,
Impostors to true fear, would well become
A woman's story, at a winter's fire,
Authorized by her grandam.

2. Thou slave! thou wretch! thou coward!
Thou little valiant, great in villainy!
Thou ever strong upon the stronger side!
Thou fortune's champion, thou dost never fight
But when her humorous ladyship is by
To teach thee safety! Thou art perjured too,
And sooth'st up greatness. What a fool art thou,
A ramping fool; to brag, and stamp, and sweat,
Upon my party! thou cold-blooded slave!

3. God! thou art mighty! At thy footstool bound,
Lie, gazing to thee, Chance, and Life, and Death;
Nor in the angel circle flaming round,
Nor in the million worlds that blaze beneath,
Is one that can withstand thy wrath's hot breath.
Woe, in thy frown: in thy smile, victory:
Hear my last prayer! I ask no mortal wreath;
Let but these eyes my rescued country see,
Then take my spirit, all Omnipotent, to thee.

4. O Thou eternal One! whose presence bright
All space doth occupy, all motion guide,
Unchanged through time's all-devastating blight!
Thou only God, there is no god beside!
Being above all things, mighty One,
Whom none can comprehend and none explore;
Who fill'st existence with thyself alone,—
Embracing all, supporting, ruling o'er,—
Being whom we call God, and know no more!

QUANTITY AND QUALITY.

Quantity, in reading and speaking, means the length of time occupied in uttering a syllable or a word. Sounds and syllables vary greatly in quantity. Some are long, some short, and others intermediate between those which are long or short. Some sounds, also, may be prolonged or shortened in utterance to any desired extent. Quantity may be classified as *Long*, *Medium*, or *Short*.

DIRECTIONS FOR PRACTICE ON LONG QUANTITY.—Select some word of one syllable ending with a long vocal or a subvocal sound; pronounce it many times in succession, increasing the quantity at each repetition, until you can dwell upon it any desired length of time, without drawling, and in a natural tone.

REMARK.—Practice in accordance with this direction will enable the pupil to secure that fullness and roundness of voice which is exemplified in the hailing of a ship, "ship aho——y;" in the reply of the sailor, when, in the roar of the storm, he answers his captain, "ay——e, ay——e;" and in the command of the officer to his troops, when, amid the thunder of artillery, he gives the order, "ma——rch," or "ha——lt."

This fullness or roundness of tone is secured, by dwelling on the *vocal sound*, and indefinitely protracting it. The mouth should be opened wide, the tongue kept down, and the aperture left as round and as free for the voice as possible.

It is this artificial rotundity which, in connection with a distinct articulation, enables one who speaks in the open air, or in a very large apartment, to send his voice to the most distant point. It is a certain degree of this quality, which distinguishes declamatory or public speaking or reading from private conversation, and no one can accomplish much, as a public speaker, without cultivating it. It must be carefully distinguished from the "high tone," which is an elevation of *pitch*, and from "loudness," or "strength" of voice.

It will be observed that clearness and distinctness of utterance are secured by a proper use of the subvocals and aspirates—these sounds giving to words their *shape*, as it were; but a clear, full, and well-modulated utterance of the vocals gives to words their *fullness*.

LONG QUANTITY.

1. Liberty! Freedom! Tyranny is dead!
2. Woe, woe, to the inhabitants of Jerusalem!

3. O righteous Heaven! ere Freedom found a grave,
Why slept the sword, omnipotent to save?
Where was thine arm, O Vengeance! where thy rod,
That smote the foes of Zion and of God?

4. O sailor boy! sailor boy! never again
Shall home, love, or kindred thy wishes repay;
Unblessed and unhonored, down deep in the main,
Full many a fathom, thy frame shall decay.

5. O Lord, our Lord, how excellent is thy name in all the earth! who hast set thy glory above the heavens! When I consider thy heavens, the work of thy fingers; the moon and the stars, which thou hast ordained; what is man, that thou art mindful of him? and the son of man, that thou visitest him? For thou hast made him a little lower than the angels, and hast crowned him with glory and honor. Thou madest him to have dominion over the work of thy hands; thou hast put all things under his feet. O Lord, our Lord, how excellent is thy name in all the earth!

MEDIUM QUANTITY.

1. Between Nose and Eyes a strange contest arose;
The spectacles set them, unhappily, wrong;
The point in dispute was, as all the world knows,
To which the said spectacles ought to belong.

2. Bird of the broad and sweeping wing!
Thy home is high in heaven,
Where the wide storms their banners fling,
And the tempest clouds are driven.

3. At midnight, in his guarded tent,
The Turk lay dreaming of the hour
When Greece, her knee in suppliance bent,
Should tremble at his power.

4. On New Year's night, an old man stood at his window, and looked, with a glance of fearful despair, up the immovable, unfading heaven, and down upon the still, pure, white earth, on which no one was now so joyless and sleepless as he.

SHORT QUANTITY.

1. Quick! or he faints! stand with the cordial near!

2. Back to thy punishment, false fugitive!

3. Fret till your proud heart breaks! Must I observe you? Must I crouch beneath your testy humor?

4. Up drawbridge, grooms! what, warder, ho!
Let the portcullis fall!

5. Quick, man the lifeboat! see yon bark,
That drives before the blast!
There's a rock ahead, the fog is dark,
And the storm comes thick and fast.

6. I am at liberty, like every other man, to use my own language; and though, perhaps, I may have some ambition to please this gentleman, I shall not lay myself under any restraint, nor very solicitously copy his diction, or his mien, however matured by age or modeled by experience.

MOVEMENT.

Movement is the rapidity with which the voice moves in reading and speaking. It varies with the nature of the thought or sentiment to be expressed, and should be increased or diminished as good taste may determine. With pupils generally, the tendency is to read too fast. The result is, reading or speaking in too high a key and an unnatural style of delivery—both of which faults are difficult to be corrected when once formed. The kinds of movement are *Slow*, *Moderate*, and *Quick*.

DIRECTIONS.—Read a selection as slowly as possible, without drawling. Read it again and again, increasing the rate of movement at each reading, until it can be read no faster without the utterance becoming indistinct. Reverse this process, reading more and more slowly at each repetition, until the slowest movement is obtained.

SLOW MOVEMENT.

1. Oh that those lips had language! Life has passed
With me but roughly, since I heard them last.

2. A tremulous sigh from the gentle night wind
Through the forest leaves slowly is creeping,
While stars up above, with their glittering eyes,
Keep guard; for the army is sleeping.

3. O Lord′! have mercy upon us, miserable offenders′!

4. So live, that when thy summons comes to join
The innumerable caravan that moves
To the pale realms of shade, where each shall take
His chamber in the silent halls of death,
Thou go not, like the quarry slave at night,
Scourged to his dungeon, but, sustained and soothed
By an unfaltering trust, approach thy grave
Like one who wraps the drapery of his couch
About him, and lies down to pleasant dreams.

MODERATE MOVEMENT.

1. The good′, the brave′, the beautiful‵,
How dreamless‵ is their sleep,
Where rolls the dirge-like music′
Of the ever-tossing deep‵!
Or where the surging night winds
Pale Winter's robes have spread
Above the narrow palaces,
In the cities of the dead‵!

2. Lives of great men all remind us
We can make our lives sublime,
And, departing, leave behind us
Footprints on the sands of time.

3. Cast your eyes over this extensive country. Observe the salubrity of your climate, the variety and fertility of your soil; and see that soil intersected in every quarter by bold, navigable streams, flowing to the east and to the west, as if the finger of heaven were marking out the course of your settlements, inviting you to enterprise, and pointing the way to wealth.

QUICK MOVEMENT.

1. Awake\! arise\! or be forever fallen.

2. Merrily swinging on brier and weed,
Near to the nest of his little dame,
Over the mountain side or mead,
Robert of Lincoln is telling his name.

3. Not a word to each other; we kept the great pace—
Neck by neck, stride by stride, never changing our place;
I turned in my saddle and made its girths tight,
Then shortened each stirrup and set the pique right,
Rebuckled the check strap, chained slacker the bit,
Nor galloped less steadily Roland a whit.

4. Oh my dear uncle, you don't know the effect of a fine spring morning upon a fellow just arrived from Russia. The day looked bright, trees budding, birds singing, the park so gay, that I took a leap out of your balcony, made your deer fly before me like the wind, and chased them all around the park to get an appetite for breakfast, while you were snoring in bed, uncle.

Quality.—We notice a difference between the soft, insinuating tones of persuasion; the full, strong voice of command and decision; the harsh, irregular, and sometimes grating explosion of the sounds of passion; the plaintive notes of sorrow and pity; and the equable and unimpassioned flow of words in argumentative style. This difference consists in a variation in the *quality* of the voice by which it is adapted to the character of the thought or sentiment read or spoken. In our attempts to imitate nature, however, it is important that all *affectation* be avoided, for perfect monotony is preferable to this fault. The tones of the voice should be made to correspond with the nature of the subject, *without apparent effort.*

EXAMPLES.

Passion and Grief.

"Come back! come back!" he cried, in grief,
"Across this stormy water;
And I'll forgive your Highland chief,
My daughter! O, my daughter!"

Plaintive.

I have lived long enough: my way of life
Is fallen into the sear, the yellow leaf:
And that which should accompany old age,
As honor, love, obedience, troops of friends,
I must not look to have.

Calm.

A very great portion of this globe is covered with water, which is called sea, and is very distinct from rivers and lakes.

Fierce Anger.

Burned Marmion's swarthy cheek like fire,
And shook his very frame for ire;
And—"This to me!" he said,—
"An 't were not for thy hoary beard,
Such hand as Marmion's had not spared
To cleave the Douglas' head!

Loud and Explosive.

"Even in thy pitch of pride,
Here, in thy hold, thy vassals near,
I tell thee, thou 'rt defied!
And if thou said'st I am not peer
To any lord in Scotland here,
Lowland or Highland, far or near,
Lord Angus, thou hast lied!"

VI. GESTURE.

Gesture is that part of the speaker's manner which pertains to his attitude, to the use and carriage of his person, and the movement of his limbs in delivery.

Every person, in beginning to speak, feels the natural embarrassment resulting from his new position. The novelty of the situation destroys his self-possession, and, with the loss of that, he becomes awkward, his arms and hands hang clumsily, and now, for the first time, seem to him worse than superfluous members. This embarrassment will be overcome gradually, as the speaker becomes familiar with his position; and it is sometimes overcome at once, by a powerful exercise of the attention upon the matter of the speech. When that fills and possesses the mind, the orator is likely to take the attitude which is becoming, and, at least, easy and natural, if not graceful.

1st. The first general direction that should be given to the speaker is, that he should *stand erect and firm*, and in that posture which gives an expanded chest and full play to the organs of respiration and utterance.

2d. Let the attitude be such that it can be *shifted easily* and *gracefully*. The student will find, by trial, that no attitude is so favorable to this end as that in which the weight of the body is thrown upon one leg, leaving the other free to be advanced or thrown back, as fatigue or the proper action of delivery may require.

The student who has any regard to grace or elegance, will of course avoid all the gross faults which are so common among public speakers, such as resting one foot upon a stool or bench, or throwing the body forward upon the support of the rostrum.

3d. Next to attitude, come the movements of the person and limbs. In these, two objects are to be observed, and, if possible, combined, viz., *propriety* and *grace*. There is expression in the extended arm, the clinched hand, the open palm, and the smiting of the breast. But let no gesture be made that is not in harmony with the thought or sentiment uttered; for it is this harmony which constitutes propriety. As far as possible, let there be a correspondence between the style of action and the train of thought. Where the thought flows on calmly, let there be grace and ease in gesture and action. Where the style is sharp and abrupt, there is propriety in quick, short, and abrupt gesticulation. Especially avoid that ungraceful sawing of the air with the arms, into which an ill-regulated fervor betrays many young speakers.

What is called a *graceful manner*, can only be attained by those who have some natural advantages of person. So far as it is in the reach of study or practice, it seems to depend chiefly upon the general cultivation of manners, implying freedom from all embarrassments, and entire self-possession. The secret of acquiring a graceful style of gesture, we apprehend, lies in the habitual practice, not only when speaking but at all times, of free and graceful movements of the limbs.

There is no limb nor feature which the accomplished speaker will not employ with effect, in the course of a various and animated delivery. The arms, however, are the chief reliance of the orator in gesture; and it will not be amiss to give a hint or two in reference to their proper use.

First—It is not an uncommon fault to use one arm exclusively, and to give that a uniform movement. Such movement may, sometimes, have become habitual from one's profession or employment; but in learners, also, there is often a predisposition to this fault.

Second—It is not unusual to see a speaker use only the lower half of his arm. This always gives a stiff and constrained manner to delivery. Let the whole arm move, and let the movement be free and flowing.

Third—As a general rule, let the hand be open, with the fingers slightly curved. It then seems liberal, communicative, and candid; and, in some degree, gives that expression to the style of delivery. Of course there are passages which require the clinched hand, the pointed finger, etc., etc.; but these are used to give a particular expression.

Fourth—In the movements of the arm, study variety and the grace of curved lines.

When a gesture is made with one arm only, the *eye* should be cast in the direction of that arm; not *at* it, but *over* it.

All speakers employ, more or less, the motions of the head. In reference to that member, we make but one observation. Avoid the continuous shaking and bobbing of the head, which is so conspicuous in the action of many ambitious public speakers.

The beauty and force of all gesture consist in its timely, judicious, and natural employment, when it can serve to illustrate the meaning or give emphasis to the force of an important passage. The usual fault of young speakers is too much action. To emphasize all parts alike, is equivalent to no emphasis; and by employing forcible gestures on unimportant passages, we diminish our power to render other parts impressive.

ELOCUTION AND READING.

The business of training youth in elocution, must be commenced in childhood. The first school is the nursery. There, at least, may be formed a distinct articulation, which is the first requisite for good speaking. How rarely is it found in perfection among our orators.

"Words," says one, referring to articulation, should "be delivered out from the lips, as beautiful coins, newly issued from the mint; deeply and accurately impressed, perfectly finished; neatly struck by the proper organs, distinct, in due succession, and of due weight." How rarely do we hear a speaker whose tongue, teeth, and lips, do their office so perfectly as to answer to this beautiful description! And the common faults in articulation, it should be remembered, take their rise from the very nursery.

Grace in eloquence, in the pulpit, at the bar, can not be separated from grace in the ordinary manners, in private life, in the social circle, in the family. It can not well be superinduced upon all the other acquisitions of youth, any more than that nameless, but invaluable, quality called good breeding. Begin, therefore, the work of forming the orator with the child; not merely by teaching him to declaim, but what is of more consequence, by observing and correcting his daily manners, motions, and attitudes. You can say, when he comes into your apartment, or presents you with something, a book or letter, in an awkward and blundering manner, "Return, and enter this room again," or, "Present me that book in a different manner," or, "Put yourself in a different attitude." You can explain to him the difference between thrusting or pushing out his hand and arm, in straight lines and at acute angles, and moving them in flowing circular lines, and easy graceful action. He will readily understand you. Nothing is more true than that the motions of children are originally graceful; it is by suffering them to be perverted, that we lay the foundation of invincible awkwardness in later life.

In schools for children, it ought to be a leading object to teach the art of reading. It ought to occupy *threefold more time* than it does. The teachers of these schools should labor to improve *themselves.* They should feel that to them, for a time, are committed the future orators of the land.

It is better that a girl should return from school a first-rate *reader*, than a first-rate performer on the pianoforte. The accomplishment, in its perfection, would give more pleasure. The voice of song is not sweeter than the voice of eloquence; and there may be eloquent *readers*, as well as eloquent *speakers*. We speak of *perfection* in this art: and it is something, we must say in defense of our preference, which we have never yet seen. Let the same pains be devoted to reading, as are required to form an accomplished performer on an instrument; let us have, as the ancients had, the formers of the voice, the music masters of the *reading* voice; let us see years devoted to this accomplishment, and then we should be prepared to stand the comparison.

Reading is, indeed, a most intellectual accomplishment. So is music, too, in its perfection. We do by no means undervalue this noble and most delightful art, to which Socrates applied himself even in his old age. But one recommendation of the art of reading is, that it requires a constant exercise of mind. It involves, in its perfection, the whole art of criticism on language. A man may possess a fine genius without being a perfect reader; but he can not be a perfect reader without genius.

ON MODULATION.

FROM LLOYD.

'T is not enough the *voice*\ be sound and clear′,
'T is *modulation*′ that must charm the ear.
When desperate heroes grieve with tedious moan,
And whine their sorrows in a seesaw tone,
The same soft sounds of unimpassioned woes,
Can only make the yawning hearers doze.

The voice all modes of passion can express
That marks the proper word with proper stress:
But none emphatic can that speaker call,
Who lays an *equal* emphasis on *all.*

Some o'er the tongue the labored measure roll,
Slow and deliberate as the parting toll;
Point every stop, mark every pause so strong,
Their words like stage processions stalk along.

All affectation but creates disgust;
And e'en in *speaking*, we may seem too just.
In vain for *them*\ the pleasing measure flows,
Whose recitation runs it all to prose;
Repeating what the poet sets not down,
The verb disjointing from its favorite noun,
While pause, and break, and repetition join
To make a discord in each tuneful line\.

Some\ placid natures fill the allotted scene
With lifeless drawls, insipid and serene;
While *others*′ thunder every couplet o'er,
And almost crack your ears with rant and roar;
More nature oft, and finer strokes are shown
In the low whisper than tempestuous tone;
And Hamlet's hollow voice and fixed amaze,
More powerful terror to the mind conveys
Than he, who, swollen with impetuous rage,
Bullies the bulky phantom of the stage.

He who, in earnest, studies o'er his part,
Will find true nature cling about his heart.
The modes of grief are not included all
In the white handkerchief and mournful drawl:
A single *look*\ more marks the internal woe,
Than all the windings of the lengthened *Oh*\!

McGUFFEY'S

SIXTH READER.

McGUFFEY'S

SIXTH READER.

SELECTIONS FOR READING.

I. ANECDOTE OF THE DUKE OF NEWCASTLE.

A LAUGHABLE story was circulated during the administration of the old Duke of Newcastle, and retailed to the public in various forms. This nobleman, with many good points, was remarkable for being profuse of his promises on all occasions, and valued himself particularly on being able to anticipate the words or the wants of the various persons who attended his levees, before they uttered a word. This sometimes led him into ridiculous embarrassments; and it was this proneness to lavish promises, which gave occasion for the following anecdote:

At the election of a certain borough in Cornwall, where the opposite interests were almost equally poised, a single vote was of the highest importance. This object the Duke, by well applied argument and personal application, at length attained; and the gentleman he recommended, gained the election. In the warmth of gratitude, his grace poured forth acknowledgments and promises without ceasing, on the fortunate possessor of the casting vote; called him his best and dearest friend; protested, that he should consider himself as forever indebted to him; and that he would serve him by night or by day.

The Cornish voter, who was an honest fellow, and would not have thought himself entitled to any reward, but for such a torrent of acknowledgments, thanked the Duke for his kindness, and told him the supervisor of excise was old and infirm, and, if he would have the goodness to recommend his son-in-law to the commissioners, in case of the old man's death, he should think himself and his family bound to render his grace every assistance in their power, on any future occasion.

"My dear friend, why do you ask for such a trifling employment?" exclaimed his grace; "your relative shall have it the moment the place is vacant, if you will but call my attention to it."

"But how shall I get admitted to you, my lord? For in London, I understand, it is a very difficult business to get a sight of you great folks, though you are so kind and complaisant to us in the country."

"The instant the man dies," replied the Duke, "set out posthaste for London; drive directly to my house, and, be it by night or by day, thunder at the door; I will leave word with my porter to show you upstairs directly; and the employment shall be disposed of according to your wishes."

The parties separated; the Duke drove to a friend's house in the neighborhood, without a wish or desire to see his new acquaintance till that day seven years; but the memory of the Cornish elector, not being burdened with such a variety of objects, was more retentive. The supervisor died a few months after, and the Duke's humble friend, relying on the word of a peer, was conveyed to London posthaste, and ascended with alacrity the steps of that nobleman's palace.

The reader should be informed, that just at this time,

no less a person than the King of Spain was expected hourly to depart this life,—an event in which the minister of Great Britain was particularly concerned; and the Duke of Newcastle, on the very night that the proprietor of the decisive vote arrived at his door, had sat up anxiously expecting dispatches from Madrid. Wearied by official business and agitated spirits, he retired to rest, having previously given particular instructions to his porter not to go to bed, as he expected every minute a messenger with advices of the greatest importance, and desired that he might be shown upstairs, the moment of his arrival.

His grace was sound asleep; and the porter, settled for the night in his armchair, had already commenced a sonorous nap, when the vigorous arm of the Cornish voter roused him from his slumbers. To his first question, "Is the Duke at home?" the porter replied, "Yes, and in bed; but has left particular orders that, come when you will, you are to go up to him directly."

"Bless him, for a worthy and honest gentleman," cried our applicant for the vacant post, smiling and nodding with approbation at the prime minister's kindness, "how punctual his grace is; I knew he would not deceive me; let me hear no more of lords and dukes not keeping their words; I verily believe they are as honest, and mean as well as any other folks." Having ascended the stairs as he was speaking, he was ushered into the Duke's bed-chamber.

"Is he dead?" exclaimed his grace, rubbing his eyes, and scarcely awakened from dreaming of the King of Spain, "Is he dead?"

"Yes, my lord," replied the eager expectant, delighted to find the election promise, with all its circumstances, so fresh in the nobleman's memory.

"When did he die?"

"The day before yesterday, exactly at half past one o'clock, after being confined three weeks to his bed, and taking *a power of doctor's stuff*; and I hope your grace will be as good as your word, and let my son-in-law succeed him."

The Duke, by this time perfectly awake, was staggered at the impossibility of receiving intelligence from Madrid in so short a space of time; and perplexed at the absurdity of a king's messenger applying for his son-in-law to succeed the King of Spain: "Is the man drunk, or mad? Where are your dispatches?" exclaimed his grace, hastily drawing back his curtain; where, instead of a royal courier, he recognized at the bedside, the fat, good-humored countenance of his friend from Cornwall, making low bows, with hat in hand, and "hoping my lord would not forget the gracious promise he was so good as to make, in favor of his son-in-law, at the last election."

Vexed at so untimely a disturbance, and disappointed of news from Spain, the Duke frowned for a moment; but chagrin soon gave way to mirth, at so singular and ridiculous a combination of circumstances, and, yielding to the impulse, he sunk upon the bed in a violent fit of laughter, which was communicated in a moment to the attendants.

The relater of this little narrative, concludes, with observing, "Although the Duke of Newcastle could not place the relative of his old acquaintance on the throne of His Catholic Majesty, he advanced him to a post *not less honorable*—he made him an exciseman."

—*Blackwood's Magazine.*

NOTES.—**Duke of Newcastle.**—Thomas Holles Pelham (b. 1693, d. 1768), one of the chief ministers of state in the reign of George II. of England.

Cornwall.—A county forming the extreme southwestern part of England.

King of Spain.—Ferdinand VI. was then the king of Spain. He died in 1759.

His Catholic Majesty, a title applied to the kings of Spain; first given to Alfonso I. by Pope Gregory III. in 739.

II. THE NEEDLE.

THE gay belles of fashion may boast of excelling
In waltz or cotillon, at whist or quadrille;
And seek admiration by vauntingly telling
Of drawing, and painting, and musical skill:
But give me the fair one, in country or city,
Whose home and its duties are dear to her heart,
Who cheerfully warbles some rustical ditty,
While plying the needle with exquisite art:
The bright little needle, the swift-flying needle,
The needle directed by beauty and art.

If Love have a potent, a magical token,
A talisman, ever resistless and true,
A charm that is never evaded or broken,
A witchery certain the heart to subdue,
'T is this; and his armory never has furnished
So keen and unerring, or polished a dart;
Let beauty direct it, so polished and burnished,
And oh! it is certain of touching the heart:
The bright little needle, the swift-flying needle,
The needle directed by beauty and art.

Be wise, then, ye maidens, nor seek admiration,
By dressing for conquest, and flirting with all;
You never, whate'er be your fortune or station,
Appear half so lovely at rout or at ball,
As gayly convened at the work-covered table,
Each cheerfully active, playing her part,
Beguiling the task with a song or a fable,
And plying the needle with exquisite art:
The bright little needle, the swift-flying needle,
The needle directed by beauty and art.

—*Samuel Woodworth.*

III. DAWN.

Edward Everett, 1794-1865. He was born at Dorchester, Mass., now a part of Boston, and graduated from Harvard College with the highest honors of his class, at the age of seventeen. While yet in college, he had quite a reputation as a brilliant writer. Before he was twenty years of age, he was settled as pastor over the Brattle Street Church, in Boston, and at once became famous as an eloquent preacher. In 1814, he was elected Professor of Greek Literature in his *Alma Mater;* and, in order to prepare himself for the duties of his office, he entered on an extended course of travel in Europe. He edited the "North American Review," in addition to the labors of his professorship, after he returned to America.

In 1825, Mr. Everett was elected to Congress, and held his seat in the House for ten years. He was Governor of his native state from 1835 to 1839. In 1841, he was appointed Minister to England. On his return, in 1846, he was chosen President of Harvard University, and held the office for three years. In 1852, he was appointed Secretary of State. February 22, 1856, he delivered, in Boston, his celebrated lecture on Washington. This lecture was afterwards delivered in most of the principal cities and towns in the United States. The proceeds were devoted to the purchase of Mt. Vernon. In 1860, he was a candidate for the Vice Presidency of the United States. He is celebrated as an elegant and forcible writer, and a chaste orator.

This extract, a wonderful piece of word painting, is a portion of an address on the "Uses of Astronomy," delivered at the inauguration of the Dudley Observatory, at Albany, N. Y. Note the careful use of words, and the strong figures in the third and fourth paragraphs.

I HAD occasion, a few weeks since, to take the early train from Providence to Boston; and for this purpose rose at two o'clock in the morning. Everything around was wrapped in darkness and hushed in silence, broken only by what seemed at that hour the unearthly clank and rush of the train. It was a mild, serene, midsummer's night,—the sky was without a cloud, the winds were whist. The moon, then in the last quarter, had just risen, and the stars shone with a spectral luster but little affected by her presence.

Jupiter, two hours high, was the herald of the day; the Pleiades, just above the horizon, shed their sweet influence in the east; Lyra sparkled near the zenith; Andromeda veiled her newly-discovered glories from the naked eye in the south; the steady Pointers, far beneath the pole, looked meekly up from the depths of the north to their sovereign.

Such was the glorious spectacle as I entered the train. As we proceeded, the timid approach of twilight became more perceptible; the intense blue of the sky began to soften; the smaller stars, like little children, went first to rest; the sister beams of the Pleiades soon melted together; but the bright constellations of the west and north remained unchanged. Steadily the wondrous transfiguration went on. Hands of angels, hidden from mortal eyes, shifted the scenery of the heavens; the glories of night dissolved into the glories of the dawn.

The blue sky now turned more softly gray; the great watch stars shut up their holy eyes; the east began to kindle. Faint streaks of purple soon blushed along the sky; the whole celestial concave was filled with the inflowing tides of the morning light, which came pouring down from above in one great ocean of radiance; till at length, as we reached the Blue Hills, a flash of purple fire blazed out from above the horizon, and turned the dewy teardrops of flower and leaf into rubies and diamonds. In a few seconds, the everlasting gates of the morning were thrown wide open, and the lord of day, arrayed in glories too severe for the gaze of man, began his state.

I do not wonder at the superstition of the ancient Magians, who, in the morning of the world, went up to the hilltops of Central Asia, and, ignorant of the true God, adored the most glorious work of his hand. But I am filled with amazement, when I am told, that, in this enlightened age and in the heart of the Christian world, there are persons who can witness this daily manifestation of the power and wisdom of the Creator, and yet say in their hearts, "There is no God."

NOTES.—**Jupiter**, the largest planet of the solar system, and, next to Venus, the brightest. **Pleiades** (pro. plē′ya-dēz), a group of seven small stars in the constellation of Taurus.

Lyra, Androm′eda, two brilliant constellations in the northern part of the heavens. **Pointers,** two stars of the group called the Dipper, in the Great Bear. These stars and the Polar Star are nearly in the same straight line.

Blue Hills, hills about seven hundred feet high, southwest of Boston, Massachusetts.

Magians, Persian worshipers of fire and the sun, as representatives of the Supreme Being.

IV. DESCRIPTION OF A STORM.

Benjamin Disraeli, 1805–1881, was of Jewish descent. His ancestors were driven out of Spain by the Inquisition, and went to Venice. In 1748, his grandfather came to England. His father was Isaac Disraeli, well known as a literary man. Benjamin was born in London, and received his early education under his father. He afterwards studied for a lawyer, but soon gave up his profession for literature. His first novel, "Vivian Grey," appeared when the author was twenty-one years of age; it received much attention. After several defeats he succeeded in an election to Parliament, and took his seat in that body, in the first year of Victoria's reign. On his first attempt to speak in Parliament, the House refused to hear him. It is said that, as he sat down, he remarked that the time would come when they would hear him. In 1849, he became the leader of the Conservative party in the House. During the administration of W. E. Gladstone, Mr. Disraeli was leader of the opposition. In 1868, he became prime minister, holding the office for a short time. In 1874, he was again appointed to the same office, where he remained until 1880. His wife was made Viscountess of Beaconsfield in 1868. After her death, the title of Earl of Beaconsfield was conferred on Disraeli. He ranked among the most eminent statesmen of the age, but always devoted a portion of his time to literature. "Lothair," a novel, was published in 1870.

* * * THEY looked round on every side, and hope gave way before the scene of desolation. Immense branches were shivered from the largest trees; small ones were entirely stripped of their leaves; the long grass was bowed to the earth; the waters were whirled in eddies out of the little rivulets; birds, leaving their nests to seek shelter in the crevices of the rocks, unable to stem the driving air, flapped their wings and fell upon the earth; the frightened animals of the plain, almost suffocated by the impetuosity

of the wind, sought safety and found destruction; some of the largest trees were torn up by the roots; the sluices of the mountains were filled, and innumerable torrents rushed down the before empty gullies. The heavens now open, and the lightning and thunder contend with the horrors of the wind.

In a moment, all was again hushed. Dead silence succeeded the bellow of the thunder, the roar of the wind, the rush of the waters, the moaning of the beasts, the screaming of the birds. Nothing was heard save the plash of the agitated lake, as it beat up against the black rocks which girt it in.

Again, greater darkness enveloped the trembling earth. Anon, the heavens were rent with lightning, which nothing could have quenched but the descending deluge. Cataracts poured down from the lowering firmament. For an instant, the horses dashed madly forward; beast and rider blinded and stifled by the gushing rain, and gasping for breath. Shelter was nowhere. The quivering beasts reared, and snorted, and sank upon their knees, dismounting their riders.

He had scarcely spoken, when there burst forth a terrific noise, they knew not what; a rush, they could not understand; a vibration which shook them on their horses. Every terror sank before the roar of the cataract. It seemed that the mighty mountain, unable to support its weight of waters, shook to the foundation. A lake had burst upon its summit, and the cataract became a falling ocean. The source of the great deep appeared to be discharging itself over the range of mountains; the great gray peak tottered on its foundation!—It shook!—it fell! and buried in its ruins the castle, the village, and the bridge!

V. AFTER THE THUNDERSTORM.

James Thomson, 1700-1748, the son of a clergyman, was born in Scotland. He studied at the University of Edinburgh, and intended to follow the profession of his father, but never entered upon the duties of the sacred office. In 1724 he went to London, where he spent most of his subsequent life. He had shown some poetical talent when a boy; and, in 1826, he published "Winter," a part of a longer poem, entitled "The Seasons," the best known of all his works. He also wrote several plays for the stage; none of them, however, achieved any great success. In the last year of his life, he published his "Castle of Indolence,' the most famous of his works excepting "The Seasons." Thomson was heavy and dull in his personal appearance, and was indolent in his habits. The moral tone of his writings is always good. This extract is from "The Seasons."

As from the face of heaven the shattered clouds
Tumultuous rove, the interminable sky
Sublimer swells, and o'er the world expands
A purer azure.

Through the lightened air
A higher luster and a clearer calm,
Diffusive, tremble; while, as if in sign
Of danger past, a glittering robe of joy,
Set off abundant by the yellow ray,
Invests the fields; and nature smiles revived.

'T is beauty all, and grateful song around,
Joined to the low of kine, and numerous bleat
Of flocks thick-nibbling through the clovered vale:
And shall the hymn be marred by thankless man,
Most favored; who, with voice articulate,
Should lead the chorus of this lower world?

Shall man, so soon forgetful of the Hand
That hushed the thunder, and serenes the sky,
Extinguished feel that spark the tempest waked,
That sense of powers exceeding far his own,
Ere yet his feeble heart has lost its fears?

VI. HOUSE CLEANING.

Francis Hopkinson, 1737-1791. He was the son of an Englishman; was born in Philadelphia, and was educated at the college of that city, now the University of Pennsylvania. He represented New Jersey in the Congress of 1776, and was one of the signers of the Declaration of Independence. He was one of the most sensible and elegant writers of his time, and distinguished himself both in prose and verse. His lighter writings abound in humor and keen satire; his more solid writings are marked by clearness and good sense. His pen did much to forward the cause of American independence. His "Essay on Whitewashing," from which the following extract is taken, was mistaken for the composition of Dr. Franklin, and published among his writings. It was originally in the form of "A Letter from a Gentleman in America to his Friend in Europe, on Whitewashing."

THERE is no season of the year in which the lady may not, if she pleases, claim her privilege; but the latter end of May is generally fixed upon for the purpose. The attentive husband may judge, by certain prognostics, when the storm is at hand. If the lady grows uncommonly fretful, finds fault with the servants, is discontented with the children, and complains much of the nastiness of everything about her, these are symptoms which ought not to be neglected, yet they sometimes go off without any further effect.

But if, when the husband rises in the morning, he should observe in the yard a wheelbarrow with a quantity of lime in it, or should see certain buckets filled with a solution of lime in water, there is no time for hesitation. He immediately locks up the apartment or closet where his papers and private property are kept, and, putting the key into his pocket, betakes himself to flight. A husband, however beloved, becomes a perfect nuisance during this season of female rage. His authority is superseded, his commission suspended, and the very scullion who cleans the brasses in the kitchen becomes of more importance than he. He has nothing for it but to abdicate for a time, and run from an evil which he can neither prevent nor mollify.

The husband gone, the ceremony begins. The walls are stripped of their furniture—paintings, prints, and looking-glasses lie huddled in heaps about the floors; the curtains are torn from their testers, the beds crammed into windows, chairs and tables, bedsteads and cradles, crowd the yard, and the garden fence bends beneath the weight of carpets, blankets, cloth cloaks, old coats, under petticoats, and ragged breeches. Here may be seen the lumber of the kitchen, forming a dark and confused mass for the foreground of the picture; gridirons and frying pans, rusty shovels and broken tongs, joint stools, and the fractured remains of rush-bottomed chairs. There a closet has disgorged its bowels—riveted plates and dishes, halves of china bowls, cracked tumblers, broken wineglasses, phials of forgotten physic, papers of unknown powders, seeds and dried herbs, tops of teapots, and stoppers of departed decanters—from the rag hole in the garret, to the rat hole in the cellar, no place escapes unrummaged. It would seem as if the day of general doom had come, and the utensils of the house were dragged forth to judgment.

In this tempest, the words of King Lear unavoidably present themselves, and might, with little alteration, be made strictly applicable.

> "Let the great gods,
> That keep this dreadful pother o'er our heads,
> Find out their enemies now. Tremble, thou wretch,
> That hast within thee undivulged crimes
> Unwhipp'd of justice.
> Close pent-up guilts,
> Rive your concealing continents, and cry
> These dreadful summoners grace."

This ceremony completed, and the house thoroughly evacuated, the next operation is to smear the walls and ceilings with brushes dipped into a solution of lime, called whitewash; to pour buckets of water over every floor; and scratch all the partitions and wainscots with hard brushes, charged with soft soap and stonecutters' sand.

The windows by no means escape the general deluge A servant scrambles out upon the penthouse, at the risk of her neck, and, with a mug in her hand and a bucket within reach, dashes innumerable gallons of water against the glass panes, to the great annoyance of passengers in the street.

I have been told that an action at law was once brought against one of these water nymphs, by a person who had a new suit of clothes spoiled by this operation: but after long argument, it was determined that no damages could be awarded; inasmuch as the defendant was in the exercise of a legal right, and not answerable for the consequences. And so the poor gentleman was doubly nonsuited; for he lost both his suit of clothes and his suit at law.

These smearings and scratchings, these washings and dashings, being duly performed, the next ceremonial is to cleanse and replace the distracted furniture. You may have seen a house raising, or a ship launch—recollect, if you can, the hurry, bustle, confusion, and noise of such a scene, and you will have some idea of this cleansing match. The misfortune is, that the sole object is to make things *clean.* It matters not how many useful, ornamental, or valuable articles suffer mutilation or death under the operation. A mahogany chair and a carved frame undergo the same discipline; they are to be made *clean* at all events; but their preservation is not worthy of attention.

For instance: a fine large engraving is laid flat upon the floor; a number of smaller prints are piled upon it, until the superincumbent weight cracks the lower glass—but this is of no importance. A valuable picture is placed leaning against the sharp corner of a table; others are made to lean against that, till the pressure of the whole forces the corner of the table through the canvas of the first. The frame and glass of a fine print are to be cleaned; the spirit and oil used on this occasion are suffered to leak through and deface the engraving—no matter. If the glass is clean and the frame shines, it is sufficient—the rest is not worthy of consideration. An able arithmetician hath made a calculation, founded on long experience, and proved that the losses and destruction incident to two whitewashings are equal to one removal, and three removals equal to one fire.

This cleansing frolic over, matters begin to resume their pristine appearance: the storm abates, and all would be well again; but it is impossible that so great a convulsion in so small a community should pass over without producing some consequences. For two or three weeks after the operation, the family are usually afflicted with sore eyes, sore throats, or severe colds, occasioned by exhalations from wet floors and damp walls.

I know a gentleman here who is fond of accounting for everything in a philosophical way. He considers this, what I call a *custom*, as a real periodical *disease* peculiar to the climate. His train of reasoning is whimsical and ingenious, but I am not at leisure to give you the detail. The result was, that he found the distemper to be incurable; but after much study, he thought he had discovered a method to divert the evil he could not subdue. For this purpose, he caused a small building, about twelve feet square, to be erected in his garden, and furnished with some ordinary chairs and tables, and a few prints of the cheapest sort. His hope was, that when the whitewashing frenzy seized the females of his family, they might repair to this apartment, and scrub, and scour, and smear to their hearts' content; and so spend the violence of the disease in this outpost, whilst he enjoyed himself in quiet at headquarters. But the experiment did not answer his expectation. It was impossible it should, since a principal part of the gratification consists in the lady's having an uncontrolled right to torment her husband at least once in every year; to turn him out of doors, and take the reins of government into her own hands.

There is a much better contrivance than this of the philosopher's; which is, to cover the walls of the house with paper. This is generally done. And though it does not abolish, it at least shortens the period of female dominion. This paper is decorated with various fancies; and made so ornamental that the women have admitted the fashion without perceiving the design.

There is also another alleviation to the husband's distress. He generally has the sole use of a small room or closet for his books and papers, the key of which he is allowed to keep. This is considered as a privileged place, even in the whitewashing season, and stands like the land of Goshen amidst the plagues of Egypt. But then he must be extremely cautious, and ever upon his guard; for, should he inadvertently go abroad and leave the key in his door, the housemaid, who is always on the watch for such an opportunity, immediately enters in triumph with buckets, brooms, and brushes—takes possession of the premises, and forthwith puts all his books and papers "to rights," to his utter confusion, and sometimes serious detriment.

NOTES.—**Lear.**—The reference is to Shakespeare's tragedy, Act III, Scene 2.

Goshen.—The portion of Egypt settled by Jacob and his family. In the Bible, Exodus viii, 22, Goshen was exempted from the plague of the flies.

The teacher should ascertain that the pupils note the satire and humor of this selection.

This letter was written about a hundred years ago. What word in the first paragraph that would probably not be used by an elegant writer of the present day? Note the words that indicate changes in domestic customs; such as *testers, joint stools, wainscots, house raising.*

VII. SCHEMES OF LIFE OFTEN ILLUSORY.

Samuel Johnson, 1709-1784. This truly remarkable man was the son of a bookseller and stationer; he was born in Lichfield, Staffordshire, England. He entered Pembroke College, Oxford, in 1728; but, at the end of three years, his poverty compelled him to leave without taking his degree. In 1736, he married Mrs. Porter, a widow of little culture, much older than himself, but possessed of some property. The marriage seems to have been a happy one, nevertheless; and, on the death of his wife, in 1752, Johnson mourned for her most sincerely. Soon after his marriage, he opened a private school, but obtained only three pupils, one of whom was David Garrick, afterward the celebrated actor. In 1737, he removed to London, where he lived for most of the remainder of his life. Here he entered upon literary work, in which he continued, and from which he derived his chief support, although at times it was but a meager one. His "Vanity of Human Wishes" was sold for ten guineas. His great Dictionary, the first one of the English language worthy of mention, brought him £1575, and occupied his time for seven years. Most of the money he received for the work went to pay his six amanuenses. The other most famous of his numerous literary works are "The Rambler," "Rasselas," "The Lives of the English Poets," and his edition of Shakespeare. In person, Johnson was heavy and awkward; he was the victim of scrofula in his youth, and of dropsy in his old age. In manner, he was boorish and overbearing; but his great powers and his wisdom caused his company to be sought by many eminent men of his time.

OMAR, the son of Hassan, had passed seventy-five years in honor and prosperity. The favor of three successive caliphs had filled his house with gold and silver; and whenever he appeared, the benedictions of the people proclaimed his passage.

Terrestrial happiness is of short continuance. The brightness of the flame is wasting its fuel; the fragrant flower is passing away in its own odors. The vigor of Omar began to fail; the curls of beauty fell from his head; strength departed from his hands, and agility from his feet. He gave back to the caliph the keys of trust, and the seals of secrecy; and sought no other pleasure for the remainder of life than the converse of the wise and the gratitude of the good.

The powers of his mind were yet unimpaired. His chamber was filled by visitants, eager to catch the dictates of experience, and officious to pay the tribute of admiration. Caleb, the son of the viceroy of Egypt, entered every day early, and retired late. He was beautiful and eloquent: Omar admired his wit, and loved his docility.

"Tell me," said Caleb, "thou to whose voice nations have listened, and whose wisdom is known to the extremities of Asia, tell me, how I may resemble Omar the prudent? The arts by which thou hast gained power and preserved it, are to thee no longer necessary or useful; impart to me the secret of thy conduct, and teach me the plan upon which thy wisdom has built thy fortune."

"Young man," said Omar, "it is of little use to form plans of life. When I took my first survey of the world, in my twentieth year, having considered the various conditions of mankind, in the hour of solitude I said thus to myself, leaning against a cedar which spread its branches over my head: 'Seventy years are allowed to man; I have yet fifty remaining.

"'Ten years I will allot to the attainment of knowledge, and ten I will pass in foreign countries; I shall be learned, and therefore I shall be honored; every city will shout at my arrival, and every student will solicit my friendship. Twenty years thus passed will store my mind with images which I shall be busy through the rest of my life in combining and comparing. I shall revel in inexhaustible accumulations of intellectual riches; I shall find new pleasures for every moment, and shall never more be weary of myself.

"'I will not, however, deviate too far from the beaten track of life; but will try what can be found in female delicacy. I will marry a wife as beautiful as the houries, and wise as Zobeide; and with her I will live twenty years within the suburbs of Bagdad, in every pleasure that wealth can purchase, and fancy can invent.

"'I will then retire to a rural dwelling, pass my days in obscurity and contemplation; and lie silently down on the bed of death. Through my life it shall be my settled resolution, that I will never depend on the smile of princes; that I will never stand exposed to the artifices of courts; I will never pant for public honors, nor disturb my quiet with the affairs of state.' Such was my scheme of life, which I impressed indelibly upon my memory.

"The first part of my ensuing time was to be spent in search of knowledge, and I know not how I was diverted from my design. I had no visible impediments without, nor any ungovernable passion within. I regarded knowledge as the highest honor, and the most engaging pleasure; yet day stole upon day, and month glided after month, till I found that seven years of the first ten had vanished, and left nothing behind them.

"I now postponed my purpose of traveling; for why should I go abroad, while so much remained to be learned at home? I immured myself for four years, and studied the laws of the empire. The fame of my skill reached the judges: I was found able to speak upon doubtful questions, and I was commanded to stand at the footstool of the caliph. I was heard with attention; I was consulted with confidence, and the love of praise fastened on my heart.

"I still wished to see distant countries; listened with rapture to the relations of travelers, and resolved some time to ask my dismission, that I might feast my soul with novelty; but my presence was always necessary, and the stream of business hurried me along. Sometimes, I was afraid lest I should be charged with ingratitude; but I still proposed to travel, and therefore would not confine myself by marriage.

"In my fiftieth year, I began to suspect that the time of my traveling was past; and thought it best to lay hold on the felicity yet in my power, and indulge myself in domestic pleasures. But, at fifty, no man easily finds a woman beautiful as the houries, and wise as Zobeide. I inquired and rejected, consulted and deliberated, till the sixty-second year made me ashamed of wishing to marry I had now nothing left but retirement; and for retirement I never found a time, till disease forced me from public employment.

"Such was my scheme, and such has been its consequence. With an insatiable thirst for knowledge, I trifled away the years of improvement; with a restless desire of seeing different countries, I have always resided in the same city; with the highest expectation of connubial felicity, I have lived unmarried; and with an unalterable resolution of contemplative retirement, I am going to die within the walls of Bagdad."

NOTES.—**Bag dȧd.′**—A large city of Asiatic Turkey, on the river Tigris. In the ninth century, it was the greatest center of Moslem power and learning.

Zobeide (Zo-bād′).—A lady of Bagdad, whose story is given in the "Three Calendars" of the "Arabian Nights."

In this selection the form of an allegory is used to express a general truth.

VIII. THE BRAVE OLD OAK.

Henry Fothergill Chorley, 1808-1872. He is known chiefly as a musical critic and author; for thirty-eight years he was connected with the "London Athenæum." His books are mostly novels.

A SONG to the oak, the brave old oak,
Who hath ruled in the greenwood long;
Here 's health and renown to his broad green crown,
And his fifty arms so strong.
There 's fear in his frown, when the sun goes down,
And the fire in the west fades out;
And he showeth his might on a wild midnight,
When the storms through his branches shout.

In the days of old, when the spring with cold
  Had brightened his branches gray,
Through the grass at his feet, crept maidens sweet,
  To gather the dews of May.
And on that day, to the rebec gay
  They frolicked with lovesome swains;
They are gone, they are dead, in the churchyard laid,
  But the tree—it still remains.

He saw rare times when the Christmas chimes
  Were a merry sound to hear,
When the Squire's wide hall and the cottage small
  Were filled with good English cheer.
Now gold hath the sway we all obey,
  And a ruthless king is he;
But he never shall send our ancient friend
  To be tossed on the stormy sea.

Then here's to the oak, the brave old oak,
  Who stands in his pride alone;
And still flourish he, a hale green tree,
  When a hundred years are gone.

IX. THE ARTIST SURPRISED.

It may not be known to all the admirers of the genius of Albert Dürer, that that famous engraver was endowed with a "better half," so peevish in temper, that she was the torment not only of his own life, but also of his pupils and domestics. Some of the former were cunning enough to purchase peace for themselves by conciliating the common tyrant, but woe to those unwilling or unable to offer aught in propitiation. Even the wiser ones were spared only by having their offenses visited upon a scapegoat.

This unfortunate individual was Samuel Duhobret, a disciple whom Dürer had admitted into his school out of charity. He was employed in painting signs and the coarser tapestry then used in Germany. He was about forty years of age, little, ugly, and humpbacked; he was the butt of every ill joke among his fellow disciples, and was picked out as an object of especial dislike by Madame Dürer. But he bore all with patience, and ate, without complaint, the scanty crusts given him every day for dinner, while his companions often fared sumptuously.

Poor Samuel had not a spice of envy or malice in his heart. He would, at any time, have toiled half the night to assist or serve those who were wont oftenest to laugh at him, or abuse him loudest for his stupidity. True, he had not the qualities of social humor or wit, but he was an example of indefatigable industry. He came to his studies every morning at daybreak, and remained at work until sunset. Then he retired into his lonely chamber, and wrought for his own amusement.

Duhobret labored three years in this way, giving himself no time for exercise or recreation. He said nothing to a single human being of the paintings he had produced in the solitude of his cell, by the light of his lamp. But his bodily energies wasted and declined under incessant toil. There was none sufficiently interested in the poor artist, to mark the feverish hue of his wrinkled cheek, or the increasing attenuation of his misshapen frame.

None observed that the uninviting pittance set aside for his midday repast, remained for several days untouched. Samuel made his appearance regularly as ever, and bore with the same meekness the gibes of his fellow-pupils, or the taunts of Madame Dürer, and worked with the same untiring assiduity, though his hands would sometimes tremble, and his eyes become suffused, a weakness probably owing to the excessive use he had made of them.

One morning, Duhobret was missing at the scene of his daily labors. His absence created much remark, and many were the jokes passed upon the occasion. One surmised this, and another that, as the cause of the phenomenon; and it was finally agreed that the poor fellow must have worked himself into an absolute skeleton, and taken his final stand in the glass frame of some apothecary, or been blown away by a puff of wind, while his door happened to stand open. No one thought of going to his lodgings to look after him or his remains.

Meanwhile, the object of their mirth was tossing on a bed of sickness. Disease, which had been slowly sapping the foundations of his strength, burned in every vein; his eyes rolled and flashed in delirium; his lips, usually so silent, muttered wild and incoherent words. In his days of health, poor Duhobret had his dreams, as all artists, rich or poor, will sometimes have. He had thought that the fruit of many years' labor, disposed of to advantage, might procure him enough to live, in an economical way, for the rest of his life. He never anticipated fame or fortune; the height of his ambition or hope was, to possess a tenement large enough to shelter him from the inclemencies of the weather, with means enough to purchase one comfortable meal per day.

Now, alas! however, even that one hope had deserted him. He thought himself dying, and thought it hard to die without one to look kindly upon him, without the words of comfort that might soothe his passage to another world. He fancied his bed surrounded by fiendish faces, grinning at his sufferings, and taunting his inability to summon power to disperse them. At length the apparition faded away, and the patient sunk into an exhausted slumber.

He awoke unrefreshed; it was the fifth day he had lain there neglected. His mouth was parched; he turned over, and feebly stretched out his hand toward the earthen pitcher, from which, since the first day of his illness, he had quenched his thirst. Alas! it was empty! Samuel lay for a few moments thinking what he should do. He knew he must die of want if he remained there alone; but to whom could he apply for aid?

An idea seemed, at last, to strike him. He arose slowly, and with difficulty, from the bed, went to the other side of the room, and took up the picture he had painted last. He resolved to carry it to the shop of a salesman, and hoped to obtain for it sufficient to furnish him with the necessaries of life for a week longer. Despair lent him strength to walk, and to carry his burden. On his way, he passed a house, about which there was a crowd. He drew nigh, asked what was going on, and received for an answer, that there was to be a sale of many specimens of art, collected by an amateur in the course of thirty years. It has often happened that collections made with infinite pains by the proprietor, have been sold without mercy or discrimination after his death.

Something whispered to the weary Duhobret, that here would be the market for his picture. It was a long way yet to the house of the picture dealer, and he made up his mind at once. He worked his way through the crowd, dragged himself up the steps, and, after many inquiries, found the auctioneer. That personage was a busy man, with a handful of papers; he was inclined to notice somewhat roughly the interruption of the lean, sallow hunchback, imploring as were his gesture and language.

"What do you call your picture?" at length, said he, carefully looking at it.

"It is a view of the Abbey of Newburg, with its village and the surrounding landscape," replied the eager and trembling artist.

The auctioneer again scanned it contemptuously, and asked what it was worth. "Oh, that is what you please; whatever it will bring," answered Duhobret.

"Hem! it is too odd to please, I should think; I can promise you no more than three thalers."

Poor Samuel sighed deeply. He had spent on that piece the nights of many months. But he was starving now; and the pitiful sum offered would give bread for a few days. He nodded his head to the auctioneer, and retiring took his seat in a corner.

The sale began. After some paintings and engravings had been disposed of, Samuel's was exhibited. "Who bids at three thalers? Who bids?" was the cry. Duhobret listened eagerly, but none answered. "Will it find a purchaser?" said he despondingly, to himself. Still there was a dead silence. He dared not look up; for it seemed to him that all the people were laughing at the folly of the artist, who could be insane enough to offer so worthless a piece at a public sale.

"What will become of me?" was his mental inquiry. "That work is certainly my best;" and he ventured to steal another glance. "Does it not seem that the wind actually stirs those boughs and moves those leaves! How transparent is the water! What life breathes in the animals that quench their thirst at that spring! How that steeple shines! How beautiful are those clustering trees!" This was the last expiring throb of an artist's vanity. The ominous silence continued, and Samuel, sick at heart, buried his face in his hands.

"Twenty-one thalers!" murmured a faint voice, just as the auctioneer was about to knock down the picture. The stupefied painter gave a start of joy. He raised his head and looked to see from whose lips those blessed words had come. It was the picture dealer, to whom he had first thought of applying.

"Fifty thalers," cried a sonorous voice. This time a tall man in black was the speaker. There was a silence of hushed expectation. "One hundred thalers," at length thundered the picture dealer.

"Three hundred!" "Five hundred!" "One thousand!" Another profound silence, and the crowd pressed around the two opponents, who stood opposite each other with eager and angry looks.

"Two thousand thalers!" cried the picture dealer, and glanced around him triumphantly, when he saw his adversary hesitate. "Ten thousand!" vociferated the tall man, his face crimson with rage, and his hands clinched convulsively. The dealer grew paler; his frame shook with agitation; he made two or three efforts, and at last cried out "Twenty thousand!"

His tall opponent was not to be vanquished. He bid forty thousand. The dealer stopped; the other laughed a low laugh of insolent triumph, and a murmur of admiration was heard in the crowd. It was too much for the dealer; he felt his peace was at stake. "Fifty thousand!" exclaimed he in desperation. It was the tall man's turn to hesitate. Again the whole crowd were breathless. At length, tossing his arms in defiance, he shouted "One hundred thousand!" The crestfallen picture dealer withdrew; the tall man victoriously bore away the prize.

How was it, meanwhile, with Duhobret, while this exciting scene was going on? He was hardly master of his senses. He rubbed his eyes repeatedly, and murmured to himself, "After such a dream, my misery will seem more cruel!" When the contest ceased, he rose up bewildered, and went about asking first one, then another, the price of the picture just sold. It seemed that his apprehension could not at once be enlarged to so vast a conception.

The possessor was proceeding homeward, when a decrepit, lame, and humpbacked invalid, tottering along by the aid of a stick, presented himself before him. He threw him a piece of money, and waved his hand as dispensing with his thanks. "May it please your honor," said the supposed beggar, "I am the painter of that picture!" and again he rubbed his eyes.

The tall man was Count Dunkelsback, one of the richest noblemen in Germany. He stopped, took out his pocket-book, tore out a leaf, and wrote on it a few lines. "Take it, friend," said he; "it is a check for your money. Adieu."

Duhobret finally persuaded himself that it was not a dream. He became the master of a castle, sold it, and resolved to live luxuriously for the rest of his life, and to cultivate painting as a pastime. But, alas, for the vanity of human expectation! He had borne privation and toil; prosperity was too much for him, as was proved soon after, when an indigestion carried him off. His picture remained long in the cabinet of Count Dunkelsback, and afterward passed into the possession of the King of Bavaria.

NOTES.—**Albert Dürer** (b. 1471, d. 1528) lived at Nuremburg, Germany. He was eminent as a painter, and as an engraver on copper and wood. He was one of the first artists who studied anatomy and perspective. His influence on art is clearly felt even at the present day.

Newburg, or Neuburg, is on the Danube, fifty miles south of Nuremburg. Bergen Abbey was north of the village.

X. PICTURES OF MEMORY.

Alice Cary, 1820-1871, was born near Cincinnati. One of her ancestors was among the "Pilgrim Fathers," and the first instructor of Latin at Plymouth, Mass. Miss Cary commenced her literary career at her western home, and, in 1849, published a volume of poems, the joint work of her younger sister, Phœbe, and herself. In 1850, she moved to New York. Two of her sisters joined her there, and they supported themselves by their literary labor. Their home became a noted resort for their literary and artistic friends. Miss Cary was the author of eleven volumes, besides many articles contributed to periodicals. Her poetry is marked with great sweetness and pathos. Some of her prose works are much admired, especially her "Clovernook Children."

AMONG the beautiful pictures
That hang on Memory's wall,

Is one of a dim old forest,
  That seemeth best of all;
Not for its gnarled oaks olden,
  Dark with the mistletoe;
Not for the violets golden,
  That sprinkle the vale below;
Not for the milk-white lilies,
  That lean from the fragrant hedge,
Coquetting all day with the sunbeams,
  And stealing their golden edge;
Not for the vines on the upland,
  Where the bright red berries rest,
Nor the pinks, nor the pale, sweet cowslip,
  It seemeth to me the best.

I once had a little brother,
  With eyes that were dark and deep;
In the lap of that dim old forest,
  He lieth in peace asleep:
Light as the down of the thistle,
  Free as the winds that blow,
We roved there the beautiful summers,
  The summers of long ago;
But his feet on the hills grew weary,
  And, one of the autumn eves,
I made for my little brother,
  A bed of the yellow leaves.

Sweetly his pale arms folded
  My neck in a meek embrace,
As the light of immortal beauty
  Silently covered his face;
And when the arrows of sunset
  Lodged in the tree tops bright,
He fell, in his saintlike beauty,
  Asleep by the gates of light.

Therefore, of all the pictures
  That hang on Memory's wall,
The one of the dim old forest
  Seemeth the best of all.

XI. THE MORNING ORATORIO.

Wilson Flagg, 1805-1884, was born in Beverly, Mass. He pursued his academical course in Andover, at Phillips Academy, and entered Harvard College, but did not graduate. His chief works are: "Studies in the Field and Forest," "The Woods and Byways of New England," and "The Birds and Seasons of New England."

NATURE, for the delight of waking eyes, has arrayed the morning heavens in the loveliest hues of beauty. Fearing to dazzle by an excess of delight, she first announces day by a faint and glimmering twilight, then sheds a purple tint over the brows of the rising morn, and infuses a transparent ruddiness throughout the atmosphere. As daylight widens, successive groups of mottled and rosy-bosomed clouds assemble on the gilded sphere, and, crowned with wreaths of fickle rainbows, spread a mirrored flush over hill, grove, and lake, and every village spire is burnished with their splendor.

At length, through crimsoned vapors, we behold the sun's broad disk, rising with a countenance so serene that every eye may view him ere he arrays himself in his meridian brightness. Not many people who live in towns are aware of the pleasure attending a ramble near the woods and orchards at daybreak in the early part of summer. The drowsiness we feel on rising from our beds is gradually dispelled by the clear and healthful breezes of early day, and we soon experience an unusual amount of vigor and elasticity.

During the night, the stillness of all things is the circumstance that most powerfully attracts our notice, rendering us peculiarly sensitive to every accidental sound that meets the ear. In the morning, at this time of year, on the contrary, we are overpowered by the vocal and multitudinous chorus of the feathered tribe. If you would hear the commencement of this grand anthem of nature, you must rise at the very first appearance of dawn, before the twilight has formed a complete semicircle above the eastern porch of heaven.

The first note that proceeds from the little warbling host, is the shrill chirp of the hairbird,—occasionally vocal at all hours on a warm summer night. This strain, which is a continued trilling sound, is repeated with diminishing intervals, until it becomes almost incessant. But ere the hairbird has uttered many notes, a single robin begins to warble from a neighboring orchard, soon followed by others, increasing in numbers until, by the time the eastern sky is flushed with crimson, every male robin in the country round is singing with fervor.

It would be difficult to note the exact order in which the different birds successively begin their parts in this performance; but the bluebird, whose song is only a short, mellow warble, is heard nearly at the same time with the robin, and the song sparrow joins them soon after with his brief but finely modulated strain. The different species follow rapidly, one after another, in the chorus, until the whole welkin rings with their matin hymn of gladness.

I have often wondered that the almost simultaneous utterance of so many different notes should produce no discords, and that they should result in such complete harmony. In this multitudinous confusion of voices, no two notes are confounded, and none has sufficient duration to grate harshly with a dissimilar sound. Though each performer sings only a few strains and then makes a pause, the whole multitude succeed one another with such rapidity that we hear an uninterrupted flow of music until the broad light of day invites them to other employments.

When there is just light enough to distinguish the birds,

we may observe, here and there, a single swallow perched on the roof of a barn or shed, repeating two twittering notes incessantly, with a quick turn and a hop at every note he utters. It would seem to be the design of the bird to attract the attention of his mate, and this motion seems to be made to assist her in discovering his position. As soon as the light has tempted him to fly abroad, this twittering strain is uttered more like a continued song, as he flits rapidly through the air.

But at this later moment the purple martins have commenced their more melodious chattering, so loud as to attract for a while the most of our attention. There is not a sound in nature so cheering and animating as the song of the purple martin, and none so well calculated to drive away melancholy. Though not one of the earliest voices to be heard, the chorus is perceptibly more loud and effective when this bird has united with the choir.

When the flush of the morning has brightened into vermilion, and the place from which the sun is soon to emerge has attained a dazzling brilliancy, the robins are already less tuneful. They are now becoming busy in collecting food for their morning repast, and one by one they leave the trees, and may be seen hopping upon the tilled ground, in quest of the worms and insects that have crept out during the night from their subterranean retreats.

But as the robins grow silent, the bobolinks begin their vocal revelries; and to a fanciful mind it might seem that the robins had gradually resigned their part in the performance to the bobolinks, not one of which is heard until some of the former have concluded their songs. The little hairbird still continues his almost incessant chirping, the first to begin and the last to quit the performance. Though the voice of this bird is not very sweetly modulated, it blends harmoniously with the notes of other birds, and greatly increases the charming effect of the combination.

It would be tedious to name all the birds that take part

in this chorus; but we must not omit the pewee, with his melancholy ditty, occasionally heard like a short minor strain in an oratorio; nor the oriole, who is really one of the chief performers, and who, as his bright plumage flashes upon the sight, warbles forth a few notes so clear and mellow as to be heard above every other sound. Adding a pleasing variety to all this harmony, the lisping notes of the meadowlark, uttered in a shrill tone, and with a peculiar pensive modulation, are plainly audible, with short rests between each repetition.

There is a little brown sparrow, resembling the hairbird, save a general tint of russet in his plumage, that may be heard distinctly among the warbling host. He is rarely seen in cultivated grounds, but frequents the wild pastures, and is the bird that warbles so sweetly at midsummer, when the whortleberries are ripe, and the fields are beautifully spangled with red lilies.

There is no confusion in the notes of his song, which consists of one syllable rapidly repeated, but increasing in rapidity and rising to a higher key towards the conclusion. He sometimes prolongs his strain, when his notes are observed to rise and fall in succession. These plaintive and expressive notes are very loud and constantly uttered, during the hour that precedes the rising of the sun. A dozen warblers of this species, singing in concert, and distributed in different parts of the field, form, perhaps, the most delightful part of the woodland oratorio to which we have listened.

At sunrise hardly a robin can be heard in the whole neighborhood, and the character of the performance has completely changed during the last half hour. The first part was more melodious and tranquilizing, the last is more brilliant and animating. The grass finches, the vireos, the wrens, and the linnets have joined their voices to the chorus, and the bobolinks are loudest in their song. But the notes of the birds in general are not so incessant as before sunrise. One by one they discontinue their **lays**, until at high noon the bobolink and the warbling fly-catcher are almost the only vocalists to be heard in the fields.

XII. SHORT SELECTIONS IN POETRY.

I. THE CLOUD.

A CLOUD lay cradled near the setting sun,
  A gleam of crimson tinged its braided snow;
Long had I watched the glory moving on,
  O'er the still radiance of the lake below:
  Tranquil its spirit seemed, and floated slow,
E'en in its very motion there was rest,
  While every breath of eve that chanced to blow,
Wafted the traveler to the beauteous west.
Emblem, methought, of the departed soul,
  To whose white robe the gleam of bliss is given,
And by the breath of mercy made to roll
  Right onward to the golden gate of heaven,
While to the eye of faith it peaceful lies,
And tells to man his glorious destinies.

John Wilson.

II. MY MIND.

My mind to me a kingdom is;
  Such perfect joy therein I find,
As far exceeds all earthly bliss
  That God or nature hath assigned;
Though much I want that most would have,
Yet still my mind forbids to crave.

NOTE.—This is the first stanza of a poem by William Byrd (b. 1543, d. 1623), an English composer of music.

III. A GOOD NAME.

Good name, in man or woman, dear my lord,
Is the immediate jewel of their souls.
Who steals my purse, steals trash; 't is something, nothing;
'T was mine, 't is his, and has been slave to thousands;
But he that filches from me my good name,
Robs me of that which not enriches him,
And makes me poor indeed.

Shakespeare.—Othello, Act III, Scene III.

IV. SUNRISE.

But yonder comes the powerful king of day,
Rejoicing in the east. The lessening cloud,
The kindling azure, and the mountain's brow
Illumed with liquid gold, his near approach
Betoken glad. Lo! now apparent all,
Aslant the dew-bright earth and colored air
He looks in boundless majesty abroad,
And sheds the shining day that, burnished, plays
On rocks, and hills, and towers, and wandering streams,
High gleaming from afar.

Thomson.

V. OLD AGE AND DEATH.

Edmund Waller, 1605-1687, an English poet, was a cousin of John Hampden, and related to Oliver Cromwell. He was educated at Eton and Cambridge. Waller was for many years a member of Parliament. He took part in the civil war, and was detected in a treasonable plot. Several years of his life were spent in exile in France. After the Restoration he came into favor at court. His poetry is celebrated for smoothness and sweetness, but is disfigured by affected conceits.

The seas are quiet when the winds give o'er;
So calm are we when passions are no more.
For then we know how vain it was to boast
Of fleeting things, too certain to be lost.
Clouds of affection from our younger eyes
Conceal that emptiness which age descries.

The soul's dark cottage, battered and decayed,
Lets in new light through chinks that time has made:
Stronger by weakness, wiser men become,
As they draw near to their eternal home.
Leaving the old, both worlds at once they view,
That stand upon the threshold of the new.

VI. MILTON.

John Dryden, 1631-1703, was a noted English writer, who was made poet laureate by James II. On the expulsion of James, and the accession of William and Mary, Dryden lost his offices and pensions, and was compelled to earn his bread by literary work. It was during these last years of his life that his best work was done. His "Ode for St. Cecilia's Day" is one of his most celebrated poems. His prose writings are specimens of good, strong English.

Three poets, in three distant ages born,
Greece, Italy, and England did adorn;
The first in loftiness of thought surpassed,
The next in majesty, in both the last.
The force of nature could no further go;
To make a third she joined the other two.

NOTE.—The two poets referred to, other than Milton, are Homer and Dante.

XIII. DEATH OF LITTLE NELL.

Charles Dickens, 1812-1870, one of the greatest novelists of modern times, was born in Portsmouth, but spent nearly all his life in London. His father was a conscientious man, but lacked capacity for getting a livelihood. In consequence, the boy's youth was much darkened by poverty. It has been supposed that he pictured his father in the character of "Micawber." He began his active life as a lawyer's apprentice; but soon left this employment to become a reporter. This occupation he followed from 1831 to 1836. His first book was entitled "Sketches of London Society, by Boz." This was followed, in 1837, by the "Pickwick Papers," a work which suddenly brought much fame to the author. His other works followed with great rapidity, and his last was unfinished at the time of his death. He was buried in Westminster Abbey. Mr. Dickens visited America in 1842, and again in 1867. During his last visit, he read his works in public, in the principal cities of the United States.

The resources of Dickens's genius seemed exhaustless. He copied no author, imitated none, but relied entirely on his own powers. He excelled especially in humor and pathos. He gathered materials for his works by the most careful and faithful observation. And he painted his characters with a fidelity so true to their different individualities that, although they sometimes have a quaint grotesqueness bordering on caricature, they stand before the memory as living realities. He was particularly successful in the delineation of the joys and griefs of childhood. "Little Nell" and little "Paul Dombey" are known, and have been loved and wept over, in almost every household where the English language is read. His writings present very vividly the wants and sufferings of the poor, and have a tendency to prompt to kindness and benevolence. His works have not escaped criticism. It has been said that "his good characters act from impulse, not from principle," and that he shows "a tricksy spirit of fantastic exaggeration." It has also been said that his novels sometimes lack skillful plot, and that he seems to speak approvingly of conviviality and dissipation. "The Old Curiosity Shop," from which the following extract is taken, was published in 1840.

SHE was dead. No sleep so beautiful and calm, so free from trace of pain, so fair to look upon. She seemed a creature fresh from the hand of God, and waiting for the breath of life; not one who had lived, and suffered death. Her couch was dressed with here and there some winter berries and green leaves, gathered in a spot she had been used to favor. "When I die, put near me something that has loved the light, and had the sky above it always." These were her words.

She was dead. Dear, gentle, patient, noble Nell was dead. Her little bird, a poor, slight thing the pressure of a finger would have crushed, was stirring nimbly in its cage, and the strong heart of its child mistress was mute and motionless forever! Where were the traces of her early cares, her sufferings, and fatigues? All gone. Sorrow was dead, indeed, in her; but peace and perfect happiness were born, imaged in her tranquil beauty and profound repose.

And still her former self lay there, unaltered in this change. Yes! the old fireside had smiled upon that same sweet face; it had passed, like a dream, through haunts of misery and care; at the door of the poor schoolmaster on the summer evening, before the furnace fire upon the cold wet night, at the still bedside of the dying boy, there had been the same mild and lovely look. So shall we know the angels, in their majesty, after death.

The old man held one languid arm in his, and had the small hand tight folded to his breast for warmth. It was the hand she had stretched out to him with her last smile; the hand that had led him on through all their wanderings. Ever and anon he pressed it to his lips; then hugged it to his breast again, murmuring that it was warmer now, and, as he said it, he looked in agony to those who stood around, as if imploring them to help her.

She was dead, and past all help, or need of help. The ancient rooms she had seemed to fill with life, even while her own was waning fast, the garden she had tended, the eyes she had gladdened, the noiseless haunts of many a thoughtful hour, the paths she had trodden, as it were, but yesterday, could know her no more.

"It is not," said the schoolmaster, as he bent down to kiss her on the cheek, and gave his tears free vent, "it is not in *this* world that heaven's justice ends. Think what earth is, compared with the world to which her young spirit has winged its early flight, and say, if one deliberate wish, expressed in solemn tones above this bed, could call her back to life, which of us would utter it?"

She had been dead two days. They were all about her at the time, knowing that the end was drawing on. She died soon after daybreak. They had read and talked to her in the earlier portion of the night; but, as the hours crept on, she sank to sleep. They could tell by what she faintly uttered in her dreams, that they were of her journeyings with the old man; they were of no painful scenes, but of people who had helped them, and used them kindly; for she often said "God bless you!" with great fervor.

Waking, she never wandered in her mind but once, and that was at beautiful music, which, she said, was in the air. God knows. It may have been. Opening her eyes, at last, from a very quiet sleep, she begged that they would kiss her once again. That done, she turned to the old man, with a lovely smile upon her face, such, they said, as they had never seen, and could never forget, and clung, with both her arms, about his neck. She had never murmured or complained; but, with a quiet mind, and manner quite unaltered, save that she every day became more earnest and more grateful to them, faded like the light upon the summer's evening.

The child who had been her little friend, came there, almost as soon as it was day, with an offering of dried flowers, which he begged them to lay upon her breast. He told them of his dream again, and that it was of her being restored to them, just as she used to be. He begged hard to see her: saying, that he would be very quiet, and that they need not fear his being alarmed, for he had sat alone by his young brother all day long, when he was dead, and had felt glad to be so near him. They let him have his wish; and, indeed, he kept his word, and was, in his childish way, a lesson to them all.

Up to that time, the old man had not spoken once, except to her, or stirred from the bedside. But, when he saw her little favorite, he was moved as they had not seen him yet, and made as though he would have him come nearer. Then, pointing to the bed, he burst into tears for the first time, and they who stood by, knowing that the sight of this child had done him good, left them alone together.

Soothing him with his artless talk of her, the child persuaded him to take some rest, to walk abroad, to do almost as he desired him. And, when the day came, on which they must remove her, in her earthly shape, from earthly eyes forever, he led him away, that he might not know when she was taken from him. They were to gather fresh leaves and berries for her bed.

And now the bell, the bell she had so often heard by night and day, and listened to with solemn pleasure, almost as a living voice, rung its remorseless toll for her, so young, so beautiful, so good. Decrepit age, and vigorous life, and blooming youth, and helpless infancy,—on crutches, in the pride of health and strength, in the full blush of promise, in the mere dawn of life, gathered round her. Old men were there, whose eyes were dim and senses failing, grandmothers, who might have died ten years ago, and still been old, the deaf, the blind, the lame, the palsied, the living dead, in many shapes and forms, to see the closing of that early grave.

Along the crowded path they bore her now, pure as the newly fallen snow that covered it, whose day on earth had been as fleeting. Under that porch, where she had sat when heaven, in its mercy, brought her to that peaceful spot, she passed again, and the old church received her in its quiet shade.

XIV. VANITY OF LIFE.

Johann Gottfried von Herder, 1744-1803, an eminent German poet, preacher, and philosopher, was born in Mohrungen, and died in Weimar. His published works comprise sixty volumes. This selection is from his "Hebrew Poetry."

MAN, born of woman,
Is of a few days,
And full of trouble;
He cometh forth as a flower, and is cut down;
He fleeth also as a shadow,
And continueth not.

Upon such dost thou open thine eye,
And bring me unto judgment with thee?
Among the impure is there none pure?
Not one.

Are his days so determined?
Hast thou numbered his months,
And set fast his bounds for him
Which he can never pass?
Turn then from him that he may rest,
And enjoy, as an hireling, his day.

The tree hath hope, if it be cut down,
It becometh green again,
And new shoots are put forth.
If even the root is old in the earth,
And its stock die in the ground,
From vapor of water it will bud,
And bring forth boughs as a young plant.

But man dieth, and his power is gone;
He is taken away, and where is he?

Till the waters waste from the sea,
Till the river faileth and is dry land,
Man lieth low, and riseth not again.
Till the heavens are old, he shall not awake,
Nor be aroused from his sleep.

Oh, that thou wouldest conceal me
In the realm of departed souls!
Hide me in secret, till thy wrath be past;
Appoint me then a new term,
And remember me again.
But alas! if a man die
Shall he live again?

So long, then, as my toil endureth,
Will I wait till a change come to me.
Thou wilt call me, and I shall answer;
Thou wilt pity the work of thy hands.
Though now thou numberest my steps,
Thou shalt then not watch for my sin.
My transgression will be sealed in a bag,
Thou wilt bind up and remove my iniquity.

Yet alas! the mountain falleth and is swallowed up,
The rock is removed out of its place,
The waters hollow out the stones,
The floods overflow the dust of the earth,
And thus, thou destroyest the hope of man.

Thou contendest with him, till he faileth,
Thou changest his countenance, and sendeth him away.
Though his sons become great and happy,
Yet he knoweth it not;
If they come to shame and dishonor,
He perceiveth it not.

Note.—Compare with the translation of the same as given in the ordinary version of the Bible. Job xiv.

XV. A POLITICAL PAUSE.

Charles James Fox, 1749-1806, a famous English orator and statesman, was the son of Hon. Henry Fox, afterward Lord Holland; he was also a lineal descendant of Charles II. of England and of Henry IV. of France. He received his education at Westminster, Eton, and Oxford, but left the University without graduating. He was first elected to Parliament before he was twenty years old. During the American Revolution, he favored the colonies; later, he was a friend and fellow-partisan both with Burke and Wilberforce. Burke said of him, "He is the most brilliant and successful debater the world ever saw." In his later years, Mr. Fox was as remarkable for carelessness in dress and personal appearance, as he had been for the opposite in his youth. He possessed many pleasing traits of character, but his morals were not commendable; he was a gambler and a spendthrift. Yet he exercised a powerful influence on the politics of his times. This extract is from a speech delivered during a truce in the long war between England and France.

"BUT we must pause," says the honorable gentleman. What! must the bowels of Great Britain be torn out, her best blood spilt, her treasures wasted, that you may make an experiment? Put yourselves — Oh! that you *would* put yourselves on the field of battle, and learn to judge of the sort of horrors you excite. In former wars, a man might at least have some feeling, some interest, that served to balance in his mind the impressions which a scene of carnage and death must inflict.

But if a man were present now at the field of slaughter, and were to inquire for what they were fighting — "Fighting!" would be the answer; "they are not *fighting;* they are *pausing.*" "Why is that man expiring? Why is that other writhing with agony? What means this implacable fury?" The answer must be, "You are quite wrong, sir, you deceive yourself,—they are not fighting,—do not disturb them,— they are merely pausing! This man is not expiring with agony,— that man is not dead,— he is only pausing! Bless you, sir, they are not angry with one another; they have now no cause of quarrel; but their country thinks that there should be a pause. All that you see is nothing like fighting,— there is no harm, nor cruelty, nor bloodshed in it; it is nothing more than a political pause. It is merely to try an experiment — to see whether Bonaparte will not behave himself better than heretofore; and, in the meantime, we have agreed to a pause, in pure friendship!"

And is this the way that you are to show yourselves the advocates of order? You take up a system calculated to uncivilize the world, to destroy order, to trample on religion, to stifle in the heart not merely the generosity of noble sentiment, but the affections of social nature; and in the prosecution of this system, you spread terror and devastation all around you.

NOTE.—In this lesson, the influence of a *negative* in determining the rising inflection, is noticeable. See Rule V, p. 24.

XVI. MY EXPERIENCE IN ELOCUTION.

John Neal, 1793-1876, a brilliant but eccentric American writer, was born in Portland, Maine. He went into business, when quite young, in company with John Pierpont, the well-known poet. They soon failed, and Mr. Neal then turned his attention to the study of law. He practiced his profession somewhat, but devoted most of his time to literature. For a time he resided in England, where he wrote for "Blackwood's Magazine" and other periodicals. His writings were produced with great rapidity, and with a purposed disregard of what is known as "classical English."

IN the academy I attended, elocution was taught in a way I shall never forget—never! We had a yearly exhibition, and the favorites of the preceptor were allowed to speak a piece; and a pretty time they had of it. Somehow I was never a favorite with any of my teachers after the first two or three days; and, as I went barefooted, I dare say it was thought unseemly, or perhaps cruel, to expose me upon the platform. And then, as I had no particular aptitude for public speaking, and no relish for what was called oratory, it was never my luck to be called up.

Among my schoolmates, however, was one—a very amiable, shy boy—to whom was assigned, at the first exhibition I attended, that passage in Pope's Homer beginning with,

"Aurora, now, fair daughter of the dawn!"

This the poor boy gave with so much emphasis and discretion, that, to me, it sounded like "O roarer!" and I was wicked enough, out of sheer envy, I dare say, to call him

"O roarer!"—a nickname which clung to him for a long while, though no human being ever deserved it less; for in speech and action both, he was quiet, reserved, and sensitive.

My next experience in elocution was still more disheartening, so that I never had a chance of showing what I was capable of in that way till I set up for myself. Master Moody, my next instructor, was thought to have uncommon qualifications for teaching oratory. He was a large, handsome, heavy man, over six feet high; and having understood that the first, second, and third prerequisite in oratory was *action*, the boys he put in training were encouraged to most vehement and obstreperous manifestations. Let me give an example, and one that weighed heavily on my conscience for many years after the poor man passed away.

Among his pupils were two boys, brothers, who were thought highly gifted in elocution. The master, who was evidently of that opinion, had a habit of parading them on all occasions before visitors and strangers; though one had lost his upper front teeth and lisped badly, and the other had the voice of a penny trumpet. Week after week these boys went through the quarrel of Brutus and Cassius, for the benefit of myself and others, to see if their example would not provoke us to a generous competition for all the honors.

How it operated on the other boys in after life I can not say; but the effect on me was decidedly unwholesome —discouraging, indeed,—until I was old enough to judge for myself, and to carry into operation a system of my own.

On coming to the passage,—

> "Be ready, gods, with all your thunderbolts;
> Dash him to pieces!"—

the elder of the boys gave it after the following fashion:

"Be ready, godths, with all your thunderbolths,—dath him in pietheth!"—bringing his right fist down into his left palm with all his strength, and his lifted foot upon the platform, which was built like a sounding-board, so that the master himself, who had suggested the action and obliged the poor boy to rehearse it over and over again, appeared to be utterly carried away by the magnificent demonstration; while to me—so deficient was I in rhetorical taste—it sounded like a crash of broken crockery, intermingled with chicken peeps.

I never got over it; and to this day can not endure stamping, nor even tapping of the foot, nor clapping the hands together, nor thumping the table for illustration; having an idea that such noises are not oratory, and that untranslatable sounds are not language.

My next essay was of a somewhat different kind. I took the field in person, being in my nineteenth year, well proportioned, and already beginning to have a sincere relish for poetry, if not for declamation. I had always been a great reader; and in the course of my foraging depredations I had met with "The Mariner's Dream" and "The Lake of the Dismal Swamp," both of which I had committed to memory before I knew it.

And one day, happening to be alone with my sister, and newly rigged out in a student's gown, such as the lads at Brunswick sported when they came to show off among their old companions, I proposed to astonish her by rehearsing these two poems in appropriate costume. Being very proud of her brother, and very obliging, she consented at once,—upon condition that our dear mother, who had never seen anything of the sort, should be invited to make one of the audience.

On the whole, I rather think that I succeeded in astonishing both. I well remember their looks of amazement—for they had never seen anything better or worse in all their lives, and were no judges of acting—as I swept to and fro in that magnificent robe, with outstretched arms and uplifted eyes, when I came to passages like the following, where an apostrophe was called for:

> "And near him the she wolf stirred the brake,
> And the copper snake breathed in his ear,
> Till he, starting, cried, from his dream awake,
> 'Oh, when shall I see the dusky lake,
> And the white canoe of my dear?'"

Or like this:

> "On beds of green sea flowers thy limbs shall be laid;
> Around thy white bones the red coral shall grow,
> Of thy fair yellow locks, threads of amber be made,
> And every part suit to thy mansion below;"—

throwing up my arms, and throwing them out in every possible direction as the spirit moved me, or the sentiment prompted; for I always encouraged my limbs and features to think for themselves, and to act for themselves, and never predetermined, never forethought, a gesture nor an intonation in my life; and should as soon think of counterfeiting another's look or step or voice, or of modulating my own by a pitch pipe (as the ancient orators did, with whom oratory was acting elocution, a branch of the dramatic art), as of adopting or imitating the gestures and tones of the most celebrated rhetorician I ever saw.

The result was rather encouraging. My mother and sister were both satisfied. At any rate, they said nothing to the contrary. Being only in my nineteenth year, what might I not be able to accomplish after a little more experience!

How little did I think, while rehearsing before my mother and sister, that anything serious would ever come of it, or that I was laying the foundations of character for life, or that I was beginning what I should not be able to finish within the next forty or fifty years following.

Yet so it was. I had broken the ice without knowing it. These things were but the foreshadowing of what happened long afterward.

NOTES.—**Brunswick,** Maine, is the seat of Bowdoin College.
"**The Mariner's Dream**" is a poem by William Dimond.
"**The Lake of the Dismal Swamp**" is by Thomas Moore.

XVII. ELEGY IN A COUNTRY CHURCHYARD.

Thomas Gray, 1716-1771, is often spoken of as "the author of the Elegy,"—this simple yet highly finished and beautiful poem being by far the best known of all his writings. It was finished in 1749,—seven years from the time it was commenced. Probably no short poem in the language ever deserved or received more praise. Gray was born in London; his father possessed property, but was indolent and selfish; his mother was a successful woman of business, and supported her son in college from her own earnings. The poet was educated at Eton and Cambridge; at the latter place, he resided for several years after his return from a continental tour, begun in 1739. He was small and delicate in person, refined and precise in dress and manners, and shy and retiring in disposition. He was an accomplished scholar in many fields of learning, but left comparatively little finished work in any department. He declined the honor of poet laureate; but, in 1769, was appointed Professor of History at Cambridge.

THE curfew tolls the knell of parting day,
  The lowing herd winds slowly o'er the lea,
The plowman homeward plods his weary way,
  And leaves the world to darkness and to me.

Now fades the glimmering landscape on the sight,
  And all the air a solemn stillness holds,
Save where the beetle wheels his droning flight,
  And drowsy tinklings lull the distant folds:

Save that from yonder ivy-mantled tower,
  The moping owl does to the moon complain
Of such as, wandering near her secret bower,
  Molest her ancient solitary reign.

Beneath those rugged elms, that yew tree's shade,
  Where heaves the turf in many a moldering heap,
Each in his narrow cell forever laid,
  The rude forefathers of the hamlet sleep.

The breezy call of incense-breathing morn,
  The swallow twittering from the straw-built shed,
The cock's shrill clarion, or the echoing horn,
  No more shall rouse them from their lowly bed.

For them no more the blazing hearth shall burn,
  Or busy housewife ply her evening care;
No children run to lisp their sire's return,
  Or climb his knees the envied kiss to share.

Oft did the harvest to their sickle yield,
  Their furrow oft the stubborn glebe has broke:
How jocund did they drive their team afield!
  How bowed the woods beneath their sturdy stroke!

Let not Ambition mock their useful toil,
  Their homely joys, and destiny obscure;
Nor Grandeur hear with a disdainful smile
  The short and simple annals of the poor.

The boast of heraldry, the pomp of power,
  And all that beauty, all that wealth e'er gave,
Await alike, the inevitable hour:
  The paths of glory lead but to the grave.

Nor you, ye proud, impute to these the fault,
  If Memory o'er their tomb no trophies raise;
Where, through the long-drawn aisle and fretted vault,
  The pealing anthem swells the note of praise.

Can storied urn, or animated bust,
  Back to its mansion call the fleeting breath?
Can Honor's voice provoke the silent dust,
  Or Flattery soothe the dull, cold ear of Death?

Perhaps, in this neglected spot is laid
  Some heart once pregnant with celestial fire;
Hands, that the rod of empire might have swayed,
  Or waked to ecstasy the living lyre:

But Knowledge to their eyes her ample page,
  Rich with the spoils of time, did ne'er unroll;
Chill Penury repressed their noble rage,
  And froze the genial current of the soul.

Full many a gem of purest ray serene,
  The dark, unfathomed caves of ocean bear:
Full many a flower is born to blush unseen,
  And waste its sweetness on the desert air.

Some village Hampden, that, with dauntless breast,
  The little tyrant of his fields withstood,
Some mute, inglorious Milton here may rest,
  Some Cromwell, guiltless of his country's blood.

The applause of listening senates to command,
  The threats of pain and ruin to despise,
To scatter plenty o'er a smiling land,
  And read their history in a nation's eyes,

Their lot forbade: nor, circumscribed alone
  Their growing virtues, but their crimes confined;
Forbade to wade through slaughter to a throne.
  And shut the gates of mercy on mankind,

The struggling pangs of conscious truth to hide,
  To quench the blushes of ingenuous shame,
Or heap the shrine of luxury and pride
  With incense kindled at the Muse's flame.

Far from the madding crowd's ignoble strife,
  Their sober wishes never learned to stray;
Along the cool, sequestered vale of life,
  They kept the noiseless tenor of their way.

Yet even these bones, from insult to protect,
  Some frail memorial still, erected nigh,
With uncouth rhymes and shapeless sculpture decked,
  Implores the passing tribute of a sigh.

Their name, their years, spelt by the unlettered Muse,
  The place of fame and elegy supply;
And many a holy text around she strews,
  That teach the rustic moralist to die.

For who, to dumb forgetfulness a prey,
  This pleasing, anxious being e'er resigned,
Left the warm precincts of the cheerful day,
  Nor cast one longing, lingering look behind?

On some fond breast the parting soul relies,
  Some pious drops the closing eye requires;
E'en from the tomb the voice of Nature cries,
  E'en in our ashes live their wonted fires.

For thee, who, mindful of the unhonored dead,
  Dost in these lines their artless tale relate,
If chance, by lonely contemplation led,
  Some kindred spirit shall inquire thy fate,—

Haply some hoary-headed swain may say,
  "Oft have we seen him at the peep of dawn
Brushing, with hasty step, the dews away,
  To meet the sun upon the upland lawn:

"There, at the foot of yonder nodding beech,
  That wreathes its old, fantastic roots so high,
His listless length at noontide would he stretch,
  And pore upon the brook that babbles by.

"Hard by yon wood, now smiling as in scorn,
  Muttering his wayward fancies, he would rove;
Now, drooping, woeful-wan, like one forlorn,
  Or crazed with care, or crossed in hopeless love.

"One morn, I missed him on the customed hill,
  Along the heath, and near his favorite tree:
Another came; nor yet beside the rill,
  Nor up the lawn, nor at the wood was he:

"The next, with dirges due, in sad array
  Slow through the church-way path we saw him borne:—
Approach and read (for thou canst read) the lay
  'Graved on the stone beneath yon aged thorn."

THE EPITÁPH.

Here rests his head upon the lap of Earth,
  A youth, to Fortune and to Fame unknown:
Fair Science frowned not on his humble birth,
  And Melancholy marked him for her own.

Large was his bounty, and his soul sincere,
  Heaven did a recompense as largely send:
He gave to Misery (all he had) a tear;
  He gained from Heaven ('t was all he wished) a friend.

No farther seek his merits to disclose,
  Or draw his frailties from their dread abode
(There they alike in trembling hope repose),
  The bosom of his Father, and his God.

NOTES.—**John Hampden** (b. 1594, d. 1643) was noted for his resolute resistance to the forced loans and unjust taxes imposed by Charles I. on England. He took part in the contest between King and Parliament, and was killed in a skirmish.

John Milton. See biographical notice, page 312.

Oliver Cromwell (b. 1599, d. 1658) was the leading character in the Great Rebellion in England. He was Lord Protector the last five years of his life, and in many respects the ablest ruler that England ever had.

XVIII. TACT AND TALENT.

TALENT is something, but tact is everything. Talent is serious, sober, grave, and respectable: tact is all that, and more too. It is not a sixth sense, but it is the life of all the five. It is the open eye, the quick ear, the judging taste, the keen smell, and the lively touch; it is the interpreter of all riddles, the surmounter of all difficulties, the remover of all obstacles. It is useful in all places, and at all times; it is useful in solitude, for it shows a man into the world; it is useful in society, for it shows him his way through the world.

Talent is power, tact is skill; talent is weight, tact is momentum; talent knows what to do, tact knows how to do it; talent makes a man respectable, tact will make him respected; talent is wealth, tact is ready money. For all the practical purposes, tact carries it against talent ten to one.

Take them to the theater, and put them against each other on the stage, and talent shall produce you a tragedy that shall scarcely live long enough to be condemned, while tact keeps the house in a roar, night after night, with its successful farces. There is no want of dramatic talent, there is no want of dramatic tact; but they are seldom together: so we have successful pieces which are not respectable, and respectable pieces which are not successful.

Take them to the bar, and let them shake their learned curls at each other in legal rivalry; talent sees its way clearly, but tact is first at its journey's end. Talent has many a compliment from the bench, but tact touches fees. Talent makes the world wonder that it gets on no faster, tact arouses astonishment that it gets on so fast. And the secret is, that it has no weight to carry; it makes no false steps; it hits the right nail on the head; it loses no time; it takes all hints; and, by keeping its eye on the weathercock, is ready to take advantage of every wind that blows.

Take them into the church: talent has always something worth hearing, tact is sure of abundance of hearers; talent may obtain a living, tact will make one; talent gets a good name, tact a great one; talent convinces, tact converts; talent is an honor to the profession, tact gains honor from the profession.

Take them to court: talent feels its weight, tact finds its way; talent commands, tact is obeyed; talent is honored with approbation, and tact is blessed by preferment. Place them in the senate: talent has the ear of the house, but tact wins its heart, and has its votes; talent is fit for employment, but tact is fitted for it. It has a knack of slipping into place with a sweet silence and glibness of movement, as a billiard ball insinuates itself into the pocket.

It seems to know everything, without learning anything. It has served an extemporary apprenticeship; it wants no drilling; it never ranks in the awkward squad; it has no left hand, no deaf ear, no blind side. It puts on no look of wondrous wisdom, it has no air of profundity, but plays with the details of place as dexterously as a well-taught hand flourishes over the keys of the pianoforte. It has all the air of commonplace, and all the force and power of genius.

XIX. SPEECH BEFORE THE VIRGINIA CONVENTION.

Patrick Henry, 1736-1799, was born in Hanover County, Virginia. He received instruction in Latin and mathematics from his father, but seemed to develop a greater fondness for hunting, fishing, and playing the fiddle than for study. Twice he was set up in business, and twice failed before he was twenty-four. He was then admitted to the bar after six weeks' study of the law. He got no business at first in his profession, but lived with his father-in-law. His wonderful powers of oratory first showed themselves in a celebrated case which he argued in Hanover Courthouse, his own father being the presiding magistrate. He began very awkwardly, but soon rose to a surprising height of eloquence, won his case against great odds, and was carried off in triumph by the delighted spectators. His fame was now established; business flowed in, and he was soon elected to the Virginia Legislature, He was a delegate to the Congress of 1774, and in 1775 made the prophetic speech of which the following selection is a portion. It was on his own motion that the "colony be immediately put in a state of defense." During the Revolution he was, for several years, Governor of Virginia. In 1788, he earnestly opposed the adoption of the Federal Constitution. When he died, he left a large family and an ample fortune. In person, Mr. Henry was tall and rather awkward, with a face stern and grave. When he spoke on great occasions, his awkwardness forsook him, his face lighted up, and his eyes flashed with a wonderful fire. In his life, he was good-humored, honest, and temperate. His patriotism was of the noblest type; and few men in those stormy times did better service for their country than he.

It is natural for man to indulge in the illusions of hope. We are apt to shut our eyes against a painful truth, and listen to the song of that siren till she transforms us into beasts. Is this the part of wise men, engaged in a great and arduous struggle for liberty? Are we disposed to be of the number of those, who, having eyes, see not, and having ears, hear not the things which so nearly concern their temporal salvation? For my part, whatever anguish of spirit it may cost, I am willing to know the whole truth; to know the worst, and to provide for it.

I have but one lamp by which my feet are guided; and that is the lamp of experience. I know of no way of judging of the future but by the past; and, judging by the past, I wish to know what there has been in the conduct of the British ministry for the last ten years to justify those hopes with which gentlemen have been pleased to solace themselves and the house? Is it that insidious smile with which our petition has been lately received? Trust it not: it will prove a snare to your feet. Suffer not yourselves to be betrayed with a kiss. Ask yourselves, how this gracious reception of our petition comports with those warlike preparations which cover our waters and darken our land. Are fleets and armies necessary to a work of love and reconciliation? Have we shown ourselves so unwilling to be reconciled that force must be called in to win back our love? Let us not deceive ourselves. These are the implements of war and subjugation,—the last arguments to which kings resort.

I ask, gentlemen, what means this martial array, if its purpose be not to force us into submission? Can gentlemen assign any other possible motive for it? Has Great Britain any enemy in this quarter of the world, to call for all this accumulation of navies and armies? No, she has none. They are meant for us: they can be meant for no other. They are sent over to bind and rivet upon us those chains which the British ministry have been so long forging. And what have we to oppose to them? Shall we try argument? We have been trying that for the last ten years. Have we anything new to offer upon the subject? Nothing. We have held the subject up in every light in which it was capable; but it has been all in vain.

Shall we resort to entreaty and humble supplication? What terms shall we find which have not been already exhausted? Let us not, I beseech you, deceive ourselves longer. We have done everything that could be done, to avert the storm which is now coming on. We have petitioned; we have remonstrated; we have supplicated; we have prostrated ourselves at the foot of the throne, and implored its interposition to arrest the tyrannical hands of the ministry and parliament. Our petitions have been slighted; our remonstrances have produced additional violence and insult; our supplications disregarded; and we have been spurned with contempt from the foot of the throne.

In vain, after these things, may we indulge the fond hope of peace and reconciliation. There is no longer any room for hope. If we wish to be free; if we mean to preserve inviolate those inestimable privileges for which we have been so long contending; if we mean not basely to abandon the noble struggle in which we have been so long engaged, and which we have pledged ourselves never to abandon until the glorious object of our contest shall be obtained—we must fight! I repeat it, we must fight! An appeal to arms and the God of Hosts, is all that is left us.

They tell us that we are weak; unable to cope with so formidable an adversary. But when shall we be stronger? Will it be the next week, or the next year? Will it be when we are totally disarmed, and when a British guard shall be stationed in every house? Shall we gather strength by irresolution and inaction? Shall we acquire the means of effectual resistance by lying supinely on our backs, and hugging the delusive phantom of hope, until our enemies shall have bound us hand and foot? We are not weak, if we make a proper use of those means which the God of nature hath placed in our power.

Three millions of people, armed in the holy cause of liberty, and in such a country as that which we possess, are invincible by any force which our enemy can send against us. Besides, we shall not fight our battles alone. There is a just God who presides over the destinies of nations; and who will raise up friends to fight our battles for us. The battle is not to the strong alone; it is to the vigilant, the active, the brave. Besides, we have no election. If we were base enough to desire it, it is now too late to retire from the contest. There is no retreat but in submission and slavery! Our chains are forged. Their clanking may be heard on the plains of Boston! The war is inevitable; and, let it come! I repeat it, let it come!

It is in vain to extenuate the matter. Gentlemen may cry peace, peace; but there is no peace. The war is actually begun. The next gale that sweeps from the north, will bring to our ears the clash of resounding arms! Our brethren are already in the field! Why stand we here idle? What is it that gentlemen wish? What would they have? Is life so dear, or peace so sweet, as to be purchased at the price of chains and slavery? Forbid it, Almighty God! I know not what course others may take; but as for me, give me liberty, or give me death.

Notes.—Observe, in this lesson, the all-controlling power of emphasis in determining the falling inflection. The words "see," "hear," and "my," in the first paragraph, the word "that" in the second, and "spurned" and "contempt" in the fourth paragraph, are examples of this. Let the reader remember that a high degree of emphasis is sometimes expressed by a whisper; also, that emphasis is often expressed by a pause.

It will be well to read in this connection some good history of the opening scenes of the Revolution.

XX. THE AMERICAN FLAG.

Joseph Rodman Drake, 1795-1820, was born in New York City. His father died when he was very young, and his early life was a struggle with poverty. He studied medicine, and took his degree when he was about twenty years old. From a child, he showed remarkable poetical powers, having made rhymes at the early age of five. Most of his published writings were produced during a period of less than two years. "The Culprit Fay" and the "American Flag" are best known. In disposition, Mr. Drake was gentle and kindly; and, on the occasion of his death, his intimate friend, Fitz-Greene Halleck, expressed his character in the well-known couplet:

"None knew thee but to love thee,
Nor named thee but to praise."

WHEN Freedom, from her mountain height,
Unfurled her standard to the air,
She tore the azure robe of night,
And set the stars of glory there:
She mingled with its gorgeous dyes
The milky baldric of the skies,
And striped its pure, celestial white
With streakings of the morning light;
Then, from his mansion in the sun,
She called her eagle bearer down,
And gave into his mighty hand
The symbol of her chosen land.

Majestic monarch of the cloud!
Who rear'st aloft thy regal form,
To hear the tempest trumpings loud,
And see the lightning lances driven,
When strive the warriors of the storm,
And rolls the thunder drum of heaven;—
Child of the sun! to thee 't is given
To guard the banner of the free,
To hover in the sulphur smoke,
To ward away the battle stroke,

And bid its blendings shine afar,
Like rainbows on the cloud of war,
  The harbingers of victory!

Flag of the brave! thy folds shall fly,
The sign of hope and triumph high!
When speaks the signal trumpet tone,
And the long line comes gleaming on,
Ere yet the lifeblood, warm and wet,
Has dimmed the glistening bayonet,
Each soldier's eye shall brightly turn
To where thy sky-born glories burn,
And, as his springing steps advance,
Catch war and vengeance from the glance.
And when the cannon mouthings loud
Heave in wild wreaths the battle shroud,
And gory sabers rise and fall,
Like shoots of flame on midnight's pall,
Then shall thy meteor glances glow,
And cowering foes shall sink beneath
Each gallant arm, that strikes below
That lovely messenger of death.

Flag of the seas! on ocean's wave
Thy stars shall glitter o'er the brave;
When death, careering on the gale,
Sweeps darkly round the bellied sail,
And frighted waves rush wildly back,
Before the broadside's reeling rack,
Each dying wanderer of the sea
Shall look at once to heaven and thee,
And smile to see thy splendors fly
In triumph o'er his closing eye.

Flag of the free heart's hope and home,
  By angel hands to valor given,

Thy stars have lit the welkin dome,
And all thy hues were born in heaven.
Forever float that standard sheet!
Where breathes the foe but falls before us,
With Freedom's soil beneath our feet,
And Freedom's banner streaming o'er us?

XXI. IRONICAL EULOGY ON DEBT.

DEBT is of the very highest antiquity. The first debt in the history of man is the debt of nature, and the first instinct is to put off the payment of it to the last moment. Many persons, it will be observed, following the natural procedure, would die before they would pay their debts.

Society is composed of two classes, debtors and creditors. The creditor class has been erroneously supposed the more enviable. Never was there a greater misconception; and the hold it yet maintains upon opinion is a remarkable example of the obstinacy of error, notwithstanding the plainest lessons of experience. The debtor has the sympathies of mankind. He is seldom spoken of but with expressions of tenderness and compassion—"the poor debtor!"—and "the unfortunate debtor!" On the other hand, "harsh" and "hard-hearted" are the epithets allotted to the creditor. Who ever heard the "poor creditor," the "unfortunate creditor" spoken of? No, the creditor never becomes the object of pity, unless he passes into the debtor class. A creditor may be ruined by the poor debtor, but it is not until he becomes unable to pay his own debts, that he begins to be compassionated.

A debtor is a man of mark. Many eyes are fixed upon him; many have interest in his well-being; his movements are of concern; he can not disappear unheeded; his name is in many mouths; his name is upon many books; he is a man of note—of *promissory* note; he fills the speculation of many minds; men conjecture about him, wonder about him,—wonder and conjecture whether he will pay. He is a man of consequence, for many are running after him. His door is thronged with duns. He is inquired after every hour of the day. Judges hear of him and know him. Every meal he swallows, every coat he puts upon his back, every dollar he borrows, appears before the country in some formal document. Compare his notoriety with the obscure lot of the creditor,—of the man who has nothing but claims on the world; a landlord, or fund-holder, or some such disagreeable, hard character.

The man who pays his way is unknown in his neighborhood. You ask the milkman at his door, and he can not tell his name. You ask the butcher where Mr. Payall lives, and he tells you he knows no such name, for it is not in his books. You shall ask the baker, and he will tell you there is no such person in the neighborhood. People that have his money fast in their pockets, have no thought of his person or appellation. His house only is known. No. 31 is good pay. No. 31 is ready money. Not a scrap of paper is ever made out for No. 31. It is an anonymous house; its owner pays his way to obscurity. No one knows anything about him, or heeds his movements. If a carriage be seen at his door, the neighborhood is not full of concern lest he be going to run away. If a package be removed from his house, a score of boys are not employed to watch whether it be carried to the pawnbroker. Mr. Payall fills no place in the public mind; no one has any hopes or fears about him.

The creditor always figures in the fancy as a sour, single man, with grizzled hair, a scowling countenance, and a peremptory air, who lives in a dark apartment, with musty deeds about him, and an iron safe, as impenetrable as his heart, grabbing together what he does not enjoy, and what there is no one about him to enjoy. The debtor, on the other hand, is always pictured with a wife and six fair-haired daughters, bound together in affection and misery, full of sensibility, and suffering without a fault. The creditor, it is never doubted, thrives without a merit. He has no wife and children to pity. No one ever thinks it desirable that he should have the means of living. He is a brute for insisting that he must receive, in order to pay. It is not in the imagination of man to conceive that his creditor has demands upon him which must be satisfied, and that he must do to others as others must do to him. A creditor is a personification of exaction. He is supposed to be always taking in, and never giving out.

People idly fancy that the possession of riches is desirable. What blindness! Spend and regale. Save a shilling and you lay it by for a thief. The prudent men are the men that live beyond their means. Happen what may, they are safe. They have taken time by the forelock. They have anticipated fortune. "The wealthy fool, with gold in store," has only denied himself so much enjoyment, which another will seize at his expense. Look at these people in a panic. See who are the fools then. You know them by their long faces. You may say, as one of them goes by in an agony of apprehension, "There is a stupid fellow who fancied himself rich, because he had fifty thousand dollars in bank." The history of the last ten years has taught the moral, "spend and regale." Whatever is laid up beyond the present hour, is put in jeopardy. There is no certainty but in instant enjoyment. Look at schoolboys sharing a plum cake. The knowing ones eat, as for a race; but a stupid fellow saves his portion; just nibbles a bit, and "keeps the rest for another time." Most provident blockhead! The others, when they have gobbled up their shares, set upon him, plunder him, and thrash him for crying out.

Before the terms "depreciation," "suspension," and "going into liquidation," were heard, there might have been some reason in the practice of "laying up;" but now it denotes the darkest blindness. The prudent men of the present time, are the men in debt. The tendency being to sacrifice creditors to debtors, and the debtor party acquiring daily new strength, everyone is in haste to get into the favored class. In any case, the debtor is safe. He has put his enjoyments behind him; they are safe; no turns of fortune can disturb them. The substance he has eaten up, is irrecoverable. The future can not trouble his past. He has nothing to apprehend. He has anticipated more than fortune would ever have granted him. He has tricked fortune; and his creditors—bah! who feels for creditors? What are creditors? Landlords; a pitiless and unpitiable tribe; all griping extortioners! What would become of the world of debtors, if it did not steal a march upon this rapacious class?

XXII. THE THREE WARNINGS.

Hester Lynch Thrale, 1739-1821, owes her celebrity almost wholly to her long intimacy with Dr. Samuel Johnson. This continued for twenty years, during which Johnson spent much time in her family. She was born in Caernarvonshire, Wales; her first husband was a wealthy brewer, by whom she had several children. In 1784, she married an Italian teacher of music named Piozzi. Her writings are quite numerous; the best known of her books is the "Anecdotes of Dr. Johnson;" but nothing she ever wrote is so well known as the "Three Warnings."

THE tree of deepest root is found
Least willing still to quit the ground;
'T was therefore said by ancient sages,
  That love of life increased with years
So much, that in our latter stages,
When pains grow sharp, and sickness rages,
  The greatest love of life appears.

This great affection to believe,
Which all confess, but few perceive,
If old assertions can't prevail,
Be pleased to hear a modern tale.

When sports went round, and all were gay,
On neighbor Dodson's wedding day,
Death called aside the jocund groom
With him into another room;
And looking grave, "You must," says he,
"Quit your sweet bride, and come with me."
"With you! and quit my Susan's side?
With you!" the hapless bridegroom cried:
"Young as I am, 'tis monstrous hard!
Besides, in truth, I'm not prepared."

What more he urged, I have not heard;
His reasons could not well be stronger:
So Death the poor delinquent spared,
And left to live a little longer.
Yet, calling up a serious look,
His hourglass trembled while he spoke:
"Neighbor," he said, "farewell! no more
Shall Death disturb your mirthful hour;
And further, to avoid all blame
Of cruelty upon my name,
To give you time for preparation,
And fit you for your future station,
Three several warnings you shall have
Before you're summoned to the grave:
Willing for once I'll quit my prey,
And grant a kind reprieve;
In hopes you'll have no more to say,
But, when I call again this way,
Well pleased the world will leave."

To these conditions both consented,
And parted perfectly contented.

What next the hero of our tale befell,
How long he lived, how wisely, and how well,
It boots not that the Muse should tell;
He plowed, he sowed, he bought, he sold,
Nor once perceived his growing old,
  Nor thought of Death as near;
His friends not false, his wife no shrew,
Many his gains, his children few,
He passed his hours in peace.
But, while he viewed his wealth increase,
While thus along life's dusty road,
The beaten track, content he trod,
Old Time, whose haste no mortal spares,
Uncalled, unheeded, unawares,
  Brought on his eightieth year.

And now, one night, in musing mood,
  As all alone he sate,
  The unwelcome messenger of Fate
Once more before him stood.
Half-killed with wonder and surprise,
"So soon returned!" old Dodson cries.
"So *soon* d' ye call it?" Death replies:
"Surely, my friend, you 're but in jest;
  Since I was here before,
'T is six and thirty years at least,
  And you are now fourscore."
"So much the worse!" the clown rejoined;
"To spare the aged would be kind:
Besides, you promised me *three warnings*,
Which I have looked for nights and mornings!"

"I know," cries Death, "that at the best,
I seldom am a welcome guest;
But do n't be captious, friend; at least,
I little thought that you 'd be able
To stump about your farm and stable;
Your years have run to a great length,
Yet still you seem to have your strength."

"Hold!" says the farmer, "not so fast!
I have been lame, these four years past."
"And no great wonder," Death replies,
"However, you still keep your eyes;
And surely, sir, to see one's friends,
For legs and arms would make amends."
"Perhaps," says Dodson, "so it might,
But latterly I 've lost my sight."
"This is a shocking story, faith;
But there 's some comfort still," says Death;
"Each strives your sadness to amuse;
I warrant you hear all the news."
"There 's none," cries he, "and if there were,
I 've grown so deaf, I could not hear."

"Nay, then," the specter stern rejoined,
"These are unpardonable yearnings;
If you are lame, and deaf, and blind,
You 've *had* your *three* sufficient warnings,
So, come along; no more we 'll part."
He said, and touched him with his dart:
And now old Dodson, turning pale,
Yields to his fate—so ends my tale.

XXIII. THE MEMORY OF OUR FATHERS.

Lyman Beecher, 1775-1863, a famous Congregational minister of New England, was born in New Haven, graduated from Yale College in 1797, and studied theology with Dr. Timothy Dwight. His first settlement was at East Hampton, L. I., at a salary of three hundred dollars per year. He was pastor of the church in Litchfield, Ct., from 1810 till 1826, when he removed to Boston, and took charge of the Hanover Street Church. In the religious controversies of the time, Dr. Beecher was one of the most prominent characters. From 1832 to 1842, he was President of Lane Theological Seminary, in the suburbs of Cincinnati. He then returned to Boston, where he spent most of the closing years of his long and active life. His death occurred in Brooklyn, N. Y. As a theologian, preacher, and advocate of education, temperance, and missions, Dr. Beecher occupied a very prominent place for nearly half a century. He left a large family of sons and two daughters, who are well known as among the most eminent preachers and authors in America.

We are called upon to cherish with high veneration and grateful recollections, the memory of our fathers. Both the ties of nature and the dictates of policy demand this. And surely no nation had ever less occasion to be ashamed of its ancestry, or more occasion for gratulation in that respect; for while most nations trace their origin to barbarians, the foundations of our nation were laid by civilized men, by Christians. Many of them were men of distinguished families, of powerful talents, of great learning and of preëminent wisdom, of decision of character, and of most inflexible integrity. And yet not unfrequently they have been treated as if they had no virtues; while their sins and follies have been sedulously immortalized in satirical anecdote.

The influence of such treatment of our fathers is too manifest. It creates and lets loose upon their institutions, the vandal spirit of innovation and overthrow; for after the memory of our fathers shall have been rendered contemptible, who will appreciate and sustain their institutions? "The memory of our fathers" should be the watchword of liberty throughout the land; for, imperfect as they were, the world before had not seen their like, nor will it soon, we fear, behold their like again. Such models of moral excellence, such apostles of civil and religious liberty, such shades of the illustrious dead looking down upon their descendants with approbation or reproof, according as they follow or depart from the good way, constitute a censorship inferior only to the eye of God; and to ridicule them is national suicide.

The doctrines of our fathers have been represented as gloomy, superstitious, severe, irrational, and of a licentious tendency. But when other systems shall have produced a piety as devoted, a morality as pure, a patriotism as disinterested, and a state of society as happy, as have prevailed where their doctrines have been most prevalent, it may be in season to seek an answer to this objection.

The persecutions instituted by our fathers have been the occasion of ceaseless obloquy upon their fair fame. And truly, it was a fault of no ordinary magnitude, that sometimes they did persecute. But let him whose ancestors were not ten times more guilty, cast the first stone, and the ashes of our fathers will no more be disturbed. Theirs was the fault of the age, and it will be easy to show that no class of men had, at that time, approximated so nearly to just apprehensions of religious liberty; and that it is to them that the world is now indebted for the more just and definite views which now prevail.

The superstition and bigotry of our fathers are themes on which some of their descendants, themselves far enough from superstition, if not from bigotry, have delighted to dwell. But when we look abroad, and behold the condition of the world, compared with the condition of New England, we may justly exclaim, "Would to God that the ancestors of all the nations had been not only almost, but altogether such bigots as our fathers were."

XXIV. SHORT SELECTIONS IN PROSE.

I. DRYDEN AND POPE.

DRYDEN knew more of man in his general nature, and Pope in his local manners. The notions of Dryden were formed by comprehensive speculation, those of Pope by minute attention. There is more dignity in the knowledge of Dryden, more certainty in that of Pope. The style of Dryden is capricious and varied, that of Pope cautious and uniform. Dryden obeys the motions of his own mind; Pope constrains his mind to his own rules of composition. Dryden's page is a natural field, rising into inequalities, and diversified by the varied exuberance of abundant vegetation; Pope's is the velvet lawn, shaven by the scythe, and leveled by the roller. If the flights of Dryden are higher, Pope continues longer on the wing. If, of Dryden's fire, the blaze is brighter, of Pope's the heat is more regular and constant. Dryden often surpasses expectation, and Pope never falls below it. Dryden is read with frequent astonishment, and Pope with perpetual delight.

—*Samuel Johnson.*

NOTE.—A fine example of *antithesis.* See p. 26.

II. LAS CASAS DISSUADING FROM BATTLE.

Is then the dreadful measure of your cruelty not yet complete? Battle! against whom? Against a king, in whose mild bosom your atrocious injuries, even yet, have not excited hate; but who, insulted or victorious, still sues for peace. Against a people, who never wronged the living being their Creator formed; a people, who received you as cherished guests, with eager hospitality and confiding kindness. Generously and freely did they share with you their comforts, their treasures, and their homes; you repaid them by fraud, oppression, and dishonor.

Pizarro, hear me! Hear me, chieftains! And thou, All-powerful! whose thunder can shiver into sand the adamantine rock, whose lightnings can pierce the core of the riven and quaking earth, oh let thy power give effect to thy servant's words, as thy Spirit gives courage to his will! Do not, I implore you, chieftains,—do not, I implore you, renew the foul barbarities your insatiate avarice has inflicted on this wretched, unoffending race. But hush, my sighs! fall not, ye drops of useless sorrow! heart-breaking anguish, choke not my utterance.

—*R. B. Sheridan.*

NOTE.—Examples of *series*. See p. 28.

III. ACTION AND REPOSE.

John Ruskin, 1819 ——, is a distinguished English art critic and author. From 1869 to 1884, he was Professor of the Fine Arts at Oxford University. His writings are very numerous, and are noted for their eloquent and brilliant style.

About the river of human life there is a wintry wind, though a heavenly sunshine; the iris colors its agitation, the frost fixes upon its repose. Let us beware that our rest become not the rest of stones, which, so long as they are tempest-tossed and thunderstricken, maintain their majesty; but when the stream is silent and the storm passed, suffer the grass to cover them, and are plowed into the dust.

IV. TIME AND CHANGE.

Sir Humphry Davy, 1778–1829, was an eminent chemist of England. He made many important chemical discoveries, and was the inventor of the miner's safety lamp.

Time is almost a human word, and Change entirely a human idea; in the system of nature, we should rather say progress than change. The sun appears to sink in the ocean in darkness, but it rises in another hemisphere; the ruins of a city fall, but they are often used to form more magnificent structures: even when they are destroyed so as to produce only dust, Nature asserts her empire over them; and the vegetable world rises in constant youth, in a period of annual successions, by the labors of man—providing food, vitality, and beauty—upon the wrecks of monuments which were raised for the purposes of glory, but which are now applied to objects of utility.

V. THE POET.

William Ellery Channing, 1780-1842, was a distinguished clergyman and orator. He took a leading part in the public affairs of his day, and wrote and lectured eloquently on several topics.

It is not true that the poet paints a life which does not exist. He only extracts and concentrates, as it were, life's ethereal essence, arrests and condenses its volatile fragrance, brings together its scattered beauties, and prolongs its more refined but evanescent joys; and in this he does well, for it is good to feel that life is not wholly usurped by cares for subsistence and physical gratifications, but admits, in measures which may be indefinitely enlarged, sentiments and delights worthy of a higher being.

VI. MOUNTAINS.

William Howitt, 1795-1879, was an English author. He published many books, and was associated with his wife, Mary Howitt, in the publication of many others.

There is a charm connected with mountains, so powerful that the merest mention of them, the merest sketch of their magnificent features, kindles the imagination, and carries the spirit at once into the bosom of their enchanted regions. How the mind is filled with their vast solitude! How the inward eye is fixed on their silent, their sublime, their everlasting peaks! How our hearts bound to the music of their solitary cries, to the tinkle of their gushing rills, to the sound of their cataracts! How inspiriting are the odors that breathe from the upland turf, from the rock-hung flower, from the hoary and solemn pine! How beautiful are those lights and shadows thrown abroad, and that fine, transparent haze which is diffused over the valleys and lower slopes, as over a vast, inimitable picture!

XXV. THE JOLLY OLD PEDAGOGUE.

George Arnold, 1834-1865, was born in New York City. He never attended school, but was educated at home, by his parents. His literary career occupied a period of about twelve years. In this time he wrote stories, essays, criticisms in art and literature, poems, sketches, etc., for several periodicals. Two volumes of his poems have been published since his death.

'T WAS a jolly old pedagogue, long ago,
Tall, and slender, and sallow, and dry;
His form was bent, and his gait was slow,
And his long, thin hair was white as snow,
But a wonderful twinkle shone in his eye:
And he sang every night as he went to bed,
"Let us be happy down here below;
The living should live, though the dead be dead,"
Said the jolly old pedagogue, long ago.

He taught the scholars the Rule of Three,
Reading, and writing, and history too;
He took the little ones on his knee,
For a kind old heart in his breast had he,
And the wants of the littlest child he knew.
"Learn while you're young," he often said,
"There is much to enjoy down here below;
Life for the living, and rest for the dead!"
Said the jolly old pedagogue, long ago.

With the stupidest boys, he was kind and cool,
  Speaking only in gentlest tones;
The rod was scarcely known in his school—
Whipping to him was a barbarous rule,
  And too hard work for his poor old bones;
Besides it was painful, he sometimes said:
  "We should make life pleasant down here below—
The living need charity more than the dead,"
  Said the jolly old pedagogue, long ago.

He lived in the house by the hawthorn lane,
  With roses and woodbine over the door;
His rooms were quiet, and neat, and plain,
But a spirit of comfort there held reign,
  And made him forget he was old and poor.
"I need so little," he often said;
  "And my friends and relatives here below
Won't litigate over me when I am dead,"
  Said the jolly old pedagogue, long ago.

But the pleasantest times he had of all,
  Were the sociable hours he used to pass,
With his chair tipped back to a neighbor's wall,
Making an unceremonious call,
  Over a pipe and a friendly glass:
This was the finest pleasure, he said,
  Of the many he tasted here below:
"Who has no cronies had better be dead,"
  Said the jolly old pedagogue, long ago.

The jolly old pedagogue's wrinkled face
  Melted all over in sunshiny smiles;
He stirred his glass with an old-school grace,
Chuckled, and sipped, and prattled apace,
  Till the house grew merry from cellar to tiles.

"I'm a pretty old man," he gently said,
"I've lingered a long time here below;
But my heart is fresh, if my youth is fled!"
Said the jolly old pedagogue, long ago.

He smoked his pipe in the balmy air
Every night, when the sun went down;
And the soft wind played in his silvery hair,
Leaving its tenderest kisses there,
On the jolly old pedagogue's jolly old crown;
And feeling the kisses, he smiled, and said:
"'T is a glorious world down here below;
Why wait for happiness till we are dead?"
Said this jolly old pedagogue, long ago.

He sat at his door one midsummer night,
After the sun had sunk in the west,
And the lingering beams of golden light
Made his kindly old face look warm and bright,
While the odorous night winds whispered, "Rest!"
Gently, gently, he bowed his head;
There were angels waiting for him, I know;
He was sure of his happiness, living or dead,
This jolly old pedagogue, long ago!

XXVI. THE TEACHER AND SICK SCHOLAR.

SHORTLY after the schoolmaster had arranged the forms and taken his seat behind his desk, a small white-headed boy with a sunburnt face appeared at the door, and, stopping there to make a rustic bow, came in and took his seat upon one of the forms. He then put an open book, astonishingly dog's-eared, upon his knees, and, thrusting his hands into his pockets, began counting the marbles with which they were filled; displaying, in the expression of his face, a remarkable capacity of totally abstracting his mind from the spelling on which his eyes were fixed.

Soon afterward, another white-headed little boy came straggling in, and after him, a red-headed lad, and then one with a flaxen poll, until the forms were occupied by a dozen boys, or thereabouts, with heads of every color but gray, and ranging in their ages from four years old to fourteen years or more; for the legs of the youngest were a long way from the floor, when he sat upon the form; and the eldest was a heavy, good-tempered fellow, about half a head taller than the schoolmaster.

At the top of the first form—the post of honor in the school—was the vacant place of the little sick scholar; and, at the head of the row of pegs, on which those who wore hats or caps were wont to hang them, one was empty. No boy attempted to violate the sanctity of seat or peg, but many a one looked from the empty spaces to the schoolmaster, and whispered to his idle neighbor, behind his hand.

Then began the hum of conning over lessons and getting them by heart, the whispered jest and stealthy game, and all the noise and drawl of school; and in the midst of the din, sat the poor schoolmaster, vainly attempting to fix his mind upon the duties of the day, and to forget his little sick friend. But the tedium of his office reminded him more strongly of the willing scholar, and his thoughts were rambling from his pupils—it was plain.

None knew this better than the idlest boys, who, growing bolder with impunity, waxed louder and more daring; playing "odd or even" under the master's eye; eating apples openly and without rebuke; pinching each other in sport or malice, without the least reserve; and cutting their initials in the very legs of his desk. The puzzled dunce, who stood beside it to say his lesson "off the book," looked no longer at the ceiling for forgotten words, but drew closer to the master's elbow, and boldly cast his eye upon the page; the wag of the little troop squinted and made grimaces (at the smallest boy, of course), holding no book before his face, and his approving companions knew no constraint in their delight. If the master did chance to rouse himself, and seem alive to what was going on, the noise subsided for a moment, and no eye met his but wore a studious and deeply humble look; but the instant he relapsed again, it broke out afresh, and ten times louder than before.

Oh! how some of those idle fellows longed to be outside, and how they looked at the open door and window, as if they half meditated rushing violently out, plunging into the woods, and being wild boys and savages from that time forth. What rebellious thoughts of the cool river, and some shady bathing place, beneath willow trees with branches dipping in the water, kept tempting and urging that sturdy boy, who, with his shirt collar unbuttoned, and flung back as far as it could go, sat fanning his flushed face with a spelling book, wishing himself a whale, or a minnow, or a fly, or anything but a boy at school, on that hot, broiling day.

Heat! ask that other boy, whose seat being nearest to the door, gave him opportunities of gliding out into the garden, and driving his companions to madness, by dipping his face into the bucket of the well, and then rolling on the grass,—ask him if there was ever such a day as that, when even the bees were diving deep down into the cups of the flowers, and stopping there, as if they had made up their minds to retire from business, and be manufacturers of honey no more. The day was made for laziness, and lying on one's back in green places, and staring at the sky, till its brightness forced the gazer to shut his eyes and go to sleep. And was this a time to be poring over musty books in a dark room, slighted by the very sun itself? Monstrous!

The lessons over, writing time began. This was a more quiet time; for the master would come and look over the writer's shoulder, and mildly tell him to observe how such a letter was turned up, in such a copy on the wall, which had been written by their sick companion, and bid him take it as a model. Then he would stop and tell them what the sick child had said last night, and how he had longed to be among them once again; and such was the poor schoolmaster's gentle and affectionate manner, that the boys seemed quite remorseful that they had worried him so much, and were absolutely quiet; eating no apples, cutting no names, and making no grimaces for full two minutes afterward.

"I think, boys," said the schoolmaster, when the clock struck twelve, "that I shall give you an extra half holiday this afternoon." At this intelligence, the boys, led on and headed by the tall boy, raised a great shout, in the midst of which the master was seen to speak, but could not be heard. As he held up his hand, however, in token of his wish that they should be silent, they were considerate enough to leave off, as soon as the longest-winded among them were quite out of breath. "You must promise me, first," said the schoolmaster, "that you 'll not be noisy, or at least, if you are, that you 'll go away first, out of the village, I mean. I 'm sure you would n't disturb your old playmate and companion."

There was a general murmur (and perhaps a very sincere one, for they were but boys) in the negative; and the tall boy, perhaps as sincerely as any of them, called those about him to witness, that he had only shouted in a whisper. "Then pray do n't forget, there 's my dear scholars," said the schoolmaster, "what I have asked you, and do it as a favor to me. Be as happy as you can, and do n't be unmindful that you are blessed with health. Good-by, all."

"Thank 'ee, sir," and "Good-by, sir," were said a great many times in a great variety of voices, and the boys went out very slowly and softly. But there was the sun shining and there were birds singing, as the sun only shines and the birds only sing on holidays and half holidays; there were the trees waving to all free boys to climb, and nestle among their leafy branches; the hay, entreating them to come and scatter it to the pure air; the green corn, gently beckoning toward wood and stream; the smooth ground, rendered smoother still by blending lights and shadows, inviting to runs and leaps, and long walks, nobody knows whither. It was more than boy could bear, and with a joyous whoop, the whole cluster took to their heels, and spread themselves about, shouting and laughing as they went. "'T is natural, thank Heaven!" said the poor schoolmaster, looking after them; "I am very glad they did n't mind me."

Toward night, the schoolmaster walked over to the cottage where his little friend lay sick. Knocking gently at the cottage door, it was opened without loss of time. He entered a room where a group of women were gathered about one who was wringing her hands and crying bitterly. "O dame!" said the schoolmaster, drawing near her chair, "is it so bad as this?" Without replying, she pointed to another room, which the schoolmaster immediately entered; and there lay his little friend, half-dressed, stretched upon a bed.

He was a very young boy; quite a little child. His hair still hung in curls about his face, and his eyes were very bright; but their light was of heaven, not of earth. The schoolmaster took a seat beside him, and, stooping over the pillow, whispered his name. The boy sprung up, stroked his face with his hand, and threw his wasted arms around his neck, crying, that he was his dear, kind friend. "I hope I always was. I meant to be, God knows," said the poor schoolmaster. "You remember my garden, Henry?" whispered the old man, anxious to rouse him, for a dullness seemed gathering upon the child, "and how pleasant it used to be in the evening time? You must make haste to visit it again, for I think the very flowers have missed you, and are less gay than they used to be. You will come soon, very soon now, won't you?"

The boy smiled faintly—so very, very faintly—and put his hand upon his friend's gray head. He moved his lips too, but no voice came from them,—no, not a sound. In the silence that ensued, the hum of distant voices, borne upon the evening air, came floating through the open window. "What's that?" said the sick child, opening his eyes. "The boys at play, upon the green." He took a handkerchief from his pillow, and tried to wave it above his head. But the feeble arm dropped powerless down. "Shall I do it?" said the schoolmaster. "Please wave it at the window," was the faint reply. "Tie it to the lattice. Some of them may see it there. Perhaps they'll think of me, and look this way."

He raised his head and glanced from the fluttering signal to his idle bat, that lay, with slate, and book, and other boyish property, upon the table in the room. And then he laid him softly down once more, and again clasped his little arms around the old man's neck. The two old friends and companions—for such they were, though they were man and child—held each other in a long embrace, and then the little scholar turned his face to the wall and fell asleep.

* * * * * * * * *

The poor schoolmaster sat in the same place, holding the small, cold hand in his, and chafing it. It was but the hand of a dead child. He felt that; and yet he chafed it still, and could not lay it down.

From "The Old Curiosity Shop," by Dickens.

XXVII. THE SNOW SHOWER.

William Cullen Bryant, 1794-1878, was the son of Peter Bryant, a physician of Cummington, Massachusetts. Amid the beautiful scenery of this remote country town, the poet was born; and here he passed his early youth. At the age of sixteen, Bryant entered Williams College, but was honorably dismissed at the end of two years. He then entered on the study of law, and was admitted to the bar at the age of twenty-one. He practiced his profession, with much success, for about nine years. In 1826, he removed to New York, and became connected with the "Evening Post," a connection which continued to the time of his death. For more than thirty of the last years of his life, Mr. Bryant made his home near Roslyn, Long Island, where he occupied an "old-time mansion," which he bought, fitted up, and surrounded in accordance with his excellent rural taste. A poem of his, written at the age of ten years, was published in the "County Gazette," and two poems of considerable length were published in book form, when the author was only fourteen. "Thanatopsis," perhaps the best known of all his poems, was written when he was but nineteen. But, notwithstanding his precocity, his powers continued to a remarkable age. His excellent translations of the "Iliad" and the "Odyssey," together with some of his best poems, were accomplished after the poet had passed the age of seventy. Mr. Bryant visited Europe several times; and, in 1849, he continued his travels into Egypt and Syria. Abroad, he was received with many marks of distinction; and he added much to his extensive knowledge by studying the literature of the countries he visited.

All his poems exhibit a peculiar love, and a careful study, of nature; and his language, both in prose and poetry, is always chaste, elegant, and correct. His mind was well-balanced; and his personal character was one to be admired, loved, and imitated.

STAND here by my side and turn, I pray,
On the lake below thy gentle eyes;
The clouds hang over it, heavy and gray,
And dark and silent the water lies;
And out of that frozen mist the snow
In wavering flakes begins to flow;
Flake after flake
They sink in the dark and silent lake.

See how in a living swarm they come
From the chambers beyond that misty veil;

Some hover in air awhile, and some
  Rush prone from the sky like summer hail.
All, dropping swiftly, or settling slow,
Meet, and are still in the depths below;
          Flake after flake
Dissolved in the dark and silent lake.

Here delicate snow stars, out of the cloud,
  Come floating downward in airy play,
Like spangles dropped from the glistening crowd
  That whiten by night the Milky Way;
There broader and burlier masses fall;
The sullen water buries them all,—
          Flake after flake,—
All drowned in the dark and silent lake.

And some, as on tender wings they glide
  From their chilly birth cloud, dim and gray.
Are joined in their fall, and, side by side,
  Come clinging along their unsteady way;
As friend with friend, or husband with wife,
Makes hand in hand the passage of life;
          Each mated flake
Soon sinks in the dark and silent lake.

Lo! while we are gazing, in swifter haste
  Stream down the snows, till the air is white,
As, myriads by myriads madly chased,
  They fling themselves from their shadowy height.
The fair, frail creatures of middle sky,
What speed they make, with their grave so nigh;
          Flake after flake
To lie in the dark and silent lake.

I see in thy gentle eyes a tear;
  They turn to me in sorrowful thought;
Thou thinkest of friends, the good and dear,
  Who were for a time, and now are not;
Like these fair children of cloud and frost,
That glisten a moment and then are lost,—
            Flake after flake,—
All lost in the dark and silent lake.

Yet look again, for the clouds divide;
  A gleam of blue on the water lies;
And far away, on the mountain side,
  A sunbeam falls from the opening skies.
But the hurrying host that flew between
The cloud and the water no more is seen;
            Flake after flake
At rest in the dark and silent lake.

XXVIII. CHARACTER OF NAPOLEON BONAPARTE.

Charles Phillips, 1787-1859, an eminent barrister and orator, was born in Sligo, Ireland, and died in London. He gained much of his reputation as an advocate in criminal cases. In his youth he published some verses; later in life he became the author of several works, chiefly of biography.

He is fallen! We may now pause before that splendid prodigy, which towered among us like some ancient ruin, whose power terrified the glance its magnificence attracted. Grand, gloomy, and peculiar, he sat upon the throne a sceptered hermit, wrapt in the solitude of his own originality. A mind, bold, independent, and decisive; a will, despotic in its dictates; an energy that distanced expedition; and a conscience, pliable to every touch of interest, marked the outlines of this extraordinary character—the most extraordinary, perhaps, that in the annals of this world ever rose, or reigned, or fell.

Flung into life in the midst of a revolution that quickened every energy of a people who acknowledged no superior, he commenced his course, a stranger by birth, and a scholar by charity. With no friend but his sword, and no fortune but his talents, he rushed into the lists where rank, and wealth, and genius had arrayed themselves, and competition fled from him, as from the glance of destiny.

He knew no motive but interest; acknowledged no criterion but success; he worshiped no God but ambition; and, with an eastern devotion, he knelt at the shrine of his idolatry. Subsidiary to this, there was no creed that he did not profess, there was no opinion that he did not promulgate: in the hope of a dynasty, he upheld the crescent; for the sake of a divorce, he bowed before the cross; the orphan of St. Louis, he became the adopted child of the Republic; and, with a parricidal ingratitude, on the ruins both of the throne and the tribune, he reared the throne of his despotism. A professed Catholic, he imprisoned the Pope; a pretended patriot, he impoverished the country; and in the name of Brutus, he grasped without remorse, and wore without shame, the diadem of the Cæsars.

The whole continent trembled at beholding the audacity of his designs, and the miracle of their execution. Skepticism bowed to the prodigies of his performance; romance assumed the air of history; nor was there aught too incredible for belief, or too fanciful for expectation, when the world saw a subaltern of Corsica waving his imperial flag over her most ancient capitals. All the visions of antiquity became commonplace in his contemplation: kings were his people; nations were his outposts; and he disposed of courts, and crowns, and camps, and churches, and cabinets, as if they were the titular dignitaries of the chessboard! Amid all these changes, he stood immutable as adamant. It mattered little whether in the field, or in the drawing-room; with the mob, or the levee; wearing the Jacobin bonnet, or the iron crown; banishing a Braganza, or espousing a Hapsburg; dictating peace on a raft to the Czar of Russia, or contemplating defeat at the gallows of Leipsic; he was still the same military despot.

In this wonderful combination, his affectations of literature must not be omitted. The jailer of the press, he affected the patronage of letters; the proscriber of books, he encouraged philosophy; the persecutor of authors, and the murderer of printers, he yet pretended to the protection of learning; the assassin of Palm, the silencer of De Staël, and the denouncer of Kotzebue, he was the friend of David, the benefactor of De Lille, and sent his academic prize to the philosopher of England.

Such a medley of contradictions, and, at the same time, such an individual consistency, were never united in the same character. A royalist, a republican, and an emperor; a Mohammedan, a Catholic, and a patron of the synagogue; a subaltern and a sovereign; a traitor and a tyrant; a Christian and an infidel; he was, through all his vicissitudes, the same stern, impatient, inflexible original; the same mysterious, incomprehensible self; the man without a model, and without a shadow.

NOTES.—**St. Louis** (b. 1215, d. 1270), a wise and pious king of France, known as Louis IX. Napoleon was appointed to the Military School at Brienne, by Louis XVI. **Brutus,** Lucius Junius, abolished the royal office at Rome (509 B. C.), and ruled as consul for two years.

Jacobin Bonnet.—The Jacobins were a powerful political club during the first French Revolution. A peculiar bonnet or hat was their badge. **Braganza,** the name of the royal family of Portugal. Maria of Portugal, and her father, Charles IV. of Spain, were both expelled by Napoleon. **Hapsburg,** the name of the royal family of Austria. Napoleon's second wife was Maria Louisa, the daughter of the Emperor. **Czar.**—The treaty of Tilsit was agreed to between Bonaparte and the Czar Alexander on the river Memel. **Leipsic.**—Napoleon was defeated by the allied forces, in October, 1813, at this city.

Palm, a German publisher, shot, in 1806, by order of Napoleon, for publishing a pamphlet against him. **De Staël** (pro. De Stäl), a celebrated French authoress, banished from Paris, in 1802, by Napoleon. **Kotzebue,** an eminent German dramatist. **David,** the leading historical painter of his times in France. **De Lille,** an eminent French poet and professor.

XXIX. NAPOLEON AT REST.

John Pierpont, 1785-1866, was born in Litchfield, Connecticut, and graduated from Yale College in 1804. The next four years he spent as a private tutor in the family of Col. William Allston, of South Carolina. On his return, he studied law in the law school of his native town. He entered upon practice, but soon left the law for mercantile pursuits, in which he was unsuccessful. Having studied theology at Cambridge, in 1819 he was ordained pastor of the Hollis Street Unitarian Church, in Boston, where he continued nearly twenty years. He afterwards preached four years for a church in Troy, New York, and then removed to Medford, Massachusetts. At the age of seventy-six, he became chaplain of a Massachusetts regiment; but, on account of infirmity, was soon obliged to give up the position. Mr. Pierpont published a series of school readers, which enjoyed a well-deserved popularity for many years.

His poetry is smooth, musical, and vigorous. Most of his pieces were written for special occasions.

His falchion flashed along the Nile;
His hosts he led through Alpine snows;
O'er Moscow's towers, that blazed the while,
His eagle flag unrolled,—and froze.
Here sleeps he now, alone! Not one
Of all the kings, whose crowns he gave,
Bends o'er his dust;—nor wife nor son
Has ever seen or sought his grave.

Behind this seagirt rock, the star,
That led him on from crown to crown,
Has sunk; and nations from afar
Gazed as it faded and went down.
High is his couch;—the ocean flood,
Far, far below, by storms is curled;
As round him heaved, while high he stood,
A stormy and unstable world.

Alone he sleeps! The mountain cloud,
That night hangs round him, and the breath
Of morning scatters, is the shroud
That wraps the conqueror's clay in death.
Pause here! The far-off world, at last,
Breathes free; the hand that shook its thrones,
And to the earth its miters cast,
Lies powerless now beneath these stones.

Hark! comes there from the pyramids,
And from Siberian wastes of snow,
And Europe's hills, a voice that bids
The world he awed to mourn him? No:
The only, the perpetual dirge
That's heard there is the sea bird's cry,—
The mournful murmur of the surge,—
The cloud's deep voice, the wind's low sigh.

NOTE.—**Seagirt rock,** the island of St. Helena, is in the Atlantic Ocean, nearly midway between Africa and South America. Napoleon was confined on this island six years, until 1821, when he died and was buried there. In 1841, his remains were removed to Paris.

XXX. WAR.

Charles Sumner, 1811-1874, was born in Boston. He studied at the Latin school in his native city, graduated from Harvard University at the age of nineteen, studied law at the same institution, and was admitted to practice in 1834. He at once took a prominent position in his profession, lectured to the law classes at Cambridge for several successive years, wrote and edited several standard law books, and might have had a professorship in the law school, had he desired it. In his famous address on "The True Grandeur of Nations," delivered July 4, 1845, before the municipal authorities of Boston, he took strong grounds against war among nations. In 1851 he was elected to the United States Senate, and continued in that position till his death. As a jurist, as a statesman, as an orator, and as a profound and scholarly writer, Mr. Sumner stands high in the estimation of his countrymen. In physical appearance, Mr. Sumner was grand and imposing; men often turned to gaze after him, as he passed along the streets of his native city.

I NEED not dwell now on the waste and cruelty of war. These stare us wildly in the face, like lurid meteor lights, as we travel the page of history. We see the desolation and death that pursue its demoniac footsteps. We look upon sacked towns, upon ravaged territories, upon violated homes; we behold all the sweet charities of life changed to wormwood and gall. Our soul is penetrated by the sharp moan of mothers, sisters, and daughters—of fathers, brothers, and sons, who, in the bitterness of their bereavement, refuse to be comforted. Our eyes rest at last upon one of these fair fields, where Nature, in her abundance, spreads her cloth of gold, spacious and apt for the entertainment of mighty multitudes—or, perhaps, from the curious subtlety of its position, like the carpet in the Arabian tale, seeming to contract so as to be covered by a few only, or to dilate so as to receive an innumerable host. Here, under a bright sun, such as shone at Austerlitz or Buena Vista—amidst the peaceful harmonies of nature—on the Sabbath of peace—we behold bands of brothers, children of a common Father, heirs to a common happiness, struggling together in the deadly fight, with the madness of fallen spirits, seeking with murderous weapons the lives of brothers who have never injured them or their kindred. The havoc rages. The ground is soaked with their commingling blood. The air is rent by their commingling cries. Horse and rider are stretched together on the earth. More revolting than the mangled victims, than the gashed limbs, than the lifeless trunks, than the spattering brains, are the lawless passions which sweep, tempest-like, through the fiendish tumult.

Horror-struck, we ask, wherefore this hateful contest? The melancholy, but truthful answer comes, that this is the *established* method of determining justice between nations!

The scene changes. Far away on the distant pathway of the ocean two ships approach each other, with white canvas broadly spread to receive the flying gales. They are proudly built. All of human art has been lavished in their graceful proportions, and in their well compacted sides, while they look in their dimensions like floating happy islands on the sea. A numerous crew, with costly appliances of comfort, hives in their secure shelter. Surely these two travelers shall meet in joy and friendship; the flag at the masthead shall give the signal of friendship; the happy sailors shall cluster in the rigging, and even on the yardarms, to look each other in the face, while the exhilarating voices of both crews shall mingle in accents of gladness uncontrollable. It is not so. Not as brothers, not as friends, not as wayfarers of the common ocean, do they come together; but as enemies.

The gentle vessels now bristle fiercely with death-dealing instruments. On their spacious decks, aloft on all their masts, flashes the deadly musketry. From their sides spout cataracts of flame, amidst the pealing thunders of a fatal artillery. They, who had escaped "the dreadful touch of merchant-marring rocks"—who had sped on their long and solitary way unharmed by wind or wave—whom the hurricane had spared—in whose favor storms and seas had intermitted their immitigable war—now at last fall by the hand of each other. The same spectacle of horror greets us from both ships. On their decks, reddened with blood, the murderers of St. Bartholomew and of the Sicilian Vespers, with the fires of Smithfield, seem to break forth anew, and to concentrate their rage. Each has now become a swimming Golgotha. At length, these vessels—such pageants of the sea—once so stately—so proudly built—but now rudely shattered by cannon balls—with shivered masts and ragged sails—exist only as unmanageable wrecks, weltering on the uncertain waves, whose temporary lull of peace is now their only safety. In amazement at this strange, unnatural contest—away from country and home—where there is no country or home to defend—we ask again, wherefore this dismal duel? Again the melancholy but truthful answer promptly comes, that this is the *established* method of determining justice between nations.

NOTES.—**Austerlitz,** a small town in Austria, seventy miles north from Vienna. It is noted as the site of a battle, in December, 1805, between the allied Austrian and Russian armies, and the French under Napoleon. The latter were victorious. **Buena Vista,** a small hamlet in eastern Mexico, where, in 1847, five thousand Americans, under Gen. Taylor, defeated twenty thousand Mexicans, under Gen. Santa Anna.

Dreadful touch.—Quoted from Merchant of Venice, Act III, Scene II.

St. Bartholomew.—A terrible massacre took place in France, on St. Bartholomew's day, August 24, 1572. It has been estimated that twenty thousand persons perished.

Sicilian Vespers, a revolt and uprising against the French in Sicily, March 30, 1282, at the hour of vespers.

Smithfield, a portion of London noted as a place for execution during the sixteenth and seventeenth centuries.

XXXI. SPEECH OF WALPOLE IN REPROOF OF MR. PITT.

Sir Robert Walpole, 1676-1745, was educated at Eton and Cambridge. He entered Parliament in 1700, and soon became a good debater and skillful tactician. He was prime minister of Great Britain from 1721 to 1742, in the reigns of George I. and George II. He was an able statesman; but has been accused of employing corruption or bribery on a large scale, to control Parliament and accomplish his purposes.

I WAS unwilling to interrupt the course of this debate, while it was carried on with calmness and decency, by men who do not suffer the ardor of opposition to cloud their reason, or transport them to such expressions as the dignity of this assembly does not admit.

I have hitherto deferred answering the gentleman, who declaimed against the bill with such fluency and rhetoric, and such vehemence of gesture; who charged the advocates for the expedients now proposed, with having no regard to any interests but their own, and with making laws only to consume paper, and threatened them with the defection of their adherents, and the loss of their influence, upon this new discovery of their folly and ignorance. Nor, do I now answer him for any other purpose than to remind him how little the clamor of rage and petulancy of invective contribute to the end for which this assembly is called together; how little the discovery of truth is promoted, and the security of the nation established, by pompous diction and theatrical emotion.

Formidable sounds and furious declamation, confident assertions and lofty periods, may affect the young and inexperienced; and perhaps the gentleman may have contracted his habits of oratory by conversing more with those of his own age than with such as have more opportunities of acquiring knowledge, and more successful methods of communicating their sentiments. If the heat of temper would permit him to attend to those whose age and long acquaintance with business give them an indisputable right to deference and superiority, he would learn in time to reason, rather than declaim; and to prefer justness of argument and an accurate knowledge of facts, to sounding epithets and splendid superlatives, which may disturb the imagination for a moment, but leave no lasting impression upon the mind. He would learn, that to accuse and prove are very different; and that reproaches, unsupported by evidence, affect only the character of him that utters them.

Excursions of fancy and flights of oratory are indeed pardonable in young men, but in no other; and it would surely contribute more, even to the purpose for which some gentlemen appear to speak (that of depreciating the conduct of the administration), to prove the inconveniences and injustice of this bill, than barely to assert them, with whatever magnificence of language, or appearance of zeal honesty, or compassion.

XXXII. PITT'S REPLY TO SIR ROBERT WALPOLE.

William Pitt, 1708-1778, one of the ablest statesmen and orators of his time, was born in Cornwall, and educated at Eton and Oxford. He entered Parliament in 1735, and became a formidable opponent of the ministry of Sir Robert Walpole. He gained great reputation by his wise and vigorous management of military affairs in the last years of the reign of George II. He opposed the "Stamp Act" with great earnestness, as well as the course of the ministry in the early years of the American Revolution. In 1778, he rose from a sick bed to make his celebrated speech, in the House of Lords, in opposition to a motion to acknowledge the independence of America. At its close, he fell in an apoplectic fit, and was borne home to die in a few weeks afterward. He was buried in Westminster Abbey. Mr. Pitt possessed a fine personal presence and a powerful voice; he was very popular with the people, and is often called the "Great Commoner." He was created "Earl of Chatham" in 1766.

THE atrocious crime of being a young man, which the honorable gentleman has, with such spirit and decency, charged upon me, I shall neither attempt to palliate nor deny; but content myself with hoping that I may be one of those whose follies cease with their youth, and not of that number who are ignorant in spite of experience. Whether youth can be imputed to a man as a reproach, I will not assume the province of determining; but surely age may become justly contemptible, if the opportunities which it brings have passed away without improvement, and vice appears to prevail when the passions have subsided. The wretch, who, after having seen the consequences of a thousand errors, continues still to blunder, and whose age has only added obstinacy to stupidity, is surely the object either of abhorrence or contempt, and deserves not that his gray hairs should secure him from insult. Much more is he to be abhorred, who, as he has advanced in age, has receded from virtue, and become more wicked—with less temptation; who prostitutes himself for money which he can not enjoy, and spends the remains of his life in the ruin of his country.

But youth is not my only crime; I am accused of acting a theatrical part. A theatrical part may either imply some peculiarity of gesture, or a dissimulation of my real sentiments, and an adoption of the opinions and language of another man. In the first sense, the charge is too trifling to be confuted, and deserves only to be mentioned that it may be despised. I am at liberty, like every other man, to use my own language; and though, perhaps, I may have some ambition to please this gentleman, I shall not lay myself under any restraint, nor very solicitously copy his diction or his mien, however matured by age, or modeled by experience.

But, if any man shall, by charging me with theatrical behavior, imply that I utter any sentiments but my own, I shall treat him as a calumniator and a villain; nor shall any protection shelter him from the treatment he deserves. I shall, on such an occasion, without scruple, trample upon all those forms with which wealth and dignity intrench themselves, nor shall anything but age restrain my resentment; age,—which always brings one privilege, that of being insolent and supercilious, without punishment.

But, with regard to those whom I have offended, I am of opinion that, if I had acted a borrowed part, I should have avoided their censure: the heat that offended them was the ardor of conviction, and that zeal for the service of my country which neither hope nor fear shall influence me to suppress. I will not sit unconcerned while my liberty is invaded, nor look in silence upon public robbery. I will exert my endeavors, at whatever hazard, to repel the aggressor, and drag the thief to justice, whoever may protect him in his villainies, and whoever may partake of his plunder.

XXXIII. CHARACTER OF MR. PITT.

Henry Grattan, 1750–1820, an Irish orator and statesman, was born at Dublin, and graduated from Trinity College, in his native city. By his admiration of Mr. Pitt, the first Earl of Chatham, he was led to turn his attention to oratory. In personal appearance, he was unprepossessing; but his private character was without a blemish.

THE secretary stood alone. Modern degeneracy had not reached him. Original and unaccommodating, the features of his character had the hardihood of antiquity. His august mind overawed majesty itself. No state chicanery, no narrow system of vicious politics, no idle contest for ministerial victories, sank him to the vulgar level of the great; but overbearing, persuasive, and impracticable, his object was England, his ambition was fame.

Without dividing, he destroyed party; without corrupting, he made a venal age unanimous. France sunk beneath him. With one hand he smote the house of Bourbon, and wielded in the other the democracy of England. The sight of his mind was infinite; and his schemes were to effect, not England, not the present age only, but Europe and posterity. Wonderful were the means by which those schemes were accomplished; always seasonable, always adequate, the suggestion of an understanding animated by ardor and enlightened by prophecy.

The ordinary feelings which make life amiable and indolent, were unknown to him. No domestic difficulties, no domestic weakness, reached him; but, aloof from the sordid occurrences of life, and unsullied by its intercourse, he came occasionally into our system, to counsel and decide. A character so exalted, so strenuous, so various, so authoritative, astonished a corrupt age, and the treasury trembled at the name of Pitt, through all classes of venality. Corruption imagined, indeed, that she had found defects in this statesman, and talked much of the inconsistency of his glory, and much of the ruin of his victories; but the history of his country, and the calamities of the enemy, answered and refuted her.

Nor were his political his only talents. His eloquence was an era in the senate; peculiar and spontaneous; familiarly expressing gigantic sentiments and instructive wisdom; not like the torrent of Demosthenes, or the splendid conflagration of Tully; it resembled sometimes the thunder, and sometimes the music of the spheres. He did not conduct the understanding through the painful subtilty of argumentation, nor was he ever on the rack of exertion; but rather lightened upon the subject, and reached the point by the flashings of the mind, which, like those of the eye, were felt, but could not be followed.

Upon the whole, there was in this man something that could create, subvert, or reform; an understanding, a spirit, and an eloquence, to summon mankind to society, or to break the bonds of slavery asunder, and to rule the wildness of free minds with unbounded authority; something that could establish or overwhelm empires, and strike a blow in the world that should resound through the universe.

Notes.—**Demosthenes** (b. 385, d. 322, B. C.) was the son of a cutler at Athens, Greece. By diligent study and unremitting toil, he became the greatest orator that ever lived.

Tully, Marcus Tullius Cicero (b. 106, d. 43, B. C.), was the most remarkable of Roman orators. He held the highest office of the Republic.

XXXIV. THE SOLDIER'S REST.

Sir Walter Scott, 1771-1832, the great Scotch poet and novelist, was born in Edinburgh. Being a feeble child, he was sent to reside on his grandfather's estate in the south of Scotland. Here he spent several years, and gained much knowledge of the traditions of border warfare, as well as of the tales and ballads pertaining to it. He was also a great reader of romances in his youth. In 1779 he returned to Edinburgh, and became a pupil in the high school. Four years later, he entered the university; but, in neither school nor college, was he distinguished for scholarship. In 1797 he was admitted to the practice of law,—a profession which he soon forsook for literature. His first poems appeared in 1802. The "Lay of the Last Minstrel" was published in 1805, "Marmion" in 1808, and "The Lady of the Lake" in 1810. Several poems of less power followed. In 1814 "Waverley," his first novel, made its appearance, but the author was unknown for some time. Numerous other novels followed with great rapidity, the author reaping a rich harvest both in fame and money. In 1811 he purchased an estate near the Tweed, to which he gave the name of Abbotsford. In enlarging his estate and building a costly house, he spent vast sums of money. This, together with the failure of his publishers in 1826, involved him very heavily in debt. But he set to work with almost superhuman effort to pay his debts by the labors of his pen. In about four years, he had paid more than $300,000; but the effort was too much for his strength, and hastened his death.

In person, Scott was tall, and apparently robust, except a slight lameness with which he was affected from childhood. He was kindly in disposition, hospitable in manner, fond of outdoor pursuits and of animals, especially dogs. He wrote with astonishing rapidity, and always in the early morning. At his death, he left two sons and two daughters. A magnificent monument to his memory has been erected in the city of his birth. The following selection is from "The Lady of the Lake."

SOLDIER, rest! thy warfare o'er,
  Sleep the sleep that knows not breaking;
Dream of battlefields no more,
  Days of danger, nights of waking.
In our isle's enchanted hall,
  Hands unseen thy couch are strewing,
Fairy strains of music fall,
  Every sense in slumber dewing.
Soldier, rest! thy warfare o'er,
Dream of battlefields no more;
Sleep the sleep that knows not breaking,
Morn of toil, nor night of waking.

No rude sound shall reach thine ear,
  Armor's clang, or war steed champing,
Trump nor pibroch summon here
  Mustering clan, or squadron tramping.
Yet the lark's shrill fife may come,
  At the daybreak from the fallow,
And the bittern sound his drum,
  Booming from the sedgy shallow.
Ruder sounds shall none be near,
Guards nor warders challenge here,
Here 's no war steed's neigh and champing,
Shouting clans or squadrons stamping.

Huntsman, rest! thy chase is done;
  While our slumb'rous spells assail ye,
Dream not, with the rising sun,
  Bugles here shall sound reveille.
Sleep! the deer is in his den;
  Sleep! thy hounds are by thee lying;
Sleep! nor dream in yonder glen,
  How thy gallant steed lay dying.

Huntsman, rest; thy chase is done,
Think not of the rising sun,
For at dawning to assail ye,
Here no bugle sounds reveille.

NOTES—**Pibroch** (pro. pē′brŏk). This is a wild, irregular species of music, peculiar to the Highlands of Scotland. It is performed on a bagpipe, and adapted to excite or assuage passion, and particularly to rouse a martial spirit among troops going to battle.

Reveille (pro. re-vāl′yā) is an awakening call at daybreak. In the army it is usually sounded on the drum.

XXXV. HENRY V. TO HIS TROOPS.

William Shakespeare, 1564–1616, was born at Stratford-upon-Avon. By many (perhaps most) critics, Shakespeare is regarded as the greatest poet the world has ever produced; one calls him, "The most illustrious of the sons of men." And yet it is a curious fact that less is really known of his life and personal characteristics than is known of almost any other famous name in history. Over one hundred years ago, a writer said, "All that is known with any degree of certainty concerning Shakespeare is—that he was born at Stratford-upon-Avon—married and had children there—went to London, where he commenced acting, and wrote poems and plays—returned to Stratford, made his will, died, and was buried." All the research of the last one hundred years has added but very little to this meager record. He was married, very young, to Anne Hathaway, a woman eight years his senior; was joint proprietor of Blackfriar's Theater in 1589, and seems to have accumulated property, and retired three or four years before his death. He was buried in Stratford Church, where a monument has been erected to his memory; he also has a monument in "Poet's Corner" of Westminster Abbey. His family soon became extinct. From all we can learn, he seems to have been highly respected and esteemed by his cotemporaries.

His works consist chiefly of plays and sonnets. His writings show an astonishing knowledge of human nature, expressed in language wonderful for its point and beauty. His style is chaste and pure, judged by the standard of his times, although expressions may sometimes be found that would not be considered proper in a modern writer. It has been argued by some that Shakespeare did not write the works imputed to him; but this theory seems to have little to support it. This extract is from King Henry V., Act III, Scene I

ONCE more unto the breach, dear friends, once more;
Or close the wall up with our English dead.
In peace there 's nothing so becomes a man
As modest stillness and humility:
But when the blast of war blows in our ears,
Then imitate the action of the tiger;
Stiffen the sinews, summon up the blood,
Disguise fair nature with hard-favored rage;
Then lend the eye a terrible aspect;
Let it pry through the portage of the head
Like the brass cannon; let the brow o'erwhelm it
As fearfully as doth a galled rock
O'erhang and jutty his confounded base,
Swilled with the wild and wasteful ocean.

Now set the teeth, and stretch the nostril wide,
Hold hard the breath, and bend up every spirit
To its full height! On, on, you noblest English,
Whose blood is fet from fathers of war proof!
Fathers, that, like so many Alexanders,
Have, in these parts, from morn till even, fought,
And sheathed their swords for lack of argument;
Be copy now to men of grosser blood,
And teach them how to war.

And you, good yeomen,
Whose limbs were made in England, show us here
The mettle of your pasture; let us swear
That you are worth your breeding, which I doubt not;
For there is none of you so mean and base,
That hath not noble luster in your eyes.
I see you stand like greyhounds in the slips,
Straining upon the start. The game 's afoot;
Follow your spirit: and, upon this charge,
Cry—"God for Harry, England, and St. George!"

NOTES.—**Henry V.** (1388–1422) was king of England for nine years. During this reign almost continuous war raged in France, to the throne of which Henry laid claim. The battle of Agincourt took place in his reign.

Fet is the old form of *fetched.*

Alexanders.—Alexander the Great (356–323 B. C.) was king of Macedonia, and the celebrated conqueror of Persia, India, and the greater part of the world as then known.

XXXVI. SPEECH OF PAUL ON MARS HILL.

THEN Paul stood in the midst of Mars Hill, and said, Ye men of Athens! I perceive that in all things ye are too superstitious. For as I passed by, and beheld your devotions, I found an altar with this inscription, TO THE UNKNOWN GOD. Whom therefore ye ignorantly worship, him declare I unto you. God that made the world and all things therein (seeing that he is Lord of heaven and earth) dwelleth not in temples made with hands; neither is worshiped with men's hands, as though he needed anything, seeing he giveth to all life, and breath, and all things; and hath made of one blood all nations of men for to dwell on all the face of the earth, and hath determined the times before appointed, and the bounds of their habitation; that they should seek the Lord, if haply they might feel after him, and find him, though he be not far from everyone of us: for in him we live, and move, and have our being; as certain also of your own poets have said, For we are also his offspring. Forasmuch then as we are the offspring of God, we ought not to think that the Godhead is like unto gold, or silver, or stone, graven by art and man's device. And the times of this ignorance God winked at; but now commandeth all men everywhere to repent: because he hath appointed a day, in the which he will judge the world in righteousness by that Man whom he hath ordained; whereof he hatb given assurance unto all men, in that he hath raised him from the dead. And when they heard of the resurrection of the dead, some mocked: and others said, We will hear thee again of this matter. So Paul departed from among them. Howbeit certain men clave unto him, and believed; among the which was Dionysius the Areopagite, and a woman named Damaris, and others with them.

—*Bible.*

NOTES.—At the time this oration was delivered (50 A. D.), **Athens** still held the place she had occupied for centuries, as the center of the enlightened and refined world.

Mars Hill, or the Areopagus, was an eminence in the city. made famous as the place where the court, also called Areopagus, held its sittings.

Dionysius, surnamed Areopageita, from being a member of this court, was an eminent Greek scholar, who, after his conversion to Christianity by St. Paul, was installed, by the latter, as the first bishop of Athens. He afterwards suffered martyrdom.

XXXVII. GOD IS EVERYWHERE.

OH! show me where is He,
The high and holy One,
To whom thou bend'st the knee,
And prayest, "Thy will be done!"
I hear thy song of praise,
And lo! no form is near:
Thine eyes I see thee raise,
But where doth God appear?
Oh! teach me who is God, and where his glories shine,
That I may kneel and pray, and call thy Father mine.

"Gaze on that arch above:
The glittering vault admire.
Who taught those orbs to move?
Who lit their ceaseless fire?
Who guides the moon to run
In silence through the skies?
Who bids that dawning sun
In strength and beauty rise?
There view immensity! behold! my God is there:
The sun, the moon, the stars, his majesty declare.

"See where the mountains rise:
Where thundering torrents foam;
Where, veiled in towering skies,
The eagle makes his home:
Where savage nature dwells,
My God is present, too:
Through all her wildest dells
His footsteps I pursue:
He reared those giant cliffs, supplies that dashing stream,
Provides the daily food which stills the wild bird's scream.

"Look on that world of waves,
Where finny nations glide;
Within whose deep, dark caves
The ocean monsters hide:
His power is sovereign there,
To raise, to quell the storm;
The depths his bounty share,
Where sport the scaly swarm:
Tempests and calms obey the same almighty voice,
Which rules the earth and skies, and bids far worlds rejoice."

—*Joseph Hutton.*

XXXVIII. LAFAYETTE AND ROBERT RAIKES.

Thomas S. Grimké, 1786-1834, an eminent lawyer and scholar, was born in Charleston, South Carolina, graduated at Yale in 1807, and died of cholera near Columbus, Ohio. He descended from a Huguenot family that was exiled from France by the revocation of the edict of Nantes. He gained considerable reputation as a politician, but is best known as an advocate of peace, Sunday schools, and the Bible. He was a man of deep feeling, earnest purpose, and pure life. Some of his views were very radical and very peculiar. He proposed sweeping reforms in English orthography, and disapproved of the classics and of pure mathematics in any scheme of general education. The following is an extract from an address delivered at a Sunday-school celebration.

It is but a few years since we beheld the most singular and memorable pageant in the annals of time. It was a pageant more sublime and affecting than the progress of Elizabeth through England after the defeat of the Armada; than the return of Francis I. from a Spanish prison to his own beautiful France; than the daring and rapid march of the conqueror at Austerlitz from Frejus to Paris. It was a pageant, indeed, rivaled only in the elements of the grand and the pathetic, by the journey of our own Washington through the different states. Need I say that I allude to the visit of Lafayette to America?

But Lafayette returned to the land of the dead, rather than of the living. How many who had fought with him in the war of '76, had died in arms, and lay buried in the grave of the soldier or the sailor! How many who had survived the perils of battle, on the land and the ocean, had expired on the deathbed of peace, in the arms of mother, sister, daughter, wife! Those who survived to celebrate with him the jubilee of 1825, were stricken in years, and hoary-headed; many of them infirm in health; many the victims of poverty, or misfortune, or affliction. And, how venerable that patriotic company; how sublime their gathering through all the land; how joyful their welcome, how affecting their farewell to that beloved stranger!

But the pageant has fled, and the very materials that

gave it such depths of interest are rapidly perishing: and a humble, perhaps a nameless grave, shall hold the last soldier of the Revolution. And shall they ever meet again? Shall the patriots and soldiers of '76, the "Immortal Band," as history styles them, meet again in the amaranthine bowers of spotless purity, of perfect bliss, of eternal glory? Shall theirs be the Christian's heaven, the kingdom of the Redeemer? The heathen points to his fabulous Elysium as the paradise of the soldier and the sage. But the Christian bows down with tears and sighs, for he knows that not many of the patriots, and statesmen, and warriors of Christian lands are the disciples of Jesus.

But we turn from Lafayette, the favorite of the old and the new world, to the peaceful benevolence, the unambitious achievements of Robert Raikes. Let us imagine him to have been still alive, and to have visited our land, to celebrate this day with us. No national ships would have been offered to bear him, a nation's guest, in the pride of the star-spangled banner, from the bright shores of the rising, to the brighter shores of the setting sun. No cannon would have hailed him in the stern language of the battlefield, the fortunate champion of Freedom, in Europe and America. No martial music would have welcomed him in notes of rapture, as they rolled along the Atlantic, and echoed through the valley of the Mississippi. No military procession would have heralded his way through crowded streets, thickset with the banner and the plume, the glittering saber and the polished bayonet. No cities would have called forth beauty and fashion, wealth and rank, to honor him in the ballroom and theater. No states would have escorted him from boundary to boundary, nor have sent their chief magistrate to do him homage. No national liberality would have allotted to him a nobleman's domain and princely treasure. No national gratitude would have hailed him in the capitol itself, the nation's guest, because the nation's benefactor; and have consecrated a battle ship, in memory of his wounds and his gallantry.

Not such would have been the reception of Robert Raikes, in the land of the Pilgrims and of Penn, of the Catholic, the Cavalier, and the Huguenot. And who does not rejoice that it would be impossible thus to welcome this primitive Christian, the founder of Sunday schools? His heralds would be the preachers of the Gospel, and the eminent in piety, benevolence, and zeal. His procession would number in its ranks the messengers of the Cross and the disciples of the Savior, Sunday-school teachers and white-robed scholars. The temples of the Most High would be the scenes of his triumph. Homage and gratitude to him, would be anthems of praise and thanksgiving to God.

Parents would honor him as more than a brother; children would reverence him as more than a father. The faltering words of age, the firm and sober voice of manhood, the silvery notes of youth, would bless him as a Christian patron. The wise and the good would acknowledge him everywhere as a national benefactor, as a patriot even to a land of strangers. He would have come a messenger of peace to a land of peace. No images of camps, and sieges, and battles; no agonies of the dying and the wounded; no shouts of victory, or processions of triumph, would mingle with the recollections of the multitude who welcomed him. They would mourn over no common dangers, trials, and calamities; for the road of duty has been to them the path of pleasantness, the way of peace. Their memory of the past would be rich in gratitude to God, and love to man; their enjoyment of the present would be a prelude to heavenly bliss; their prospects of the future, bright and glorious as faith and hope. * * *

Such was the reception of Lafayette, the warrior; such would be that of Robert Raikes, the Howard of the Christian church. And which is the nobler benefactor, patriot, and philanthropist? Mankind may admire and extol Lafayette more than the founder of the Sunday schools; but religion, philanthropy, and enlightened common sense must ever esteem Robert Raikes the superior of Lafayette. His are the virtues, the services, the sacrifices of a more enduring and exalted order of being. His counsels and triumphs belong less to time than to eternity.

The fame of Lafayette is of this world; the glory of Robert Raikes is of the Redeemer's everlasting kingdom. Lafayette lived chiefly for his own age, and chiefly for his and our country; but Robert Raikes has lived for all ages and all countries. Perhaps the historian and biographer may never interweave his name in the tapestry of national or individual renown. But the records of every single church honor him as a patron; the records of the universal Church, on earth as in heaven, bless him as a benefactor.

The time may come when the name of Lafayette will be forgotten; or when the star of his fame, no longer glittering in the zenith, shall be seen, pale and glimmering, on the verge of the horizon. But the name of Robert Raikes shall never be forgotten; and the lambent flame of his glory is that eternal fire which rushed down from heaven to devour the sacrifice of Elijah. Let mortals then admire and imitate Lafayette more than Robert Raikes. But the just made perfect, and the ministering spirits around the throne of God, have welcomed him as a fellow-servant of the same Lord; as a fellow-laborer in the same glorious cause of man's redemption; as a coheir of the same precious promises and eternal rewards.

NOTES.—**Armada,** the great fleet sent out in 1588, by Philip II. of Spain, for the conquest of England, was defeated in the Channel by the English and Dutch fleets. After the victory, Queen Elizabeth made a triumphal journey through the kingdom.

Francis I. (b. 1494, d. 1547), King of France, was taken prisoner at the battle of Pavia, and confined at Madrid, Spain, nearly a year.

Austerlitz.—See Note on p. 150.

Lafayette (b. 1757, d. 1834), a French marquis, who served as major general in the Revolutionary War in America, which terminated in 1783. Lafayette revisited this country in 1824, and was received throughout the land with the greatest enthusiasm.

Robert Raikes (b. 1735, d. 1811), an English printer and philanthropist, noted as the founder of Sunday schools.

Howard, John (b. 1726, d. 1790), a celebrated English philanthropist, who spent much of his life in the endeavor to reform the condition of prisons in Europe.

XXXIX. FALL OF CARDINAL WOLSEY.

Wolsey. FAREWELL! a long farewell, to all my greatness!
This is the state of man: to-day he puts forth
The tender leaves of hopes; to-morrow blossoms,
And bears his blushing honors thick upon him;
The third day comes a frost, a killing frost,
And, when he thinks, good, easy man, full surely
His greatness is a-ripening, nips his root,
And then he falls, as I do. I have ventured,
Like little, wanton boys that swim on bladders,
This many summers in a sea of glory,
But far beyond my depth: my high-blown pride
At length broke under me, and now has left me,
Weary and old with service, to the mercy
Of a rude stream, that must forever hide me.
Vain pomp and glory of this world, I hate ye:
I feel my heart new open'd. Oh, how wretched
Is that poor man that hangs on princes' favors!
There is, betwixt that smile we would aspire to,

That sweet aspect of princes, and their ruin,
More pangs and fears than wars or women have:
And when he falls, he falls like Lucifer,
Never to hope again.

Enter CROMWELL, *and stands amazed.*

Why, how now, Cromwell!

Crom. I have no power to speak, sir.

Wol. What, amazed
At my misfortunes? Can thy spirit wonder,
A great man should decline? Nay, an you weep,
I am fall'n indeed.

Crom. How does your grace?

Wol. Why, well;
Never so truly happy, my good Cromwell.
I know myself now; and I feel within me
A peace above all earthly dignities,
A still and quiet conscience. The king has cured me,
I humbly thank his grace; and from these shoulders,
These ruin'd pillars, out of pity, taken
A load would sink a navy,—too much honor:
Oh, 't is a burthen, Cromwell, 't is a burthen,
Too heavy for a man that hopes for heaven!

Crom. I am glad your grace has made that right use of it.

Wol. I hope I have: I am able now, methinks,
Out of a fortitude of soul I feel,
To endure more miseries, and greater far,
Than my weak-hearted enemies dare offer.
What news abroad?

Crom. The heaviest, and the worst,
Is your displeasure with the king.

Wol. God bless him!

Crom. The next is, that Sir Thomas More is chosen
Lord chancellor in your place.

Wol. That's somewhat sudden:
But he's a learned man. May he continue
Long in his highness' favor, and do justice
For truth's sake and his conscience; that his bones,
When he has run his course, and sleeps in blessings,
May have a tomb of orphans' tears wept on 'em!
What more?

Crom. That Cranmer is return'd with welcome,
Install'd lord archbishop of Canterbury.

Wol. That's news indeed!

Crom. Last, that the Lady Anne,
Whom the king hath in secrecy long married,
This day was viewed in open as his queen,
Going to chapel; and the voice is now
Only about her coronation.

Wol. There was the weight that pull'd me down. O
Cromwell,
The king has gone beyond me: all my glories
In that one woman I have lost forever:
No sun shall ever usher forth mine honors,
Or gild again the noble troops that waited
Upon my smiles. Go, get thee from me, Cromwell;
I am a poor, fall'n man, unworthy now
To be thy lord and master: seek the king;
That sun, I pray, may never set! I have told him
What and how true thou art: he will advance thee;
Some little memory of me will stir him—
I know his noble nature—not to let
Thy hopeful service perish, too: good Cromwell,
Neglect him not; make use now, and provide
For thine own future safety.

Crom. O my lord,
Must I, then, leave you? Must I needs forego
So good, so noble, and so true a master?
Bear witness, all that have not hearts of iron,
With what a sorrow Cromwell leaves his lord.
The king shall have my service; but my prayers
Forever and forever shall be yours.

Wol. Cromwell, I did not think to shed a tear
In all my miseries; but thou hast forced me,
Out of thy honest truth, to play the woman.
Let's dry our eyes: and thus far hear me, Cromwell;
And, when I am forgotten, as I shall be,
And sleep in dull, cold marble, where no mention
Of me more must be heard of, say, I taught thee;
Say, Wolsey, that once trod the ways of glory,
And sounded all the depths and shoals of honor,
Found thee a way, out of his wreck, to rise in;
A sure and safe one, though thy master missed it.
Mark but my fall, and that that ruin'd me.
Cromwell, I charge thee, fling away ambition:
By that sin fell the angels; how can man, then,
The image of his Maker, hope to win by it?
Love thyself last: cherish those hearts that hate thee;
Corruption wins not more than honesty.
Still in thy right hand carry gentle peace,
To silence envious tongues. Be just, and fear not:
Let all the ends thou aim'st at be thy country's,
Thy God's, and truth's; then, if thou fall'st, O
Cromwell,
Thou fall'st a blessed martyr! Serve the king;
And,—prithee, lead me in:
There, take an inventory of all I have,
To the last penny; 't is the king's: my robe,
And my integrity to Heaven, is all
I dare now call mine own. O Cromwell, Cromwell!

Had I but served my God with half the zeal
I served my king, He would not in mine age
Have left me naked to mine enemies.

Crom. Good sir, have patience.

Wol. So I have. Farewell
The hopes of court! my hopes in Heaven do dwell.

Shakespeare.—Henry VIII, Act iii, Scene ii.

NOTES.—**Wolsey,** Cardinal Thomas (b. 1471, d. 1530), was for several years the favored minister of Henry VIII. of England. He acquired great wealth and power. In 1522, he was one of the candidates for the Papal throne. In 1529, he was disgraced at the English court and arrested.

Cromwell, Thomas (b. 1490, d. 1540), was Wolsey's servant. After Wolsey's death, he became secretary to Henry VIII., and towards the close of his life was made Earl of Essex.

XL. THE PHILOSOPHER.

John P. Kennedy, 1795-1870. This gentleman, eminent in American politics and literature, was born in Baltimore, graduated at the College of Baltimore, and died in the same city. He served several years in the Legislature of his native state, and three terms in the United States House of Representatives. He was Secretary of the Navy during a part of President Fillmore's administration, and was active in sending out the famous Japan expedition, and Dr. Kane's expedition in search of Sir John Franklin. Mr. Kennedy wrote several novels, as well as political and other papers. His writings are marked by ease and freshness. The following extract is from "Swallow Barn," a series of sketches of early Virginia.

FROM the house at Swallow Barn there is to be seen, at no great distance, a clump of trees, and in the midst of these a humble building is discernible, that seems to court the shade in which it is modestly embowered. It is an old structure built of logs. Its figure is a cube, with a roof rising from all sides to a point, and surmounted by a wooden weathercock, which somewhat resembles a fish and somewhat a fowl.

This little edifice is a rustic shrine devoted to Cadmus, and is under the dominion of parson Chub. He is a plump, rosy old gentleman, rather short and thickset, with the blood vessels meandering over his face like rivulets,—a pair of prominent blue eyes, and a head of silky hair not unlike the covering of a white spaniel. He may be said to be a man of jolly dimensions, with an evident taste for good living, sometimes sloven in his attire, for his coat—which is not of the newest—is decorated with sundry spots that are scattered over it in constellations. Besides this, he wears an immense cravat, which, as it is wreathed around his short neck, forms a bowl beneath his chin, and—as Ned says—gives the parson's head the appearance of that of John the Baptist upon a charger, as it is sometimes represented in the children's picture books. His beard is grizzled with silver stubble, which the parson reaps about twice a week—if the weather be fair.

Mr. Chub is a philosopher after the order of Socrates. He was an emigrant from the Emerald Isle, where he suffered much tribulation in the disturbances, as they are mildly called, of his much-enduring country. But the old gentleman has weathered the storm without losing a jot of that broad, healthy benevolence with which Nature has enveloped his heart, and whose ensign she has hoisted in his face. The early part of his life had been easy and prosperous, until the rebellion of 1798 stimulated his republicanism into a fever, and drove the full-blooded hero headlong into a quarrel, and put him, in spite of his peaceful profession, to standing by his pike in behalf of his principles. By this unhappy boiling over of the caldron of his valor, he fell under the ban of the ministers, and tested his share of government mercy. His house was burnt over his head, his horses and hounds (for, by all accounts, he was a perfect Actæon) were "confiscate to the state," and he was forced to fly. This brought him to America in no very compromising mood with royalty.

Here his fortunes appear to have been various, and he was tossed to and fro by the battledoor of fate, until he found a snug harbor at Swallow Barn; where, some years ago, he sat down in that quiet repose which a worried and badgered patriot is best fitted to enjoy.

He is a good scholar, and, having confined his readings entirely to the learning of the ancients, his republicanism is somewhat after the Grecian mold. He has never read any politics of later date than the time of the Emperor Constantine, not even a newspaper,—so that he may be said to have been contemporary with Æschines rather than Lord Castlereagh—until that eventful epoch of his life when his blazing rooftree awakened him from his anachronistical dream. This notable interruption, however, gave him but a feeble insight into the moderns, and he soon relapsed to Thucydides and Livy, with some such glimmerings of the American Revolution upon his remembrance as most readers have of the exploits of the first Brutus.

The old gentleman had a learned passion for folios. He had been a long time urging Meriwether to make some additions to his collections of literature, and descanted upon the value of some of the ancient authors as foundations, both moral and physical, to the library. Frank gave way to the argument, partly to gratify the parson, and partly from the proposition itself having a smack that touched his fancy. The matter was therefore committed entirely to Mr. Chub, who forthwith set out on a voyage of exploration to the north. I believe he got as far as Boston. He certainly contrived to execute his commission with a curious felicity. Some famous Elzevirs were picked up, and many other antiques that nobody but Mr. Chub would ever think of opening.

The cargo arrived at Swallow Barn in the dead of winter. During the interval between the parson's return from his expedition and the coming of the books, the reverend little schoolmaster was in a remarkably unquiet state of body, which almost prevented him from sleeping: and it is said that the sight of the long-expected treasures had the happiest effect upon him. There was ample accommodation for this new acquisition of ancient wisdom provided before its arrival, and Mr. Chub now spent a whole week in arranging the volumes on their proper shelves, having, as report affirms, altered the arrangement at least seven times during that period. Everybody wondered what the old gentleman was at, all this time; but it was discovered afterwards, that he was endeavoring to effect a distribution of the works according to a minute division of human science, which entirely failed, owing to the unlucky accident of several of his departments being without any volumes.

After this matter was settled, he regularly spent his evenings in the library. Frank Meriwether was hardly behind the parson in this fancy, and took, for a short time, to abstruse reading. They both consequently deserted the little family circle every evening after tea, and might have continued to do so all the winter but for a discovery made by Hazard.

Ned had seldom joined the two votaries of science in their philosophical retirement, and it was whispered in the family that the parson was giving Frank a quiet course of lectures in the ancient philosophy, for Meriwether was known to talk a great deal, about that time, of the old and new Academicians. But it happened upon one dreary winter night, during a tremendous snowstorm, which was banging the shutters and doors of the house so as to keep up a continual uproar, that Ned, having waited in the parlor for the philosophers until midnight, set out to invade their retreat—not doubting that he should find them deep in study. When he entered the library, both candles were burning in their sockets, with long, untrimmed wicks;

the fire was reduced to its last embers, and, in an arm-chair on one side of the table, the parson was discovered in a sound sleep over Jeremy Taylor's "Ductor Dubitantium," whilst Frank, in another chair on the opposite side, was snoring over a folio edition of Montaigne. And upon the table stood a small stone pitcher, containing a residuum of whisky punch, now grown cold. Frank started up in great consternation upon hearing Ned's footstep beside him, and, from that time, almost entirely deserted the library. Mr. Chub, however, was not so easily drawn away from the career of his humor, and still shows his hankering after his leather-coated friends.

NOTES.—**Cadmus** is said to have taught the Greeks the use of the alphabet.

Socrates (b. 469, d. 399 B. C.), a noted Athenian philosopher. **Rebellion.**—In 1798, the Irish organized and rose against the English rule. The rebellion was suppressed.

Actæon [Ak-tē′on], a fabled Greek hunter, who was changed into a stag.

Constantine, the Great (b. 272, d, 337), the first Christian emperor of Rome. He was an able general and wise legislator. In 328, he removed his capital to Byzantium, which he named Constantinople. **Æschines** [ĕs′ke-nēz] (b. 389, d. 314 B. C.), an Athenian orator, the rival of Demosthenes. **Castlereagh**, Lord (b. 1769, d. 1822), a British statesman. He was in power, and prominent in the suppression of the Rebellion. **Brutus,** see p. 145.

Elzevirs [ĕl′ze-virs̱], the name of a family of Dutch printers noted for the beauty of their workmanship. They lived from 1540 to 1680.

Academicians.—The *Old* Academy was founded by Plato, at Athens, about 380 B. C. The *New*, by Carneades, about two hundred years later.

Jeremy Taylor (b. 1613, d. 1667), an English bishop and writer. His *Ductor Dubitantium*, or "Rule of Conscience," was one of his chief works. **Montaigne**, Michel (b. 1533, d. 1592), was a celebrated French writer of peculiar characteristics. He owes his reputation entirely to his "Essais."

XLI. MARMION AND DOUGLAS.

NOT far advanced was morning day,
When Marmion did his troop array
  To Surrey's camp to ride;
He had safe conduct for his band,
Beneath the royal seal and hand,
  And Douglas gave a guide.

The train from out the castle drew,
But Marmion stopped to bid adieu:
"Though something I might plain," he said,
  "Of cold respect to stranger guest,
  Sent hither by your king's behest,
While in Tantallon's towers I staid,
  Part we in friendship from your land,
  And, noble Earl, receive my hand."
But Douglas round him drew his cloak,
Folded his arms, and thus he spoke:
  "My manors, halls, and bowers shall still
  Be open, at my sovereign's will,
    To each one whom he lists, howe'er
    Unmeet to be the owner's peer.
  My castles are my king's alone,
  From turret to foundation stone;
  The hand of Douglas is his own;
    And never shall, in friendly grasp,
    The hand of such as Marmion clasp."

Burned Marmion's swarthy cheek like fire,
And shook his very frame for ire;
  And—"This to me!" he said,—
"An 't were not for thy hoary beard,
Such hand as Marmion's had not spared
  To cleave the Douglas' head!
And, first, I tell thee, haughty peer,

C.S.R.

He who does England's message here,
Although the meanest in her state,
May well, proud Angus, be thy mate:
And, Douglas, more, I tell thee here,
Even in thy pitch of pride,
Here, in thy hold, thy vassals near,
I tell thee, thou'rt defied!
And if thou said'st I am not peer
To any lord in Scotland here,
Lowland or Highland, far or near,
Lord Angus, thou hast lied!"

On the Earl's cheek the flush of rage
O'ercame the ashen hue of age.
Fierce he broke forth,—"And dar'st thou then
To beard the lion in his den,
The Douglas in his hall?
And hop'st thou hence unscathed to go?
No, by Saint Bride of Bothwell, no!
Up drawbridge, grooms,—what, warder, ho!
Let the portcullis fall."
Lord Marmion turned,—well was his need,—
And dashed the rowels in his steed,
Like arrow through the archway sprung;
The ponderous gate behind him rung:
To pass there was such scanty room,
The bars, descending, razed his plume.

The steed along the drawbridge flies,
Just as it trembled on the rise;
Nor lighter does the swallow skim
Along the smooth lake's level brim:
And when Lord Marmion reached his band
He halts, and turns with clenchèd hand,
And shout of loud defiance pours,
And shook his gauntlet at the towers.

"Horse! horse!" the Douglas cried, "and chase!"
But soon he reined his fury's pace:
"A royal messenger he came,
Though most unworthy of the name.
Saint Mary mend my fiery mood!
Old age ne'er cools the Douglas' blood;
I thought to slay him where he stood.
'Tis pity of him, too," he cried;
"Bold he can speak, and fairly ride;
I warrant him a warrior tried."
With this his mandate he recalls,
And slowly seeks his castle halls.

—*Walter Scott.*

NOTES.—In the poem from which this extract is taken, **Marmion** is represented as an embassador sent by Henry VIII., king of England, to James IV., king of Scotland, with whom he was at war. Having finished his mission to James, Marmion was intrusted to the protection and hospitality of **Douglas**, one of the Scottish nobles. Douglas entertained him, treated him with the respect due to his office and to the honor of his sovereign, yet he despised his private character. Marmion perceived this, and took umbrage at it, though he attempted to repress his resentment, and desired to part in peace. Under these circumstances the scene, as described in this sketch, takes place.

Tantallon is the name of the Douglas castle at Bothwell, Scotland.

XLII. THE PRESENT.

Adelaide Anne Procter, 1825-1864, was the daughter of Bryan Waller Procter, known in literature as "Barry Cornwall." She is the author of several volumes of poetry, and was a contributor to "Good Words," "All the Year Round," and otner London periodicals. Her works have been republished in America.

Do not crouch to-day, and worship
The dead Past, whose life is fled

Hush your voice in tender reverence;
  Crowned he lies, but cold and dead:
For the Present reigns, our monarch,
  With an added weight of hours;
Honor her, for she is mighty!
  Honor her, for she is ours!

See the shadows of his heroes
  Girt around her cloudy throne;
Every day the ranks are strengthened
  By great hearts to him unknown;
Noble things the great Past promised,
  Holy dreams, both strange and new;
But the Present shall fulfill them;
  What he promised, she shall do.

She inherits all his treasures,
  She is heir to all his fame,
And the light that lightens round her
  Is the luster of his name;
She is wise with all his wisdom,
  Living on his grave she stands,
On her brow she bears his laurels,
  And his harvest in her hands.

Coward, can she reign and conquer
  If we thus her glory dim?
Let us fight for her as nobly
  As our fathers fought for him.
God, who crowns the dying ages,
  Bids her rule, and us obey,—
Bids us cast our lives before her,
  Bids us serve the great To-day.

XLIII. THE BAPTISM.

John Wilson, 1785-1854, a distinguished Scottish author, was born at Paisley. When fifteen years of age, he entered the University of Glasgow; but, three years later, he became a member of Magdalen College, Oxford. Here he attained eminence both as a student and as a proficient in gymnastic games and exercises. Soon after graduating, he purchased an estate near Lake Windermere, and became a companion of Wordsworth and Southey; but he soon left his estate to reside in Edinburgh. In 1817, when "Blackwood's Magazine" was established in opposition to the "Edinburgh Review," he became chief contributor to that famous periodical. In its pages, he won his chief fame as a writer. In 1820, he succeeded Dr. Thomas Brown as Professor of Moral Philosophy in the University of Edinburgh; this position he held for thirty years. His "Lights and Shadows of Scottish Life" was published in 1822. This is a collection of pathetic and beautiful tales of domestic life in Scotland. His contributions to Blackwood appeared over the pseudonym of "Christopher North," or more familiarly, "Kit North." Professor Wilson was a man of great physical power and of striking appearance. In character, he was vehement and impulsive; but his writings show that he possessed feelings of deep tenderness.

THE rite of baptism had not been performed for several months in the kirk of Lanark. It was now the hottest time of persecution; and the inhabitants of that parish found other places in which to worship God, and celebrate the ordinances of religion. It was now the Sabbath day, and a small congregation of about a hundred souls had met for divine service, in a place more magnificent than any temple that human hands had ever built to Deity. The congregation had not assembled to the toll of the bell, but each heart knew the hour and observed it; for there are a hundred sundials among the hills, woods, moors, and fields; and the shepherd and the peasant see the hours passing by them in sunshine and shadow.

The church in which they were assembled, was hewn by God's hand out of the eternal rock. A river rolled its way through a mighty chasm of cliffs, several hundred feet high, of which the one side presented enormous masses, and the other corresponding recesses, as if the great stone girdle had been rent by a convulsion. The channel was overspread with prodigious fragments of rocks or large loose stones, some of them smooth and bare, others containing soil and verdure in their rents and fissures, and here and there crowned with shrubs and trees. The eye could at once command a long-stretching vista, seemingly closed and shut up at both extremities by the coalescing cliffs. This majestic reach of river contained pools, streams, and waterfalls innumerable; and when the water was low—which was now the case, in the common drought—it was easy to walk up this scene with the calm, blue sky overhead, an utter and sublime solitude.

On looking up, the soul was bowed down by the feeling of that prodigious height of unscalable, and often overhanging, cliff. Between the channel and the summit of the far extended precipices, were perpetually flying rooks and wood pigeons, and now and then a hawk, filling the profound abyss with their wild cawing, deep murmur, or shrilly shriek. Sometimes a heron would stand erect and still, on some little stone island, or rise up like a white cloud along the black walls of the chasm, and disappear. Winged creatures alone could inhabit this region. The fox and wild cat chose more accessible haunts. Yet, here came the persecuted Christians and worshiped God, whose hand hung over their head those magnificent pillars and arches, scooped out those galleries from the solid rock, and laid at their feet the calm water, in its transparent beauty, in which they could see themselves sitting, in reflected groups, with their Bibles in their hands.

Here, upon a semicircular ledge of rocks, over a narrow chasm, of which the tiny stream played in a murmuring waterfall, and divided the congregation into two equal parts, sat about a hundred persons, all devoutly listening to their minister, who stood before them on what might be called a small, natural pulpit of living stone. Up to it there led a short flight of steps, and over it waved the canopy of a tall, graceful birch tree. The pulpit stood in the middle of the channel, directly facing the congregation, and separated from them by the clear, deep, sparkling pool, into which the scarce-heard water poured over the blackened rock. The water, as it left the pool, separated into two streams, and flowed on each side of that altar, thus placing it in an island, whose large, mossy stones were richly embowered under the golden blossoms and green tresses of the broom.

At the close of divine service, a row of maidens, all clothed in purest white, came gliding off from the congregation, and, crossing the murmuring stream on stepping stones, arranged themselves at the foot of the pulpit with those who were about to be baptized. Their devout fathers, just as though they had been in their own kirk, had been sitting there during worship, and now stood up before the minister. The baptismal water, taken from that pellucid pool, was lying, consecrated, in an appropriate receptacle, formed by the upright stones that composed one side of the pulpit, and the holy rite proceeded.

Some of the younger ones in that semicircle kept gazing down into the pool, in which the whole scene was reflected; and now and then, in spite of the grave looks and admonishing whispers of their elders, letting fall a pebble into the water, that they might judge of its depth, from the length of time that elapsed before the clear air bells lay sparkling on the agitated surface. The rite was over, and the religious service of the day closed by a psalm. The mighty rocks hemmed in the holy sound, and sent it in a more compact volume, clear, sweet, and strong, up to heaven. When the psalm ceased, an echo, like a spirit's voice, was heard dying away, high up among the magnificent architecture of the cliffs; and once more might be noticed in the silence, the reviving voice of the waterfall.

Just then, a large stone fell from the top of the cliff into the pool, a loud voice was heard, and a plaid was hung over on the point of a shepherd's staff. Their wakeful sentinel had descried danger, and this was his warning. Forthwith, the congregation rose. There were paths, dangerous to unpracticed feet, along the ledges of the rocks, leading up to several caves and places of concealment. The more active and young assisted the elder, more especially the old pastor, and the women with the infants; and many minutes had not elapsed, till not a living creature was visible in the channel of the stream, but all of them were hidden, or nearly so, in the clefts and caverns.

The shepherd who had given the alarm, had lain down again instantly in his plaid on the greensward, upon the summit of these precipices. A party of soldiers was immediately upon him, and demanded what signals he had been making, and to whom; when one of them, looking over the edge of the cliff, exclaimed, "See, see! Humphrey, we have caught the whole tabernacle of the Lord in a net at last. There they are, praising God among the stones of the river Mouse. These are the Cartland Craigs. A noble cathedral!" "Fling the lying sentinel over the cliffs. Here is a canting Covenanter for you, deceiving honest soldiers on the very Sabbath day. Over with him, over with him; out of the gallery into the pit." But the shepherd had vanished like a shadow, and, mixing with the tall, green broom and bushes, was making his unseen way toward a wood. "Satan has saved his servant; but come, my lads, follow me. I know the way down into the bed of the stream, and the steps up to Wallace's Cave. They are called, 'kittle nine stanes.' The hunt's up. We'll all be in at the death. Halloo! my boys, halloo!"

The soldiers dashed down a less precipitous part of the wooded banks, a little below the "craigs," and hurried up the channel. But when they reached the altar where the old, gray-haired minister had been seen standing, and the rocks that had been covered with people, all was silent and solitary; not a creature to be seen. "Here is a Bible, dropped by some of them," cried a soldier, and, with his foot, he spun it away into the pool. "A bonnet, a bonnet," cried another; "now for the pretty, sanctified face, that rolled its demure eyes below it." But after a few jests and oaths, the soldiers stood still, eying with a kind of mysterious dread the black and silent walls of the rocks that hemmed them in, and hearing only the small voice of the stream that sent a profounder stillness through the heart of that majestic solitude. "What if these cowardly Covenanters should tumble down upon our heads pieces of rock, from their hiding places! Advance, or retreat?"

There was no reply; for a slight fear was upon every man. Musket or bayonet could be of little use to men obliged to clamber up rocks, along slender paths, leading they know not where. And they were aware that armed men nowadays worshiped God; men of iron hearts, who feared not the glitter of the soldier's arms, neither barrel nor bayonet; men of long stride, firm step, and broad breast, who, on the open field, would have overthrown the marshaled line, and gone first and foremost, if a city had to be taken by storm.

As the soldiers were standing together irresolute, a noise came upon their ears like distant thunder, but even more appalling; and a slight current of air, as if propelled by it, passed whispering along the sweetbriers, and the broom, and the tresses of the birch trees. It came deepening, and rolling, and roaring on; and the very Cartland Craigs shook to their foundation, as if in an earthquake. "The Lord have mercy upon us! What is this?" And down fell many of the miserable wretches on their knees, and some on their faces, upon the sharp-pointed rocks. Now, it was like the sound of many myriads of chariots rolling on their iron axles down the strong channel of the torrent. The old, gray-haired minister issued from the mouth of Wallace's Cave, and said, in a loud voice, "The Lord God terrible reigneth!"

A waterspout had burst up among the moorlands, and

the river, in its power, was at hand. There it came, tumbling along into that long reach of cliffs, and, in a moment, filled it with one mass of waves. Huge, agitated clouds of foam rode on the surface of a blood-red torrent. An army must have been swept off by that flood. The soldiers perished in a moment; but high up in the cliffs, above the sweep of destruction, were the Covenanters, men, women, and children, uttering prayers to God, unheard by themselves, in the raging thunder.

NOTES.—**Lanark** is a small town in the valley of the Clyde, in Scotland. It is thirty miles southwest from Edinburgh.

Mouse River flows to the Clyde from the hills north of Lanark. **Covenanter.**—Under Charles I., the Scotch were so oppressed that they organized in resistance. The covenant was a famous paper, largely signed, in which they agreed to continue in the profession of their faith, and resist all errors.

Wallace's Cave.—William Wallace (b. 1270, d. 1305) was the foremost Scot of his times. He was declared, in the absence of the king, guardian of the kingdom. More than once was he outlawed and obliged to seek safety by concealment in the woods and caves.

XLIV. SPARROWS.

Adeline D. Train Whitney, 1824——, was born in Boston, and was educated in the school of Dr. George B. Emerson. Her father was Enoch Train, a well-known merchant of that city. At the age of nineteen, she became the wife of Mr. Seth D. Whitney. Her literary career began about 1856, since which time she has written several novels and poems; a number of them first appeared in the "Atlantic Monthly." Her writings are marked by grace and sprightliness.

LITTLE birds sit on the telegraph wires,
And chitter, and flitter, and fold their wings;
Maybe they think that, for them and their sires,
Stretched always, on purpose, those wonderful strings:
And, perhaps, the Thought that the world inspires,
Did plan for the birds, among other things.

Little birds sit on the slender lines,
And the news of the world runs under their feet,—
How value rises, and how declines,
How kings with their armies in battle meet,—
And, all the while, 'mid the soundless signs,
They chirp their small gossipings, foolish sweet.

Little things light on the lines of our lives,—
Hopes, and joys, and acts of to-day,—
And we think that for these the Lord contrives,
Nor catch what the hidden lightnings say.
Yet, from end to end, His meaning arrives,
And His word runs underneath, all the way.

Is life only wires and lightning, then,
Apart from that which about it clings?
Are the thoughts, and the works, and the prayers of men
Only sparrows that light on God's telegraph strings,
Holding a moment, and gone again?
Nay; He planned for the birds, with the larger things.

XLV. OBSERVANCE OF THE SABBATH.

Gardiner Spring, 1785-1873, was the son of Samuel Spring, D.D., who was pastor of a Congregational church in Newburyport, Massachusetts, for more than forty years. The son entered Yale College, and was valedictorian of his class in 1805. He studied law for a time; then went to Bermuda, where he taught nearly two years. On his return he completed his law studies, and practiced his profession for more than a year. In 1810, having studied theology at Andover, he was ordained as pastor of the "Brick Church" in New York City. Here he remained till his death. He was elected president of Dartmouth College, and also of Hamilton, but declined both positions. His works, embracing about twenty octavo volumes, have passed through several editions; some have been translated into foreign languages, and reprinted in Europe. As a preacher, Dr. Spring was eloquent and energetic.

THE Sabbath lies at the foundation of all true morality. Morality flows from principle. Let the principles of moral obligation become relaxed, and the practice of morality will not long survive the overthrow. No man can preserve his own morals, no parent can preserve the morals of his children, without the impressions of religious obligation.

If you can induce a community to doubt the genuineness and authenticity of the Scriptures; to question the reality and obligations of religion; to hesitate, undeciding, whether there be any such thing as virtue or vice; whether there be an eternal state of retribution beyond the grave; or whether there exists any such being as God, you have broken down the barriers of moral virtue, and hoisted the flood gates of immorality and crime. I need not say that when a people have once done this, they can no longer exist as a tranquil and happy people. Every bond that holds society together would be ruptured; fraud and treachery would take the place of confidence between man and man; the tribunals of justice would be scenes of bribery and injustice; avarice, perjury, ambition, and revenge would walk through the land, and render it more like the dwelling of savage beasts than the tranquil abode of civilized and Christianized men.

If there is an institution which opposes itself to this progress of human degeneracy, and throws a shield before the interests of moral virtue in our thoughtless and wayward world, it is the Sabbath. In the fearful struggle between virtue and vice, notwithstanding the powerful auxiliaries which wickedness finds in the bosoms of men, and in the seductions and influence of popular example, wherever the Sabbath has been suffered to live, the trembling interests of moral virtue have always been revered and sustained. One of the principal occupations of this day is to illustrate and enforce the great principles of sound morality. Where this sacred trust is preserved inviolate, you behold a nation convened one day in seven for the purpose of acquainting themselves with the best moral principles and precepts; and it can not be otherwise than that the authority of moral virtue, under such auspices, should be acknowledged and felt.

We may not, at once, perceive the effects which this weekly observance produces. Like most moral causes, it operates slowly; but it operates surely, and gradually weakens the power and breaks the yoke of profligacy and sin. No villain regards the Sabbath. No vicious family regards the Sabbath. No immoral community regards the Sabbath. The holy rest of this ever-memorable day is a barrier which is always broken down before men become giants in sin. Blackstone, in his Commentaries on the Laws of England, remarks that "a corruption of morals usually follows a profanation of the Sabbath." It is an observation of Lord Chief Justice Hale, that "of all the persons who were convicted of capital crimes, while he was on the bench, he found a few only who would not confess that they began their career of wickedness by a neglect of the duties of the Sabbath and vicious conduct on that day."

The prisons in our own land could probably tell us that they have scarcely a solitary tenant who had not broken over the restraints of the Sabbath before he was abandoned to crime. You may enact laws for the suppression of immorality, but the secret and silent power of the Sabbath constitutes a stronger shield to the vital interest of the community than any code of penal statutes that ever was enacted. The Sabbath is the keystone of the arch which sustains the temple of virtue, which, however defaced, will survive many a rude shock so long as the foundation remains firm.

The observance of the Sabbath is also most influential in securing national prosperity. The God of Heaven has said, "Them that honor me I will honor." You will not often find a notorious Sabbath breaker a permanently prosperous man; and a Sabbath-breaking community is never a happy or prosperous community. There is a multitude of unobserved influences which the Sabbath exerts upon the temporal welfare of men. It promotes the spirit of good order and harmony; it elevates the poor from want; it transforms squalid wretchedness; it imparts self-respect and elevation of character; it promotes softness and civility of manners; it brings together the rich and the poor upon one common level in the house of prayer; it purifies and strengthens the social affections, and makes the family circle the center of allurement and the source of instruction, comfort, and happiness. Like its own divine religion, "it has the promise of the life that now is and that which is to come," for men can not put themselves beyond the reach of hope and heaven so long as they treasure up this one command, "Remember the Sabbath day, to keep it holy."

NOTES.—**Sir William Blackstone** (b. 1723, d. 1780) was the son of a London silk mercer. He is celebrated as the author of the "Commentaries on the Laws of England," now universally used by law students both in England and America. He once retired from the law through failure to secure a practice, but afterwards attained the highest honors in his profession. See biographical notice on page 410.

Sir Matthew Hale (b. 1609, d. 1676), was Lord Chief Justice of England from 1671 to 1676.

XLVI. GOD'S GOODNESS TO SUCH AS FEAR HIM.

FRET not thyself because of evil doers,
Neither be thou envious against the workers of iniquity;
For they shall soon be cut down like the grass,
And wither as the green herb.
Trust in the Lord, and do good;
So shalt thou dwell in the land, and verily thou shalt be
fed.
Delight thyself also in the Lord,

And he shall give thee the desires of thine heart.
Commit thy way unto the Lord;
Trust also in him, and he shall bring it to pass.
And he shall bring forth thy righteousness as the light,
And thy judgment as the noonday.
Rest in the Lord, and wait patiently for him.

Fret not thyself because of him who prospereth in his way,
Because of the man who bringeth wicked devices to pass.
Cease from anger, and forsake wrath:
Fret not thyself in any wise to do evil,
For evil doers shall be cut off:
But those that wait upon the Lord, they shall inherit the earth.
For yet a little while, and the wicked shall not be;
Yea, thou shalt diligently consider his place, and it shall not be.
But the meek shall inherit the earth,
And shall delight themselves in the abundance of peace.

A little that a righteous man hath
Is better than the riches of many wicked;
For the arms of the wicked shall be broken,
But the Lord upholdeth the righteous.
The Lord knoweth the days of the upright,
And their inheritance shall be forever;
They shall not be ashamed in the evil time,
And in the days of famine they shall be satisfied.

But the wicked shall perish,
And the enemies of the Lord shall be as the fat of lambs;
They shall consume; into smoke shall they consume away.
The wicked borroweth, and payeth not again;
But the righteous sheweth mercy and giveth.
For such as be blessed of him shall inherit the earth.

The steps of a good man are ordered by the Lord,
And he delighteth in his way;
Though he fall, he shall not be utterly cast down;
For the Lord upholdeth him with his hand.

I have been young, and now am old,
Yet have I not seen the righteous forsaken,
Nor his seed begging bread.
He is ever merciful, and lendeth,
And his seed is blessed.

Depart from evil, and do good,
And dwell for evermore;
For the Lord loveth judgment,
And forsaketh not his saints:
They are preserved forever:
But the seed of the wicked shall be cut off.
The righteous shall inherit the land,
And dwell therein forever.
The mouth of the righteous speaketh wisdom,
And his tongue talketh of judgment;
The law of his God is in his heart;
None of his steps shall slide.
The wicked watcheth the righteous,
And seeketh to slay him.
The Lord will not leave him in his hand,
Nor condemn him when he is judged.

Wait on the Lord, and keep his way,
And he shall exalt thee to inherit the land;
When the wicked are cut off, thou shalt see it.
I have seen the wicked in great power,
And spreading himself like a green bay tree;
Yet he passed away, and, lo, he was not;
Yea, I sought him, but he could not be found.

—*From the Thirty-seventh Psalm.*

XLVII. CHARACTER OF COLUMBUS.

Washington Irving, 1783-1859. Among those whose works have enriched American literature, and have given it a place in the estimation of foreigners, no name stands higher than that of Washington Irving. He was born in the city of New York; his father was a native of Scotland, and his mother was English. He had an ordinary school education, and at the age of sixteen began the study of law. Two of his older brothers were interested in literary pursuits; and in his youth he studied the old English authors. He was also passionately fond of books of travel. At the age of nineteen, he began his literary career by writing for a paper published by his brother. In 1804 he made a voyage to the south of Europe. On his return he completed his studies in law, but never practiced his profession. "Salmagundi," his first book (partly written by others), was published in 1807. This was followed, two years later, by "Knickerbocker's History of New York." Soon after, he entered into mercantile pursuits in company with two brothers. At the close of the war with England he sailed again for Europe, and remained abroad seventeen years. During his absence he formed the acquaintance of the most eminent literary men of his time, and wrote several of his works; among them were: "The Sketch Book," "Bracebridge Hall," "Tales of a Traveler," "Life and Voyages of Columbus," and the "Conquest of Granada." On his return he made a journey west of the Mississippi, and gathered materials for several other books. From 1842 to 1846 he was Minister to Spain. On his return to America he established his residence at "Sunnyside," near Tarrytown, on the Hudson, where he passed the last years of his life. A young lady to whom he was attached having died in early life, Mr. Irving never married.

His works are marked by humor, just sentiment, and elegance and correctness of expression. They were popular both at home and abroad from the first, and their sale brought him a handsome fortune. The "Life of Washington," his last work, was completed in the same year in which he died.

Columbus was a man of great and inventive genius. The operations of his mind were energetic, but irregular; bursting forth, at times, with that irresistible force which characterizes intellect of such an order. His ambition was lofty and noble, inspiring him with high thoughts and an anxiety to distinguish himself by great achievements. He aimed at dignity and wealth in the same elevated spirit with which he sought renown; they were to rise from the territories he should discover, and be commensurate in importance.

His conduct was characterized by the grandeur of his views and the magnanimity of his spirit. Instead of ravaging the newly-found countries, like many of his cotemporary discoverers, who were intent only on immediate gain, he regarded them with the eyes of a legislator; he sought to colonize and cultivate them, to civilize the natives, to build cities, introduce the useful arts, subject everything to the control of law, order, and religion, and thus to found regular and prosperous empires. That he failed in this was the fault of the dissolute rabble which it was his misfortune to command, with whom all law was tyranny and all order oppression.

He was naturally irascible and impetuous, and keenly sensible to injury and injustice; yet the quickness of his temper was counteracted by the generosity and benevolence of his heart. The magnanimity of his nature shone forth through all the troubles of his stormy career. Though continually outraged in his dignity, braved in his authority, foiled in his plans, and endangered in his person by the seditions of turbulent and worthless men, and that, too, at times when suffering under anguish of body and anxiety of mind enough to exasperate the most patient, yet he restrained his valiant and indignant spirit, and brought himself to forbear, and reason, and even to supplicate. Nor can the reader of the story of his eventful life fail to notice how free he was from all feeling of revenge, how ready to forgive and forget on the least sign of repentance and atonement. He has been exalted for his skill in controlling others, but far greater praise is due to him for the firmness he displayed in governing himself.

His piety was genuine and fervent. Religion mingled with the whole course of his thoughts and actions, and shone forth in his most private and unstudied writings. Whenever he made any great discovery he devoutly returned thanks to God. The voice of prayer and the melody of praise rose from his ships on discovering the new world, and his first action on landing was to prostrate himself upon the earth and offer up thanksgiving. All his great enterprises were undertaken in the name of the Holy Trinity, and he partook of the holy sacrament previous to embarkation. He observed the festivals of the church in the wildest situations. The Sabbath was to him a day of sacred rest, on which he would never sail from a port unless in case of extreme necessity. The religion thus deeply seated in his soul diffused a sober dignity and a benign composure over his whole deportment; his very language was pure and guarded, and free from all gross or irreverent expressions.

A peculiar trait in his rich and varied character remains to be noticed; namely, that ardent and enthusiastic imagination which threw a magnificence over his whole course of thought. A poetical temperament is discernible throughout all his writings and in all his actions. We see it in all his descriptions of the beauties of the wild land he was discovering, in the enthusiasm with which he extolled the blandness of the temperature, the purity of the atmosphere, the fragrance of the air, "full of dew and sweetness," the verdure of the forests, the grandeur of the mountains, and the crystal purity of the running streams. It spread a glorious and golden world around him, and tinged everything with its own gorgeous colors.

With all the visionary fervor of his imagination, its fondest dreams fell short of the reality. He died in ignorance of the real grandeur of his discovery. Until his last breath, he entertained the idea that he had merely opened a new way to the old resorts of opulent commerce, and had discovered some of the wild regions of the East. What visions of glory would have broken upon his mind could he have known that he had indeed discovered a new continent equal to the old world in magnitude, and separated by two vast oceans from all the earth hitherto known by civilized man! How would his magnanimous spirit have been consoled amid the afflictions of age and the cares of penury, the neglect of a fickle public and the injustice of an ungrateful king, could he have anticipated the splendid empires which would arise in the beautiful world he had discovered, and the nations, and tongues, and languages which were to fill its land with his renown, and to revere and bless his name to the latest posterity!

NOTE.—**Christopher Columbus** (b. 1436, d. 1506) was the son of a wool comber of Genoa. At the age of fifteen he became a sailor, and in his voyages visited England, Iceland, the Guinea coast, and the Greek Isles. He was an earnest student of navigation, of cosmography, and of books of travel; thus he thoroughly prepared himself for the great undertaking which led to the discovery of America. He struggled against every discouragement for almost ten years before he could persuade a sovereign to authorize and equip his expedition.

XLVIII. "HE GIVETH HIS BELOVED SLEEP."

Elizabeth Barrett Browning, 1809-1861, was born in London, married the poet Robert Browning in 1846, and afterwards resided in Italy most of the time till her death, which occurred at Florence. She was thoroughly educated in severe and masculine studies, and began to write at a very early age. Her "Essay on Mind," a metaphysical and reflective poem, was written at the age of sixteen. She wrote very rapidly, and her friend, Miss Mitford, tells us that "Lady Geraldine's Courtship," containing ninety-three stanzas, was composed in twelve hours! She published several other long poems, "Aurora Leigh" being one of the most highly finished. Mrs. Browning is regarded as one of the most able female poets of modern times; but her writings are often obscure, and some have doubted whether she always clearly conceived what she meant to express. She had a warm sympathy with all forms of suffering and distress. "He Giveth his Beloved Sleep" is one of the most beautiful of her minor poems. The thought is an amplification of verse 2d of Psalm cxxvii.

OF all the thoughts of God that are
Borne inward unto souls afar,
  Along the Psalmist's music deep,
Now tell me if that any is,

For gift or grace, surpassing this,—
"He giveth his beloved, sleep!"

What would we give to our beloved?
The hero's heart to be unmoved,
The poet's star-tuned harp, to sweep,
The patriot's voice, to teach and rouse,
The monarch's crown, to light the brows?—
"He giveth his beloved, sleep."

What do we give to our beloved?
A little faith all undisproved,
A little dust to overweep,
And bitter memories to make
The whole earth blasted for our sake,—
"He giveth his beloved, sleep."

"Sleep soft, beloved!" we sometimes say,
But have no tune to charm away
Sad dreams that through the eyelids creep.
But never doleful dream again
Shall break his happy slumber when
"He giveth his beloved, sleep."

O earth, so full of dreary noises!
O men, with wailing in your voices!
O delvéd gold, the wailers heap!
O strife, O curse, that o'er it fall!
God strikes a silence through you all,
And "giveth his beloved, sleep."

His dews drop mutely on the hill;
His cloud above it saileth still,
Though on its slope men sow and reap.

More softly than the dew is shed,
Or cloud is floated overhead,
"He giveth his beloved, sleep."

Ay, men may wonder while they scan
A living, thinking, feeling man,
Confirmed in such a rest to keep;
But angels say—and through the word
I think their happy smile is heard—
"He giveth his beloved, sleep."

For me my heart, that erst did go
Most like a tired child at a show,
That sees through tears the mummers leap,
Would now its wearied vision close,
Would childlike on his love repose
Who "giveth his beloved, sleep."

And friends, dear friends,—when it shall be
That this low breath is gone from me,
And round my bier ye come to weep,
Let one most loving of you all
Say, "Not a tear must o'er her fall;
'He giveth his beloved, sleep.'"

XLIX. DESCRIPTION OF A SIEGE.

"The skirts of the wood seem lined with archers, although only a few are advanced from its dark shadow." "Under what banner?" asked Ivanhoe. "Under no ensign which I can observe," answered Rebecca. "A singular novelty," muttered the knight, "to advance to storm such a castle without pennon or banner displayed. Seest thou who they be that act as leaders?" "A knight clad in sable armor is the most conspicuous," said the Jewess: "he alone is armed from head to heel, and seems to assume the direction of all around him."

"Seem there no other leaders?" exclaimed the anxious inquirer. "None of mark and distinction that I can behold from this station," said Rebecca, "but doubtless the other side of the castle is also assailed. They seem, even now, preparing to advance. God of Zion protect us! What a dreadful sight! Those who advance first bear huge shields and defenses made of plank: the others follow, bending their bows as they come on. They raise their bows! God of Moses, forgive the creatures thou hast made!"

Her description was here suddenly interrupted by the signal for assault, which was given by the blast of a shrill bugle, and at once answered by a flourish of the Norman trumpets from the battlements, which, mingled with the deep and hollow clang of the kettledrums, retorted in notes of defiance the challenge of the enemy. The shouts of both parties augmented the fearful din, the assailants crying, "Saint George, for merry England!" and the Normans answering them with loud cries of "Onward, De Bracy! Front de Bœuf, to the rescue!"

"And I must lie here like a bedridden monk," exclaimed Ivanhoe, "while the game that gives me freedom or death is played out by the hand of others! Look from the window once again, kind maiden, and tell me if they yet advance to the storm." With patient courage, strengthened by the interval which she had employed in mental devotion, Rebecca again took post at the lattice, sheltering herself, however, so as not to be exposed to the arrows of the archers. "What dost thou see, Rebecca?" again demanded the wounded knight. "Nothing but the cloud of arrows flying so thick as to dazzle mine eyes, and to hide the bowmen who shoot them." "That can not endure," said Ivanhoe.

"If they press not right on, to carry the castle by force of arms, the archery may avail but little against stone walls and bulwarks. Look for the knight in dark armor, fair Rebecca, and see how he bears himself; for as the leader is, so will his followers be."

"I see him not," said Rebecca. "Foul craven!" exclaimed Ivanhoe; "does he blench from the helm when the wind blows highest?" "He blenches not! he blenches not!" said Rebecca; "I see him now: he leads a body of men close under the outer barrier of the barbacan. They pull down the piles and palisades; they hew down the barriers with axes. His high black plume floats abroad over the throng like a raven over the field of the slain. They have made a breach in the barriers, they rush in, they are thrust back! Front de Bœuf heads the defenders. I see his gigantic form above the press. They throng again to the breach, and the pass is disputed, hand to hand, and man to man. God of Jacob! it is the meeting of two fierce tides, the conflict of two oceans moved by adverse winds;" and she turned her head from the window as if unable longer to endure a sight so terrible.

Speedily recovering her self-control, Rebecca again looked forth, and almost immediately exclaimed, "Holy prophets of the law! Front de Bœuf and the Black Knight fight hand to hand on the breach, amid the roar of their followers, who watch the progress of the strife. Heaven strike with the cause of the oppressed and of the captive!" She then uttered a loud shriek, and exclaimed, "He is down! he is down!" "Who is down!" cried Ivanhoe; "for our dear Lady's sake, tell me which has fallen!" "The Black Knight," answered Rebecca, faintly; then instantly again shouted with joyful eagerness—"But no! but no! the name of the Lord of Hosts be blessed! he is on foot again, and fights as if there were twenty men's strength in his single arm—his sword is broken—he snatches an ax from a yeoman—he presses Front de Bœuf, blow on blow—the giant stoops and totters like an oak under the steel of the woodman—he falls—he falls!" "Front de Bœuf?" exclaimed Ivanhoe. "Front de Bœuf," answered the Jewess; "his men rush to the rescue, headed by the haughty Templar,—their united force compels the champion to pause—they drag Front de Bœuf within the walls."

"The assailants have won the barriers, have they not?" said Ivanhoe. "They have—they have—and they press the besieged hard upon the outer wall; some plant ladders, some swarm like bees, and endeavor to ascend upon the shoulders of each other; down go stones, beams, and trunks of trees upon their heads, and as fast as they bear the wounded to the rear, fresh men supply their places in the assault. Great God! hast thou given men thine own image that it should be thus cruelly defaced by the hands of their brethren!" "Think not of that," replied Ivanhoe; "this is no time for such thoughts. Who yield? Who push their way?"

"The ladders are thrown down," replied Rebecca, shuddering; "the soldiers lie groveling under them like crushed reptiles; the besieged have the better." "Saint George strike for us!" said the knight; "do the false yeomen give way?" "No," exclaimed Rebecca, "they bear themselves right yeomanly; the Black Knight approaches the postern with his huge ax; the thundering blows which he deals, you may hear them above all the din and shouts of the battle; stones and beams are hailed down on the brave champion; he regards them no more than if they were thistle down and feathers."

"Saint John of Acre!" said Ivanhoe, raising himself joyfully on his couch, "methought there was but one man in England that might do such a deed." "The postern gate shakes," continued Rebecca; "it crashes—it is splintered by his powerful blows—they rush in—the outwork is won! O God! they hurry the defenders from the battlements—they throw them into the moat! O men, if ye be indeed men, spare them that can resist no longer!" "The bridge—the bridge which communicates with the castle—have they won that pass?" exclaimed Ivanhoe. "No," replied Rebecca; "the Templar has destroyed the plank on which they crossed—few of the defenders escaped with him into the castle—the shrieks and cries which you hear, tell the fate of the others. Alas! I see that it is still more difficult to look upon victory than upon battle."

"What do they now, maiden?" said Ivanhoe; "look forth yet again—this is no time to faint at bloodshed." "It is over, for a time," said Rebecca; "our friends strengthen themselves within the outwork which they have mastered." "Our friends," said Ivanhoe, "will surely not abandon an enterprise so gloriously begun, and so happily attained; Oh no! I will put my faith in the good knight whose ax has rent heart of oak and bars of iron. Singular," he again muttered to himself, "if there can be two who are capable of such achievements. It is,—it must be Richard Cœur de Lion."

"Seest thou nothing else, Rebecca, by which the Black Knight may be distinguished?" "Nothing," said the Jewess, "all about him is as black as the wing of the night-raven. Nothing can I spy that can mark him further; but having once seen him put forth his strength in battle, methinks I could know him again among a thousand warriors. He rushes to the fray as if he were summoned to a banquet. There is more than mere strength; it seems as if the whole soul and spirit of the champion were given to every blow which he deals upon his enemies. God forgive him the sin of bloodshed! it is fearful, yet magnificent, to behold how the arm and heart of one man can triumph over hundreds."

—*Walter Scott.*

NOTES.—**Ivanhoe**, a wounded knight, and **Rebecca**, a Jewess, had been imprisoned in the castle of **Reginald Front de**

Bœuf. The friends of the prisoners undertake their rescue. At the request of Ivanhoe, who is unable to leave his couch, Rebecca takes her stand near a window overlooking the approach to the castle, and details to the knight the incidents of the contest as they take place. Front de Bœuf and his garrison were Normans; the besiegers, Saxons.

The **castles** of this time (twelfth century) usually consisted of a keep, or castle proper, surrounded at some distance by two walls, one within the other. Each wall was encircled on its outer side by a **moat,** or ditch, which was filled with water, and was crossed by means of a **drawbridge.** Before the main entrance of the outer wall was an outwork called the **barbacan,** which was a high wall surmounted by battlements and turrets, built to defend the gate and drawbridge. Here, also, were placed **barriers** of palisades, etc., to impede the advance of an attacking force. The **postern** gate was small, and was usually some distance from the ground; it was used for the egress of messengers during a siege.

L. MARCO BOZZARIS.

Fitz-Greene Halleck, 1790-1867, was born in Guilford, Connecticut. At the age of eighteen he entered a banking house in New York, where he remained a long time. For many years he was bookkeeper and assistant in business for John Jacob Astor. Nearly all his poems were written before he was forty years old, several of them in connection with his friend Joseph Rodman Drake. His "Young America," however, was written but a few years before his death. Mr. Halleck's poetry is carefully finished and musical; much of it is sportive, and some satirical. No one of his poems is better known than "Marco Bozzaris."

At midnight, in his guarded tent,
The Turk was dreaming of the hour
When Greece, her knee in suppliance bent,
Should tremble at his power.
In dreams, through camp and court he bore
The trophies of a conqueror;
In dreams, his song of triumph heard;

Then wore his monarch's signet ring:
Then pressed that monarch's throne—a king:
As wild his thoughts, and gay of wing,
As Eden's garden bird.

At midnight, in the forest shades,
Bozzaris ranged his Suliote band,
True as the steel of their tried blades,
Heroes in heart and hand.
There had the Persian's thousands stood,
There had the glad earth drunk their blood,
On old Platæa's day:
And now there breathed that haunted air,
The sons of sires who conquered there,
With arms to strike, and soul to dare,
As quick, as far as they.

An hour passed on—the Turk awoke;
That bright dream was his last:
He woke—to hear his sentries shriek,
"To arms! they come! the Greek! the Greek!"
He woke—to die mid flame and smoke,
And shout, and groan, and saber stroke,
And death shots falling thick and fast
As lightnings from the mountain cloud;
And heard, with voice as trumpet loud,
Bozzaris cheer his band:
"Strike—till the last armed foe expires;
Strike—for your altars and your fires;
Strike—for the green graves of your sires;
God—and your native land!"

They fought—like brave men, long and well;
They piled that ground with Moslem slain;
They conquered—but Bozzaris fell,

Bleeding at every vein.
His few surviving comrades saw
His smile, when rang their proud hurrah,
And the red field was won:
Then saw in death his eyelids close
Calmly, as to a night's repose,
Like flowers at set of sun.

Come to the bridal chamber, Death!
Come to the mother, when she feels
For the first time her firstborn's breath;
Come when the blessèd seals
That close the pestilence are broke,
And crowded cities wail its stroke;
Come in consumption's ghastly form,
The earthquake's shock, the ocean storm;
Come when the heart beats high and warm
With banquet song, and dance, and wine:
And thou art terrible — the tear,
The groan, the knell, the pall, the bier,
And all we know, or dream, or fear
Of agony, are thine.
But to the hero, when his sword
Has won the battle for the free,
Thy voice sounds like a prophet's word;
And in its hollow tones are heard
The thanks of millions yet to be.

Bozzaris! with the storied brave
Greece nurtured in her glory's time,
Rest thee — there is no prouder grave
Even in her own proud clime.
We tell thy doom without a sigh,
For thou art Freedom's, now, and Fame's.
One of the few, the immortal names,
That were not born to die.

NOTES.—**Marco Bozzaris** (b. about 1790, d. 1823) was a famous Greek patriot. His family were Suliotes, a people inhabiting the Suli Mountains, and bitter enemies of the Turks. Bozzaris was engaged in war against the latter nearly all his life, and finally fell in a night attack upon their camp near Carpenisi. This poem, a fitting tribute to his memory, has been translated into modern Greek.

Platæa was the scene of a great victory of the Greeks over the Persians in the year 479 B. C.

Moslem.—The followers of Mohammed are called Moslems.

LI. SONG OF THE GREEK BARD.

George Gordon Byron, Lord Byron, 1788-1824. This gifted poet was the son of a profligate father and of a fickle and passionate mother. He was afflicted with lameness from his birth; and, although he succeeded to his great-uncle's title at ten years of age, he inherited financial embarrassment with it. These may be some of the reasons for the morbid and wayward character of the youthful genius. It is certain that he was not lacking in affection, nor in generosity. In his college days, at Cambridge, he was willful and careless of his studies. "Hours of Idleness," his first book, appeared in 1807. It was severely treated by the "Edinburgh Review," which called forth his "English Bards and Scotch Reviewers," in 1809. Soon after, he went abroad for two years; and, on his return, published the first two cantos of "Childe Harold's Pilgrimage," a work that made him suddenly famous. He married in 1815, but separated from his wife after one year. Soured and bitter, he now left England, purposing never to return. He spent most of the next seven years in Italy, where most of his poems were written. The last year of his life was spent in Greece, aiding in her struggle for liberty against the Turks. He died at Missolonghi. As a man, Byron was impetuous, morbid and passionate. He was undoubtedly dissipated and immoral, but perhaps to a less degree than has sometimes been asserted. As a poet, he possessed noble powers, and he has written much that will last; in general, however, his poetry is not wholesome, and his fame is less than it once was.

THE isles of Greece! the isles of Greece!
Where burning Sappho loved and sung,
Where grew the arts of war and peace,—
Where Delos rose, and Phœbus sprung!
Eternal summer gilds them yet,
But all, except their sun, is set.

The Scian and the Teian muse,
  The hero's harp, the lover's lute,
Have found the fame your shores refuse;
  Their place of birth alone is mute
To sounds which echo further west
Than your sires' "Islands of the Blest."

The mountains look on Marathon,
  And Marathon looks on the sea;
And musing there an hour alone,
  I dreamed that Greece might still be free;
For, standing on the Persian's grave,
I could not deem myself a slave.

A king sat on the rocky brow
  Which looks o'er sea-born Salamis;
And ships, by thousands, lay below,
  And men in nations,—all were his!
He counted them at break of day,—
And when the sun set, where were they?

And where are they? And where art thou,
  My country? On thy voiceless shore
The heroic lay is tuneless now,—
  The heroic bosom beats no more!
And must thy lyre, so long divine,
Degenerate into hands like mine?

Must we but weep o'er days more blest?
  Must we but blush? Our fathers bled.
Earth! render back from out thy breast
  A remnant of our Spartan dead!
Of the three hundred, grant but three,
To make a new Thermopylæ!

What! silent still? and silent all?
  Ah! no;—the voices of the dead
Sound like a distant torrent's fall,
  And answer, "Let one living head,
But one, arise,—we come, we come!"
'Tis but the living who are dumb!

In vain—in vain!—strike other chords;
  Fill high the cup with Samian wine!
Leave battles to the Turkish hordes,
  And shed the blood of Scio's vine!
Hark! rising to the ignoble call,
How answers each bold Bacchanal!

You have the Pyrrhic dance as yet;
  Where is the Pyrrhic phalanx gone?
Of two such lessons, why forget
  The nobler and the manlier one?
You have the letters Cadmus gave;
Think ye he meant them for a slave?

Fill high the bowl with Samian wine!
  We will not think of themes like these!
It made Anacreon's song divine:
  He served, but served Polycrates,
A tyrant; but our masters then
Were still, at least, our countrymen.

The tyrant of the Chersonese
  Was freedom's best and bravest friend;
That tyrant was Miltiades!
  Oh that the present hour would lend
Another despot of the kind!
Such chains as his were sure to bind.

Fill high the bowl with Samian wine!
  Our virgins dance beneath the shade;
I see their glorious, black eyes shine;
  But gazing on each glowing maid,
My own the burning tear-drop laves,
To think such breasts must suckle slaves.

Place me on Sunium's marbled steep,
  Where nothing save the waves and I
May hear our mutual murmurs sweep;
  There, swanlike, let me sing and die:
A land of slaves shall ne'er be mine,—
Dash down yon cup of Samian wine!

NOTES.—**Sappho** was a Greek poetess living on the island of Lesbos, about 600 B. C. **Delos** is one of the Grecian Archipelago, and is of volcanic origin. The ancient Greeks believed that it rose from the sea at a stroke from Neptune's trident, and was moored fast to the bottom by Jupiter. It was the supposed birthplace of **Phœbus,** or Apollo. The island of Chios, or **Scios,** is one of the places which claim to be the birthplace of Homer. **Teios,** or Teos, a city in Ionia, is the birthplace of the Greek poet **Anacreon.** The **Islands of the Blest,** mentioned in ancient poetry, were imaginary islands in the west, where, it was believed, the favorites of the gods were conveyed without dying.

At **Marathon** (490 B. C.), on the east coast of Greece, 11,000 Greeks, under the generalship of **Miltiades,** routed 110,000 Persians. The island of **Salamis** lies very near the Greek coast: in the narrow channel between, the Greek fleet almost destroyed (480 B. C.) that of Xerxes, the Persian king, who witnessed the contest from a throne on the mountain side. **Thermopylæ** is a narrow mountain pass in Greece, where Leonidas, with 300 Spartans and about 1,100 other Greeks, held the entire Persian army in check until every Spartan, except one, was slain. **Samos** is one of the Grecian Archipelago, noted for its cultivation of the vine and olive.

A **Bacchanal** was a disciple of Bacchus, the god of wine. **Pyrrhus** was a Greek, and one of the greatest generals of the world. The **phalanx** was an almost invincible arrangement of troops, massed in close array, with their shields overlapping one another, and their spears projecting; this form of military tactics was peculiar to the Greeks.

Polycrates seized the island of Samos, and made himself tyrant: he was entrapped and crucified in 522 B. C. **Chersonese** is the ancient name for a peninsula. **Sunium** is the name of a promontory southeast of Athens.

LII. NORTH AMERICAN INDIANS.

Charles Sprague, 1791-1875, was born in Boston, and received his education in the public schools of that city. For sixteen years he was engaged in mercantile pursuits, as clerk and partner. In 1820 he became teller in a bank; and, from 1825, he filled the office of cashier of the Globe Bank for about forty years. In 1829 he gave his most famous poem, "Curiosity," before the Phi Beta Kappa society, in Cambridge. An active man of business all his days, he has written but little either in prose or poetry, but that little is excellent in quality, graceful, and pleasing.

The address from which this extract is taken, was delivered before the citizens of Boston, July 4th, 1825.

NOT many generations ago, where you now sit, encircled with all that exalts and embellishes civilized life, the rank thistle nodded in the wind and the wild fox dug his hole unscared. Here lived and loved another race of beings. Beneath the same sun that rolls over your head, the Indian hunter pursued the panting deer; gazing on the same moon that smiles for you, the Indian lover wooed his dusky mate. Here the wigwam blaze beamed on the tender and helpless, and the council fire glared on the wise and daring. Now they dipped their noble limbs in your sedgy lakes, and now they paddled the light canoe along your rocky shores. Here they warred; the echoing whoop, the bloody grapple, the defying death song, all were here; and when the tiger strife was over, here curled the smoke of peace.

Here, too, they worshiped; and from many a dark bosom went up a fervent prayer to the Great Spirit. He had not written his laws for them on tables of stone, but he had traced them on the tables of their hearts. The poor child of nature knew not the God of Revelation, but the God of the universe he acknowledged in everything around. He beheld him in the star that sank in beauty behind his lonely dwelling; in the sacred orb that flamed on him from his midday throne; in the flower that snapped in the morning breeze; in the lofty pine that defied a thousand whirlwinds; in the timid warbler that never left its native grove; in the fearless eagle, whose untired pinion was wet in clouds; in the worm that crawled at his feet; and in his own matchless form, glowing with a spark of that light, to whose mysterious source he bent in humble though blind adoration.

And all this has passed away. Across the ocean came a pilgrim bark, bearing the seeds of life and death. The former were sown for you; the latter sprang up in the path of the simple native. Two hundred years have changed the character of a great continent, and blotted forever from its face a whole, peculiar people. Art has usurped the bowers of nature, and the anointed children of education have been too powerful for the tribes of the ignorant. Here and there a stricken few remain; but how unlike their bold, untamable progenitors. The Indian of falcon glance and lion bearing, the theme of the touching ballad, the hero of the pathetic tale is gone, and his degraded offspring crawls upon the soil where he walked in majesty, to remind us how miserable is man when the foot of the conqueror is on his neck.

As a race they have withered from the land. Their arrows are broken, their springs are dried up, their cabins are in the dust. Their council fire has long since gone out on the shore, and their war cry is fast fading to the untrodden west. Slowly and sadly they climb the distant mountains, and read their doom in the setting sun. They are shrinking before the mighty tide which is pressing them away; they must soon hear the roar of the last wave which will settle over them forever. Ages hence, the inquisitive white man, as he stands by some growing city, will ponder on the structure of their disturbed remains, and wonder to what manner of persons they belonged. They will live only in the songs and chronicles of their exterminators. Let these be faithful to their rude virtues as men, and pay due tribute to their unhappy fate as a people.

LIII. LOCHIEL'S WARNING.

Thomas Campbell, 1777-1844, was a descendant of the famous clan of Campbells, in Kirnan, Scotland, and was born at Glasgow. At the age of thirteen he entered the university in that city, from which he graduated with distinction, especially as a Greek scholar; his translations of Greek tragedy were considered without parallel in the history of the university. During the first year after graduation, he wrote several poems of minor importance. He then removed to Edinburgh and adopted literature as his profession; here his "Pleasures of Hope" was published in 1799, and achieved immediate success. He traveled extensively on the continent, and during his absence wrote "Lochiel's Warning," "Hohenlinden," and other minor poems. In 1809 he published "Gertrude of Wyoming;" from 1820 to 1830 he edited the "New Monthly Magazine." In 1826 he was chosen lord rector of the University of Glasgow, to which office he was twice reëlected. He was active in founding the University of London. During the last years of his life he produced but little of note. He died at Boulogne, in France. During most of his life he was in straitened pecuniary circumstances, and ill-health and family afflictions cast a melancholy over his later years. His poems were written with much care, and are uniformly smooth and musical.

Seer. LOCHIEL! Lochiel! beware of the day
When the Lowlands shall meet thee in battle array!
For a field of the dead rushes red on my sight,
And the clans of Culloden are scattered in fight.
They rally, they bleed, for their kingdom and crown;
Woe, woe to the riders that trample them down!

Proud Cumberland prances, insulting the slain,
And their hoof-beaten bosoms are trod to the plain.
But hark! through the fast-flashing lightning of war,
What steed to the desert flies frantic and far?
'T is thine, O Glenullin! whose bride shall await
Like a love-lighted watch fire all night at the gate.
A steed comes at morning,—no rider is there,—
But its bridle is red with the sign of despair.
Weep, Albin! to death and captivity led!
Oh, weep! but thy tears can not number the dead:
For a merciless sword on Culloden shall wave,—
Culloden! that reeks with the blood of the brave.

Loch. Go preach to the coward, thou death-telling seer!
Or, if gory Culloden so dreadful appear,
Draw, dotard, around thy old wavering sight,
This mantle, to cover the phantoms of fright.

Seer. Ha! laugh'st thou, Lochiel, my vision to scorn?
Proud bird of the mountain thy plume shall be torn!
Say, rushed the bold eagle exultingly forth
From his home in the dark-rolling clouds of the north?
Lo! the death shot of foemen outspeeding, he rode
Companionless, bearing destruction abroad;
But down let him stoop from his havoc on high!
Ah! home let him speed, for the spoiler is nigh.
Why flames the far summit? Why shoot to the blast
Those embers, like stars from the firmament cast?
'T is the fire shower of ruin, all dreadfully driven
From his eyrie that beacons the darkness of heaven.
O crested Lochiel! the peerless in might,
Whose banners arise on the battlements' height,
Heaven's fire is around thee, to blast and to burn;
Return to thy dwelling! all lonely return!
For the blackness of ashes shall mark where it stood,
And a wild mother scream o'er her famishing brood.

Loch. False wizard, avaunt! I have marshaled my clan,
Their swords are a thousand, their bosoms are one!
They are true to the last of their blood and their
breath,
And like reapers descend to the harvest of death.
Then welcome be Cumberland's steed to the shock!
Let him dash his proud foam like a wave on the rock!
But woe to his kindred, and woe to his cause,
When Albin her claymore indignantly draws;
When her bonneted chieftains to victory crowd,
Clanronald the dauntless, and Moray the proud,
All plaided and plumed in their tartan array—

Seer. ——Lochiel, Lochiel, beware of the day!
For, dark and despairing, my sight I may seal,
But man can not cover what God would reveal:
'Tis the sunset of life gives me mystical lore,
And coming events cast their shadows before.
I tell thee, Culloden's dread echoes shall ring
With the bloodhounds that bark for thy fugitive
king.
Lo! anointed by heaven with the vials of wrath,
Behold where he flies on his desolate path!
Now, in darkness and billows, he sweeps from my
sight:
Rise, rise! ye wild tempests, and cover his flight!
'Tis finished. Their thunders are hushed on the moors;
Culloden is lost, and my country deplores.
But where is the ironbound prisoner? Where?
For the red eye of battle is shut in despair.
Say, mounts he the ocean wave, banished, forlorn,
Like a limb from his country, cast bleeding and torn?
Ah no! for a darker departure is near;
The war drum is muffled, and black is the bier;
His death bell is tolling; O mercy, dispel
Yon sight that it freezes my spirit to tell!

Life flutters convulsed in his quivering limbs,
And his blood-streaming nostril in agony swims.
Accursed be the fagots that blaze at his feet,
Where his heart shall be thrown ere it ceases to beat,
With the smoke of its ashes to poison the gale—

Loch. Down, soothless insulter! I trust not the tale:
For never shall Albin a destiny meet
So black with dishonor, so foul with retreat.
Though my perishing ranks should be strewed in
their gore,
Like ocean weeds heaped on the surf-beaten shore,
Lochiel, untainted by flight or by chains,
While the kindling of life in his bosom remains,
Shall victor exult, or in death be laid low,
With his back to the field and his feet to the foe!
And leaving in battle no blot on his name,
Look proudly to heaven from the deathbed of fame.

NOTES.—**Lochiel** was a brave and influential Highland chieftain. He espoused the cause of Charles Stuart, called the Pretender, who claimed the British throne. In the preceding piece, he is supposed to be marching with the warriors of his clan to join Charles's army. On his way he is met by a **Seer,** who having, according to the popular superstition, the gift of second-sight, or prophecy, forewarns him of the disastrous event of the enterprise, and exhorts him to return home and avoid the destruction which certainly awaits him, and which afterward fell upon him at the battle of **Culloden,** in 1746. In this battle the Highlanders were commanded by Charles in person, and the English by the Duke of **Cumberland.** The Highlanders were completely routed, and the Pretender's rebellion brought to a close. He himself shortly afterward made a narrow escape by water from the west of Scotland; hence the reference to the **fugitive king.**

Albin is the poetic name of Scotland, more particularly the Highlands. The **ironbound** prisoner refers to Lochiel.

LIV. ON HAPPINESS OF TEMPER.

Oliver Goldsmith, 1728–1774. This eccentric son of genius was an Irishman; his father was a poor curate. Goldsmith received his education at several preparatory schools, at Trinity College, Dublin, at Edinburgh, and at Leyden. He was indolent and unruly as a student, often in disgrace with his teachers; but his generosity, recklessness, and love of athletic sports made him a favorite with his fellow-students. He spent some time in wandering over the continent, often in poverty and want. In 1756 he returned to England, and soon took up his abode in London. Here he made the acquaintance and friendship of several notable men, among whom were Johnson and Sir Joshua Reynolds. "The Traveler" was published in 1764, and was soon followed by the "Vicar of Wakefield." He wrote in nearly all departments of literature, and always with purity, grace, and fluency. His fame as a poet is secured by the "Traveler" and the "Deserted Village;" as a dramatist, by "She Stoops to Conquer;" as a satirist, by the "Citizen of the World;" and as a novelist by the "Vicar of Wakefield." In his later years his writings were the source of a large income, but his gambling, careless generosity, and reckless extravagance always kept him in financial difficulty, and he died heavily in debt. His monument is in Westminster Abbey.

WRITERS of every age have endeavored to show that pleasure is in us, and not in the objects offered for our amusement. If the soul be happily disposed, everything becomes capable of affording entertainment, and distress will almost want a name. Every occurrence passes in review, like the figures of a procession; some may be awkward, others ill-dressed, but none but a fool is on that account enraged with the master of ceremonies.

I remember to have once seen a slave, in a fortification in Flanders, who appeared no way touched with his situation. He was maimed, deformed, and chained; obliged to toil from the appearance of day till nightfall, and condemned to this for life; yet, with all these circumstances of apparent wretchedness, he sang, would have danced, but that he wanted a leg, and appeared the merriest, happiest man of all the garrison. What a practical philosopher was here! A happy constitution supplied philosophy, and though seemingly destitute of wisdom he was really wise. No reading or study had contributed to disenchant the fairyland around him. Everything furnished him with an opportunity of mirth; and though some thought him, from his insensibility, a fool, he was such an idiot as philosophers should wish to imitate.

They who, like that slave, can place themselves on that side of the world in which everything appears in a pleasant light, will find something in every occurrence to excite their good humor. The most calamitous events, either to themselves or others, can bring no new affliction; the world is to them a theater, in which only comedies are acted. All the bustle of heroism, or the aspirations of ambition, seem only to heighten the absurdity of the scene, and make the humor more poignant. They feel, in short, as little anguish at their own distress, or the complaints of others, as the undertaker, though dressed in black, feels sorrow at a funeral.

Of all the men I ever read of, the famous Cardinal de Retz possessed this happiness in the highest degree. When fortune wore her angriest look, and he fell into the power of Cardinal Mazarin, his most deadly enemy, (being confined a close prisoner in the castle of Valenciennes,) he never attempted to support his distress by wisdom or philosophy, for he pretended to neither. He only laughed at himself and his persecutor, and seemed infinitely pleased at his new situation. In this mansion of distress, though denied all amusements, and even the conveniences of life, and entirely cut off from all intercourse with his friends, he still retained his good humor, laughed at the little spite of his enemies, and carried the jest so far as to write the life of his jailer.

All that the wisdom of the proud can teach, is to be stubborn or sullen under misfortunes. The Cardinal's example will teach us to be good-humored in circumstances of the highest affliction. It matters not whether our good humor be construed by others into insensibility or idiotism,—it is happiness to ourselves; and none but a fool could measure his satisfaction by what the world thinks of it.

The happiest fellow I ever knew, was of the number of those good-natured creatures that are said to do no harm to anybody but themselves. Whenever he fell into any misery, he called it "seeing life." If his head was broken by a chairman, or his pocket picked by a sharper, he comforted himself by imitating the Hibernian dialect of the one, or the more fashionable cant of the other. Nothing came amiss to him. His inattention to money matters had concerned his father to such a degree that all intercession of friends was fruitless. The old gentleman was on his deathbed. The whole family (and Dick among the number) gathered around him.

"I leave my second son, Andrew," said the expiring miser, "my whole estate, and desire him to be frugal." Andrew, in a sorrowful tone (as is usual on such occasions), prayed heaven to prolong his life and health to enjoy it himself. "I recommend Simon, my third son, to the care of his elder brother, and leave him, besides, four thousand pounds." "Ah, father!" cried Simon (in great affliction, to be sure), "may heaven give you life and health to enjoy it yourself!" At last, turning to poor Dick: "As for you, you have always been a sad dog; you'll never come to good; you'll never be rich; I leave you a shilling to buy a halter." "Ah, father!" cries Dick, without any emotion, "may heaven give you life and health to enjoy it yourself!"

NOTES.—**Cardinal de Retz,** Jean Francois Paul de Gondi (b. 1614, d. 1679), was leader of the revolt against **Jules Mazarin** (b. 1602, d. 1661), the prime minister of France during the minority of Louis XIV. This led to a war which lasted four or five years. After peace had been concluded, and Louis XIV. established on the throne, Mazarin was reinstated in power, and Cardinal de Retz was imprisoned.

Flanders, formerly part of the Netherlands, is now included in Belgium, Holland and France.

LV. THE FORTUNE TELLER.

Henry Mackenzie, 1745-1831, was born in Edinburgh, educated at the university there, and died in the same city. He was an attorney by profession, and was the associate of many famous literary men residing at that time in Edinburgh. His fame as a writer rests chiefly on two novels, "The Man of Feeling" and "The Man of the World;" both were published before the author was forty years old.

HARLEY sat down on a large stone by the wayside, to take a pebble from his shoe, when he saw, at some distance, a beggar approaching him. He had on a loose sort of coat, mended with different-colored rags, among which the blue and russet were predominant. He had a short, knotty stick in his hand, and on the top of it was stuck a ram's horn; he wore no shoes, and his stockings had entirely lost that part of them which would have covered his feet and ankles; in his face, however, was the plump appearance of good humor; he walked a good, round pace, and a crook-legged dog trotted at his heels.

"Our delicacies," said Harley to himself, "are fantastic; they are not in nature! That beggar walks over the sharpest of these stones barefooted, whilst I have lost the most delightful dream in the world from the smallest of them happening to get into my shoe." The beggar had by this time come up, and, pulling off a piece of a hat, asked charity of Harley. The dog began to beg, too. It was impossible to resist both; and, in truth, the want of shoes and stockings had made both unnecessary, for Harley had destined sixpence for him before.

The beggar, on receiving it, poured forth blessings without number; and, with a sort of smile on his countenance, said to Harley that if he wanted to have his fortune told—Harley turned his eye briskly upon the beggar; it was an unpromising look for the subject of a prediction, and silenced the prophet immediately. "I would much rather learn," said Harley, "what it is in your power to tell me. Your trade must be an entertaining one; sit down on this stone, and let me know something of your profession; I have often thought of turning fortune teller for a week or two, myself."

"Master," replied the beggar, "I like your frankness much, for I had the humor of plain dealing in me from a child; but there is no doing with it in this world,—we must do as we can; and lying is, as you call it, my profession. But I was in some sort forced to the trade, for I once dealt in telling the truth. I was a laborer, sir, and gained as much as to make me live. I never laid by, indeed, for I was reckoned a piece of a wag, and your wags, I take it, are seldom rich, Mr. Harley." "So," said Harley, "you seem to know me." "Ay, there are few folks in the country that I do n't know something of. How should I tell fortunes else?" "True,—but go on with your story; you were a laborer, you say, and a wag; your industry, I suppose, you left with your old trade; but your humor you preserved to be of use to you in your new."

"What signifies sadness, sir? A man grows lean on 't. But I was brought to my idleness by degrees; sickness first disabled me, and it went against my stomach to work, ever after. But, in truth, I was for a long time so weak that I spit blood whenever I attempted to work. I had no relation living, and I never kept a friend above a week when I was able to joke. Thus I was forced to beg my bread, and a sorry trade I have found it, Mr. Harley. I told all my misfortunes truly, but they were seldom believed; and the few who gave me a half-penny as they passed, did it with a shake of the head, and an injunction not to trouble them with a long story. In short, I found that people do n't care to give alms without some security for their money,—such as a wooden leg, or a withered arm, for example. So I changed my plan, and instead of telling my own misfortunes, began to prophesy happiness to others.

"This I found by much the better way. Folks will always listen when the tale is their own, and of many who say they do not believe in fortune telling, I have known few on whom it had not a very sensible effect. I pick up the names of their acquaintance; amours and little squabbles are easily gleaned from among servants and neighbors; and, indeed, people themselves are the best intelligencers in the world for our purpose. They dare not puzzle us for their own sakes, for everyone is anxious to hear what he wishes to believe; and they who repeat it, to laugh at it when they have done, are generally more serious than their hearers are apt to imagine. With a tolerably good memory, and some share of cunning, I succeed reasonably well as a fortune teller. With this, and showing the tricks of that dog, I make shift to pick up a livelihood.

"My trade is none of the most honest, yet people are not much cheated after all, who give a few half-pence for a prospect of happiness, which I have heard some persons say, is all a man can arrive at in this world. But I must bid you good day, sir; for I have three miles to walk before noon, to inform some boarding-school young ladies whether their husbands are to be peers of the realm or captains in the army; a question which I promised to answer them by that time."

Harley had drawn a shilling from his pocket; but Virtue bade him to consider on whom he was going to bestow it. Virtue held back his arm; but a milder form, a younger sister of Virtue's, not so severe as Virtue, nor so serious as Pity, smiled upon him; his fingers lost their compression; nor did Virtue appear to catch the money as it fell. It had no sooner reached the ground than the watchful cur (a trick he had been taught) snapped it up; and, contrary to the most approved method of stewardship, delivered it immediately into the hands of his master.

LVI. RIENZI'S ADDRESS TO THE ROMANS.

Mary Russell Mitford, 1786-1855. She was the daughter of a physician, and was born in Hampshire, England. At twenty years of age, she published three volumes of poems; and soon after entered upon literature as a lifelong occupation. She wrote tales, sketches, poems, and dramas. "Our Village" is the best known of her prose works; the book describes the daily life of a rural people, is simple but finished in style, and is marked by mingled humor and pathos. Her most noted drama is "Rienzi." Miss Mitford passed the last forty years of her life in a little cottage in Berkshire, among a simple, country people, to whom she was greatly endeared by her kindness and social virtues.

I COME not here to talk. You know too well
The story of our thraldom. We are slaves!
The bright sun rises to his course, and lights
A race of slaves! He sets, and his last beams
Fall on a slave; not such as, swept along
By the full tide of power, the conqueror led
To crimson glory and undying fame;
But base, ignoble slaves; slaves to a horde
Of petty tyrants, feudal despots, lords,
Rich in some dozen paltry villages;
Strong in some hundred spearmen; only great
In that strange spell,—a name.

Each hour, dark fraud,
Or open rapine, or protected murder,
Cries out against them. But this very day,
An honest man, my neighbor,—there he stands,—
Was struck—struck like a dog, by one who wore
The badge of Ursini; because, forsooth,
He tossed not high his ready cap in air,
Nor lifted up his voice in servile shouts,
At sight of that great ruffian! Be we men,
And suffer such dishonor? men, and wash not
The stain away in blood? Such shames are common.
I have known deeper wrongs; I that speak to ye,

I had a brother once—a gracious boy,
Full of all gentleness, of calmest hope,
Of sweet and quiet joy,—there was the look
Of heaven upon his face, which limners give
To the beloved disciple.

How I loved
That gracious boy! Younger by fifteen years,
Brother at once, and son! He left my side,
A summer bloom on his fair cheek; a smile
Parting his innocent lips. In one short hour,
That pretty, harmless boy was slain! I saw
The corse, the mangled corse, and then I cried
For vengeance! Rouse, ye Romans! rouse, ye slaves!
Have ye brave sons? Look in the next fierce brawl
To see them die. Have ye fair daughters? Look
To see them live, torn from your arms, distained,
Dishonored; and if ye dare call for justice,
Be answered by the lash.

Yet this is Rome,
That sat on her seven hills, and from her throne
Of beauty ruled the world! and we are Romans.
Why, in that elder day, to be a Roman
Was greater than a king!

And once again,—
Hear me, ye walls that echoed to the tread
Of either Brutus! Once again, I swear,
The eternal city shall be free.

NOTES.—**Rienzi** (b. about 1312, d. 1354) was the last of the Roman tribunes. In 1347 he led a successful revolt against the nobles, who by their contentions kept Rome in constant turmoil. He then assumed the title of tribune, but, after indulging in a life of reckless extravagance and pomp for a few months, he was compelled to abdicate, and fly for his life. In 1354 he was reinstated in power, but his tyranny caused his assassination the same year.

The **Ursini** were one of the noble families of Rome.

This lesson is especially adapted for drill on inflection, emphasis, and modulation.

LVII. CHARACTER OF THE PURITAN FATHERS OF NEW ENGLAND.

One of the most prominent features which distinguished our forefathers, was their determined resistance to oppression. They seemed born and brought up for the high and special purpose of showing to the world that the civil and religious rights of man — the rights of self-government, of conscience, and independent thought — are not merely things to be talked of and woven into theories, but to be adopted with the whole strength and ardor of the mind, and felt in the profoundest recesses of the heart, and carried out into the general life, and made the foundation of practical usefulness, and visible beauty, and true nobility.

Liberty, with them, was an object of too serious desire and stern resolve to be personified, allegorized, and enshrined. They made no goddess of it, as the ancients did; they had no time nor inclination for such trifling; they felt that liberty was the simple birthright of every human creature; they called it so; they claimed it as such; they reverenced and held it fast as the unalienable gift of the Creator, which was not to be surrendered to power, nor sold for wages.

It was theirs, as men; without it, they did not esteem themselves men; more than any other privilege or possession, it was essential to their happiness, for it was essential to their original nature; and therefore they preferred it above wealth, and ease, and country; and, that they might enjoy and exercise it fully, they forsook houses, and lands, and kindred, their homes, their native soil, and their fathers' graves.

They left all these; they left England, which, whatever it might have been called, was not to them a land of freedom; they launched forth on the pathless ocean, the wide, fathomless ocean, soiled not by the earth beneath, and bounded, all round and above, only by heaven; and it seemed to them like that better and sublimer freedom, which their country knew not, but of which they had the conception and image in their hearts; and, after a toilsome and painful voyage, they came to a hard and wintry coast, unfruitful and desolate, but unguarded and boundless; its calm silence interrupted not the ascent of their prayers; it had no eyes to watch, no ears to hearken, no tongues to report of them; here, again, there was an answer to their soul's desire, and they were satisfied, and gave thanks; they saw that they were free, and the desert smiled.

I am telling an old tale; but it is one which must be told when we speak of those men. It is to be added, that they transmitted their principles to their children, and that, peopled by such a race, our country was always free. So long as its inhabitants were unmolested by the mother country in the exercise of their important rights, they submitted to the form of English government; but when those rights were invaded, they spurned even the form away.

This act was the Revolution, which came of course and spontaneously, and had nothing in it of the wonderful or unforeseen. The wonder would have been if it had not occurred. It was, indeed, a happy and glorious event, but by no means unnatural; and I intend no slight to the revered actors in the Revolution when I assert that their fathers before them were as free as they — every whit as free.

The principles of the Revolution were not the suddenly acquired property of a few bosoms: they were abroad in the land in the ages before; they had always been taught, like the truths of the Bible; they had descended from father to son, down from those primitive days, when the Pilgrim, established in his simple dwelling, and seated at his blazing fire, piled high from the forest which shaded his door, repeated to his listening children the story of his wrongs and his resistance, and bade them rejoice, though the wild winds and the wild beasts were howling without, that they had nothing to fear from great men's oppression.

Here are the beginnings of the Revolution. Every settler's hearth was a school of independence; the scholars were apt, and the lessons sunk deeply; and thus it came that our country was always free; it could not be other than free.

As deeply seated as was the principle of liberty and resistance to arbitrary power in the breasts of the Puritans, it was not more so than their piety and sense of religious obligation. They were emphatically a people whose God was the Lord. Their form of government was as strictly theocratical, if direct communication be excepted, as was that of the Jews; insomuch that it would be difficult to say where there was any civil authority among them entirely distinct from ecclesiastical jurisdiction.

Whenever a few of them settled a town, they immediately gathered themselves into a church; and their elders were magistrates, and their code of laws was the Pentateuch. These were forms, it is true, but forms which faithfully indicated principles and feelings; for no people could have adopted such forms, who were not thoroughly imbued with the spirit, and bent on the practice, of religion.

God was their King; and they regarded him as truly and literally so, as if he had dwelt in a visible palace in the midst of their state. They were his devoted, resolute, humble subjects; they undertook nothing which they did not beg of him to prosper; they accomplished nothing without rendering to him the praise; they suffered nothing without carrying their sorrows to his throne; they ate nothing which they did not implore him to bless.

Their piety was not merely external; it was sincere; it had the proof of a good tree in bearing good fruit; it produced and sustained a strict morality. Their tenacious purity of manners and speech obtained for them, in the mother country, their name of Puritans, which, though given in derision, was as honorable an appellation as was ever bestowed by man on man.

That there were hypocrites among them, is not to be doubted; but they were rare. The men who voluntarily exiled themselves to an unknown coast, and endured there every toil and hardship for conscience' sake, and that they might serve God in their own manner, were not likely to set conscience at defiance, and make the service of God a mockery; they were not likely to be, neither were they, hypocrites. I do not know that it would be arrogating too much for them to say, that, on the extended surface of the globe, there was not a single community of men to be compared with them, in the respects of deep religious impressions and an exact performance of moral duty.

F. W. P. Greenwood.

NOTE.—The **Pentateuch** is the first five books of the Old Testament. The word is derived from two Greek words, πέντε (pente), five, and τεῦχος (teuchos), book.

LVIII. LANDING OF THE PILGRIM FATHERS.

Felicia Dorothea Hemans, 1794-1835, was born in Liverpool. Her father, whose name was Browne, was an Irish merchant. She spent her childhood in Wales, began to write poetry at a very early age, and was married when about eighteen to Captain Hemans. By this marriage, she became the mother of five sons; but, owing to differences of taste and disposition, her husband left her at the end of six years; and by mutual agreement they never again lived together. Mrs. Hemans now made literature a profession, and wrote much and well. In 1826 Prof. Andrews Norton brought out an edition of her poems in America, where they became popular, and have remained so.

Mrs. Hemans's poetry is smooth and graceful, frequently tinged with a shade of melancholy, but never despairing, cynical, or misanthropic. It never deals with the highest themes, nor rises to sublimity, but its influence is calculated to make the reader truer, nobler, and purer.

THE breaking waves dashed high
  On a stern and rock-bound coast,
And the woods against a stormy sky
  Their giant branches tossed;

And the heavy night hung dark,
  The hills and waters o'er,
When a band of exiles moored their bark
  On the wild New England shore.

Not as the conqueror comes,
  They, the true-hearted, came;
Not with the roll of the stirring drums,
  And the trumpet that sings of fame.

Not as the flying come,
  In silence, and in fear;—
They shook the depths of the desert gloom
  With their hymns of lofty cheer.

Amidst the storm they sang,
  And the stars heard, and the sea;
And the sounding aisles of the dim woods rang
  To the anthem of the free!

The ocean eagle soared
  From his nest by the white wave's foam;
And the rocking pines of the forest roared,—
  This was their welcome home.

There were men with hoary hair
  Amidst that pilgrim band:
Why had they come to wither there,
  Away from their childhood's land?

There was woman's fearless eye,
  Lit by her deep love's truth;
There was manhood's brow, serenely high,
  And the fiery heart of youth.

What sought they thus afar?
  Bright jewels of the mine?
The wealth of seas, the spoils of war?—
  They sought a faith's pure shrine!

Ay, call it holy ground,
  The soil where first they trod:
They have left unstained what there they found,—
  Freedom to worship God.

NOTE.—The **Pilgrim Fathers** landed at Plymouth, Mass., Dec. 11th (Old Style), 1620. The rock on which they first stepped, is in Water Street of the village, and is covered by a handsome granite canopy, surmounted by a colossal statue of Faith.

LIX. NECESSITY OF EDUCATION.

WE must educate! We must educate! or we must perish by our own prosperity. If we do not, short will be our race from the cradle to the grave. If, in our haste to be rich and mighty, we outrun our literary and religious institutions, they will never overtake us; or only come up after the battle of liberty is fought and lost, as spoils to grace the victory, and as resources of inexorable despotism for the perpetuity of our bondage.

But what will become of the West if her prosperity rushes up to such a majesty of power, while those great institutions linger which are necessary to form the mind, and the conscience, and the heart of the vast world? It must not be permitted. And yet what is done must be done quickly; for population will not wait, and commerce will not cast anchor, and manufactures will not shut off the steam, nor shut down the gate, and agriculture, pushed by millions of freemen on their fertile soil, will not withhold her corrupting abundance.

And let no man at the East quiet himself, and dream of liberty, whatever may become of the West. Our alliance of blood, and political institutions, and common interests, is such, that we can not stand aloof in the hour of her calamity, should it ever come. Her destiny is our destiny; and the day that her gallant ship goes down, our little boat sinks in the vortex!

The great experiment is now making, and from its extent and rapid filling up, is making in the West, whether the perpetuity of our republican institutions can be reconciled with universal suffrage. Without the education of the head and heart of the nation, they can not be; and the question to be decided is, can the nation, or the vast balance power of it, be so imbued with intelligence and virtue as to bring out, in laws and their administration, a perpetual self-preserving energy. We know that the work is a vast one, and of great difficulty; and yet we believe it can be done.

I am aware that our ablest patriots are looking out on the deep, vexed with storms, with great forebodings and failings of heart, for fear of the things that are coming upon us; and I perceive a spirit of impatience rising, and distrust in respect to the perpetuity of our republic; and I am sure that these fears are well founded, and am glad that they exist. It is the star of hope in our dark horizon. Fear is what we need, as the ship needs wind on a rocking sea, after a storm, to prevent foundering. But when our fear and our efforts shall correspond with our danger, the danger is past.

For it is not the impossibility of self-preservation which threatens us; nor is it the unwillingness of the nation to pay the price of the preservation, as she has paid the price of the purchase of our liberties. It is inattention and inconsideration, protracted till the crisis is past, and the things which belong to our peace are hid from our eyes. And blessed be God, that the tokens of a national waking up, the harbinger of God's mercy, are multiplying upon us!

We did not, in the darkest hour, believe that God had brought our fathers to this goodly land to lay the foundation of religious liberty, and wrought such wonders in their preservation, and raised their descendants to such heights of civil and religious liberty, only to reverse the analogy of his providence, and abandon his work.

And though there now be clouds, and the sea roaring, and men's hearts failing, we believe there is light behind the cloud, and that the imminence of our danger is intended, under the guidance of Heaven, to call forth and apply a holy, fraternal fellowship between the East and the West, which shall secure our preservation, and make the prosperity of our nation durable as time, and as abundant as the waves of the sea.

I would add, as a motive to immediate action, that if we do fail in our great experiment of self-government, our destruction will be as signal as the birthright abandoned, the mercies abused, and the provocation offered to beneficent Heaven. The descent of desolation will correspond with the past elevation.

No punishments of Heaven are so severe as those for mercies abused; and no instrumentality employed in their infliction is so dreadful as the wrath of man. No spasms are like the spasms of expiring liberty, and no wailing such as her convulsions extort.

It took Rome three hundred years to die; and our death, if we perish, will be as much more terrific as our intelligence and free institutions have given us more bone, sinew, and vitality. May God hide from me the day when the dying agonies of my country shall begin! O thou beloved land, bound together by the ties of brotherhood, and common interest, and perils! live forever—one and undivided!

—*Lyman Beecher.*

LX. RIDING ON A SNOWPLOW.

Benjamin Franklin Taylor, 1822-1887, was born at Lowville, New York, and graduated at Madison University, of which his father was president. Here he remained as resident graduate for about five years. His "Attractions of Language" was published in 1845. For many years Mr. Taylor was literary editor of the "Chicago Journal." He wrote considerably for the magazines, and was the author of many well-known fugitive pieces, both in prose and verse. He also published several books, of which "January and June," "Pictures in Camp and Field," "The World on Wheels," "Old-time Pictures and Sheaves of Rhyme," "Between the Gates," and "Songs of Yesterday," are the best known. In his later years, Mr. Taylor achieved some reputation as a lecturer. His writings are marked by an exuberant fancy.

Did you ever ride on a snowplow? Not the pet and pony of a thing that is attached to the front of an engine, sometimes, like a pilot; but a great two-storied monster of strong timbers, that runs upon wheels of its own, and that boys run after and stare at as they would after and at an elephant. You are snow-bound at Buffalo. The Lake Shore Line is piled with drifts like a surf. Two passenger trains have been half-buried for twelve hours somewhere in snowy Chautauqua. The storm howls like a congregation of Arctic bears. But the superintendent at Buffalo is determined to release his castaways, and clear the road to

Erie. He permits you to be a passenger on the great snowplow; and there it is, all ready to drive. Harnessed behind it, is a tandem team of three engines. It does not occur to you that you are going to ride on a steam drill, and so you get aboard.

It is a spacious and timbered room, with one large bull's-eye window,—an overgrown lens. The thing is a sort of Cyclops. There are ropes, and chains, and a windlass. There is a bell by which the engineer of the first engine can signal the plowman, and a cord whereby the plowman can talk back. There are two sweeps, or arms, worked by machinery, on the sides. You ask their use, and the superintendent replies, "When, in a violent shock, there is danger of the monster's upsetting, an arm is put out, on one side or the other, to keep the thing from turning a complete somersault." You get one idea, and an inkling of another. So you take out your Accident Policy for three thousand dollars, and examine it. It never mentions battles, nor duels, nor snowplows. It names "public conveyances." Is a snowplow a public conveyance? You are inclined to think it is neither that nor any other kind that you should trust yourself to, but it is too late for consideration.

You roll out of Buffalo in the teeth of the wind, and the world is turned to snow. All goes merrily. The machine strikes little drifts, and they scurry away in a cloud. The three engines breathe easily; but by and by the earth seems broken into great billows of dazzling white. The sun comes out of a cloud, and touches it up till it outsilvers Potosi. Houses lie in the trough of the sea everywhere, and it requires little imagination to think they are pitching and tossing before your eyes. A great breaker rises right in the way. The monster, with you in it, works its way up and feels of it. It is packed like a ledge of marble. Three whistles! The machine backs away and keeps backing, as a gymnast runs astern to get sea room and momentum for a big jump; as a giant swings aloft a heavy sledge, that it may come down with a heavy blow.

One whistle! You have come to a halt. Three pairs of whistles one after the other! and then, putting on all steam, you make for the drift. The superintendent locks the door, you do not quite understand why, and in a second the battle begins. The machine rocks and creaks in all its joints. There comes a tremendous shock. The cabin is as dark as midnight. The clouds of flying snow put out the day. The labored breathing of the locomotives behind you, the clouds of smoke and steam that wrap you up as in a mantle, the noonday eclipse of the sun, the surging of the ship, the rattling of chains, the creak of timbers as if the craft were aground and the sea getting out of its bed to whelm you altogether, the doubt as to what will come,—all combine to make a scene of strange excitement for a landlubber.

You have made some impression on the breaker, and again the machine backs for a fair start, and then another plunge, and shock, and twilight. And so, from deep cut to deep cut, as if the season had packed all his winter clothes upon the track, until the stalled trains are reached and passed; and then, with alternate storm and calm, and halt and shock, till the way is cleared to Erie.

It is Sunday afternoon, and Erie—"Mad Anthony Wayne's" old headquarters—has donned its Sunday clothes, and turned out by hundreds to see the great plow come in,—its first voyage over the line. The locomotives set up a crazy scream, and you draw slowly into the depot. The door opened at last, you clámber down, and gaze up at the uneasy house in which you have been living. It looks as if an avalanche had tumbled down upon it,—white as an Alpine shoulder. Your first thought is gratitude that you have made a landing alive. Your second, a resolution that, if again you ride a hammer, it will not be when three engines have hold of the handle!

Notes.—**Chautauqua** is the most western county in the state of New York; it borders on Lake Erie.

The **Cyclops** are described in Grecian mythology as giants having only one eye, which was circular, and placed in the middle of the forehead.

Cerro de **Potosi** is a mountain in Bolivia, South America, celebrated for its mineral wealth. More than five thousand mines have been opened in it; the product is chiefly silver.

"**Mad Anthony Wayne**" (b. 1745, d. 1796), so called from his bravery and apparent recklessness, was a famous American officer during the Revolution. In 1794 he conducted a successful campaign against the Indians of the Northwest, making his headquarters at Erie, Pa.

LXI. THE QUARREL OF BRUTUS AND CASSIUS.

Cas. That you have wronged me doth appear in this:
You have condemned and noted Lucius Pella
For taking bribes here of the Sardians;
Wherein my letters, praying on his side,
Because I knew the man, were slighted off.

Bru. You wronged yourself to write in such a case.

Cas. In such a time as this, it is not meet
That every nice offense should bear his comment.

Bru. Yet let me tell you, Cassius, you yourself
Are much condemned to have an itching palm,
To sell and mart your offices for gold
To undeservers.

Cas. I an itching palm!
You know that you are Brutus that speak this,
Or, by the gods, this speech were else your last.

Bru. The name of Cassius honors this corruption,
And chastisement doth therefore hide his head.

Cas. Chastisement!

Bru. Remember March, the ides of March remember!
Did not great Julius bleed for justice' sake?

What villain touched his body, that did stab,
And not for justice? What! shall one of us,
That struck the foremost man of all this world
But for supporting robbers; shall we now
Contaminate our fingers with base bribes,
And sell the mighty space of our large honors
For so much trash as may be graspèd thus?
I had rather be a dog, and bay the moon,
Than such a Roman.

Cas. Brutus, bay not me;
I'll not endure it: you forget yourself,
To hedge me in; I am a soldier, I,
Older in practice, abler than yourself
To make conditions.

Bru. Go to; you are not, Cassius.

Cas. I am.

Bru. I say you are not.

Cas. Urge me no more, I shall forget myself:
Have mind upon your health; tempt me no further.

Bru. Away, slight man!

Cas. Is't possible?

Bru. Hear me, for I will speak.
Must I give way and room to your rash choler?
Shall I be frighted when a madman stares?

Cas. O ye gods! ye gods! must I endure all this?

Bru. All this! Ay, more: fret till your proud heart break;
Go, show your slaves how choleric you are,
And make your bondmen tremble. Must I budge?
Must I observe you? Must I stand and crouch
Under your testy humor? By the gods,
You shall digest the venom of your spleen,
Though it do split you; for, from this day forth,
I'll use you for my mirth, yea for my laughter,
When you are waspish.

Cas. Is it come to this?

Bru. You say you are a better soldier:

Let it appear so; make your vaunting true,
And it shall please me well: for mine own part,
I shall be glad to learn of noble men.

Cas. You wrong me every way; you wrong me, Brutus;
I said, an elder soldier, not a better:
Did I say "better"?

Bru. If you did, I care not.

Cas. When Cæsar lived, he durst not thus have moved me.

Bru. Peace, peace! you durst not so have tempted him.

Cas. I durst not?

Bru. No.

Cas. What! durst not tempt him?

Bru. For your life, you durst not.

Cas. Do not presume too much upon my love;
I may do that I shall be sorry for.

Bru. You have done that you should be sorry for.
There is no terror, Cassius, in your threats;
For I am armed so strong in honesty
That they pass by me as the idle wind,
Which I respect not. I did send to you
For certain sums of gold, which you denied me;—
For I can raise no money by vile means:
By heaven! I had rather coin my heart,
And drop my blood for drachmas, than to wring
From the hard hands of peasants their vile trash
By any indirection.—I did send
To you for gold to pay my legions,
Which you denied me: was that done like Cassius?
Should I have answered Caius Cassius so?
When Marcus Brutus grows so covetous,
To lock such rascal counters from his friends,
Be ready, gods, with all your thunderbolts;
Dash him to pieces!

Cas. I denied you not.

Bru. You did.

Cas. I did not: he was but a fool that brought

My answer back. Brutus hath rived my heart:
A friend should bear his friend's infirmities,
But Brutus makes mine greater than they are.

Bru. I do not, till you practice them on me.

Cas. You love me not.

Bru. I do not like your faults.

Cas. A friendly eye could never see such faults.

Bru. A flatterer's would not, though they do appear
As huge as high Olympus.

Cas. Come, Antony, and young Octavius, come,
Revenge yourselves alone on Cassius,
For Cassius is aweary of the world;
Hated by one he loves; braved by his brother;
Checked like a bondman; all his faults observed,
Set in a notebook, learned, and conned by rote,
To cast into my teeth. Oh, I could weep
My spirit from mine eyes! There is my dagger,
And here my naked breast; within, a heart
Dearer than Plutus' mine, richer than gold:
If that thou be'st a Roman, take it forth:
I, that denied thee gold, will give my heart:
Strike, as thou didst at Cæsar; for, I know,
When thou didst hate him worst, thou lovedst him better
Than ever thou lovedst Cassius.

Bru. Sheathe your dagger:
Be angry when you will, it shall have scope;
Do what you will, dishonor shall be humor.
O Cassius, you are yokèd with a lamb
That carries anger as the flint bears fire;
Who, much enforcèd, shows a hasty spark,
And straight is cold again.

Cas. Hath Cassius lived
To be but mirth and laughter to his Brutus,
When grief, and blood ill-tempered, vexeth him?

Bru. When I spoke that, I was ill-tempered, too.

Cas. Do you confess so much? Give me your hand.

Bru. And my heart, too.

Cas. O Brutus!

Bru. What's the matter?

Cas. Have you not love enough to bear with me,
When that rash humor which my mother gave me
Makes me forgetful?

Bru. Yes, Cassius; and, from henceforth,
When you are over earnest with your Brutus,
He'll think your mother chides, and leave you so.

Shakespeare.—Julius Cæsar, Act iv, Scene iii.

NOTES.—**Ides** (pro. īdz) was a term used in the Roman calendar. It fell on the fifteenth day of March, May, July, and October, and on the thirteenth of other months. On the ides of March, 44 B. C., Julius Cæsar was murdered by Brutus, Cassius, and other conspirators. The populace were aroused to indignation, and the conspirators were compelled to fly.

Indirection; *i. e.*, dishonest means.

Antony and **Octavius**, who, with Lepidus, formed the triumvirate now governing Rome, were at this time marching against the forces of Brutus and Cassius.

Plutus, in ancient mythology, the god of wealth.

LXII. THE QUACK.

John Tobin, 1770-1804, a solicitor, was born at Salisbury, England, and died on shipboard near Cork. He wrote several comedies, the most popular being "The Honeymoon," from which this extract is taken; it was published in 1805.

SCENE—*The Inn. Enter* HOSTESS *followed by* LAMPEDO, *a Quack Doctor.*

Host. NAY, nay; another fortnight.

Lamp. It can't be.
The man's as well as I am: have some mercy!
He hath been here almost three weeks already.

Host. Well, then, a week.

Lamp. We may detain him a week. (*Enter* BALTHAZAR, *the patient, from behind, in his nightgown, with a drawn sword.*)
You talk now like a reasonable hostess,
That sometimes has a reckoning with her conscience.

Host. He still believes he has an inward bruise.

Lamp. I would to heaven he had! or that he'd slipped
His shoulder blade, or broke a leg or two,
(Not that I bear his person any malice,)
Or luxed an arm, or even sprained his ankle!

Host. Ay, broken anything except his neck.

Lamp. However, for a week I'll manage him,
Though he had the constitution of a horse—
A farrier should prescribe for him.

Balth. A farrier! (*Aside.*)

Lamp. To-morrow, we phlebotomize again;
Next day, my new-invented patent draught;
Then, I have some pills prepared;
On Thursday, we throw in the bark; on Friday—

Balth. (*Coming forward.*) Well, sir, on Friday—what, on Friday? Come,
Proceed.

Lamp. Discovered!

Host. Mercy, noble sir! } *They fall on their knees.*

Lamp. We crave your mercy! }

Balth. On your knees? 'tis well!
Pray! for your time is short.

Host. Nay, do not kill us.

Balth. You have been tried, condemned, and only wait
For execution. Which shall I begin with?

Lamp. The lady, by all means, sir.

Balth. Come, prepare. (*To the hostess.*)

Host. Have pity on the weakness of my sex!

Balth. Tell me, thou quaking mountain of gross flesh,
Tell me, and in a breath, how many poisons—

If you attempt it—(*To* LAMPEDO, *who is making off*
you have cooked up for me?

Host. None, as I hope for mercy!

Balth. Is not thy wine a poison?

Host. No indeed, sir;
'T is not, I own, of the first quality;
But—

Balth. What?

Host. I always give short measure, sir,
And ease my conscience that way.

Balth. Ease your conscience!
I'll ease your conscience for you.

Host. Mercy, sir!

Balth. Rise, if thou canst, and hear me.

Host. Your commands, sir?

Balth. If, in five minutes, all things are prepared
For my departure, you may yet survive.

Host. It shall be done in less.

Balth. Away, thou lumpfish. (*Exit hostess.*)

Lamp. So! now comes my turn! 't is all over with me!
There's dagger, rope, and ratsbane in his looks!

Balth. And now, thou sketch and outline of a man!
Thou thing that hast no shadow in the sun!
Thou eel in a consumption, eldest born
Of Death and Famine! thou anatomy
Of a starved pilchard!

Lamp. I do confess my leanness. I am spare,
And, therefore, spare me.

Balth. Why wouldst thou have made me
A thoroughfare, for thy whole shop to pass through?

Lamp. Man, you know, must live.

Balth. Yes: he must die, too.

Lamp. For my patients' sake!

Balth. I'll send you to the major part of them—
The window, sir, is open;—come, prepare.

Lamp. Pray consider!

MIGUEL DIAS
OPORTO
JUAN

I may hurt some one in the street.

Balth. Why, then,
I'll rattle thee to pieces in a dicebox,
Or grind thee in a coffee mill to powder,
For thou must sup with Pluto:—so, make ready!
Whilst I, with this good smallsword for a lancet,
Let thy starved spirit out (for blood thou hast none),
And nail thee to the wall, where thou shalt look
Like a dried beetle with a pin stuck through him.

Lamp. Consider my poor wife.

Balth. Thy wife!

Lamp. My wife, sir.

Balth. Hast thou dared think of matrimony, too?
Thou shadow of a man, and base as lean!

Lamp. O spare me for her sake!
I have a wife, and three angelic babes,
Who, by those looks, are well nigh fatherless.

Balth. Well, well! your wife and children shall plead for you.
Come, come; the pills! where are the pills? Produce them.

Lamp. Here is the box.

Balth. Were it Pandora's, and each single pill
Had ten diseases in it, you should take them.

Lamp. What, all?

Balth. Ay, all; and quickly, too. Come, sir, begin—
(LAMPEDO *takes one.*) That's well!—Another.

Lamp. One's a dose.

Balth. Proceed, sir.

Lamp. What will become of me?
Let me go home, and set my shop to rights,
And, like immortal Cæsar, die with decency.

Balth. Away! and thank thy lucky star I have not
Brayed thee in thine own mortar, or exposed thee
For a large specimen of the lizard genus.

Lamp. Would I were one!—for they can feed on air.

Balth. Home, sir! and be more honest.
Lamp. If I am not,
I'll be more wise, at least.

Notes.—**Pluto,** in ancient mythology, the god of the lower world.

Pandora is described in the Greek legends as the first created woman. She was sent by Jupiter to Epimetheus as a punishment, because the latter's brother, Prometheus, had stolen fire from heaven. When she arrived among men, she opened a box in which were all the evils of mankind, and everything escaped except Hope.

LXIII. RIP VAN WINKLE.

The appearance of Rip, with his long, grizzled beard, his rusty fowling piece, his uncouth dress, and an army of women and children at his heels, soon attracted the attention of the tavern politicians. They crowded around him, eying him from head to foot with great curiosity. The orator bustled up to him, and, drawing him partly aside, inquired on which side he voted. Rip stared in vacant stupidity. Another short but busy little fellow pulled him by the arm, and, rising on tiptoe, inquired in his ear "whether he was Federal or Democrat."

Rip was equally at a loss to comprehend the question; when a knowing, self-important old gentleman, in a sharp cocked hat, made his way through the crowd, putting them to the right and left with his elbows as he passed, and planting himself before Van Winkle, with one arm akimbo, the other resting on his cane, his keen eyes and sharp hat penetrating, as it were, into his very soul, demanded, in an austere tone, what brought him to the election with a gun on his shoulder, and a mob at his heels, and whether he meant to breed a riot in the village.

"Alas! gentlemen," cried Rip, somewhat dismayed, "I am a poor, quiet man, a native of the place, and a loyal subject of the king, God bless him!" Here a general shout burst from the bystanders.—"A tory! a tory! a spy! a refugee! hustle him! away with him!" It was with great difficulty that the self-important man in the cocked hat restored order; and, having a tenfold austerity of brow, demanded again of the unknown culprit, what he came there for, and whom he was seeking. The poor man humbly assured him that he meant no harm, but merely came there in search of some of his neighbors, who used to keep about the tavern. "Well, who are they? name them."

Rip bethought himself a moment, and inquired, "Where's Nicholas Vedder?" There was a silence for a little while, when an old man replied, in a thin, piping voice, "Nicholas Vedder! why he is dead and gone these eighteen years! There was a wooden tombstone in the churchyard that used to tell all about him, but that's rotten and gone too." "Where's Brom Dutcher?" "Oh, he went off to the army in the beginning of the war. Some say he was killed at the storming of Stony Point; others say he was drowned in a squall at the foot of Anthony's Nose. I don't know; he never came back again."

"Where's Van Bummel, the schoolmaster?" "He went off to the wars, too; was a great militia general, and is now in Congress." Rip's heart died away at hearing of these sad changes in his home and friends, and finding himself thus alone in the world. Every answer puzzled him, too, by treating of such enormous lapses of time, and of matters which he could not understand—war, Congress, Stony Point. He had no courage to ask after any more friends, but cried out in despair, "Does nobody here know Rip Van Winkle?"

"Oh, Rip Van Winkle!" exclaimed two or three. "Oh, to be sure! That's Rip Van Winkle yonder, leaning against the tree." Rip looked, and beheld a precise counterpart of himself as he went up the mountain; apparently as lazy, and certainly as ragged. The poor fellow was now completely confounded; he doubted his own identity, and whether he was himself or another man. In the midst of his bewilderment, the man in the cocked hat demanded who he was, and what was his name.

"God knows!" exclaimed he, at his wit's end. "I'm not myself; I'm somebody else; that's me yonder; no, that's somebody else got into my shoes. I was myself last night; but I fell asleep on the mountain, and they've changed my gun, and everything's changed, and I'm changed, and I can't tell what's my name or who I am!"

The bystanders began now to look at each other, nod, wink significantly, and tap their fingers against their foreheads. There was a whisper, also, about securing the gun, and keeping the old fellow from doing mischief, at the very suggestion of which the self-important man in the cocked hat retired with some precipitation. At this critical moment, a fresh, comely woman pressed through the throng to get a peep at the gray-bearded man. She had a chubby child in her arms, which, frightened at his looks, began to cry. "Hush, Rip!" cried she, "hush, you little fool! the old man won't hurt you."

The name of the child, the air of the mother, the tone of her voice, all awakened a train of recollections in his mind. "What is your name, my good woman?" asked he. "Judith Gardenier." "And your father's name?" "Ah, poor man! Rip Van Winkle was his name; but it's twenty years since he went away from home with his gun, and never has been heard of since; his dog came home without him; but whether he shot himself, or was carried away by the Indians, nobody can tell. I was then but a little girl."

Rip had but one question more to ask; but he put it with a faltering voice: "Where's your mother?" "Oh, she, too, died but a short time since; she broke a blood vessel in a fit of passion at a New England peddler." There was a drop of comfort, at least, in this intelligence. The honest man could contain himself no longer. He caught his daughter and her child in his arms. "I am your father!" cried he. "Young Rip Van Winkle once, old Rip Van Winkle now! Does nobody know poor Rip Van Winkle?"

All stood amazed, until an old woman, tottering out from among the crowd, put her hand to her brow, and, peering under it in his face for a moment, exclaimed, "Sure enough! it is Rip Van Winkle! it is himself! Welcome home again, old neighbor! Why, where have you been these twenty long years?" Rip's story was soon told, for the whole twenty years had been to him but as one night.

To make a long story short, the company broke up and returned to the more important concerns of the election. Rip's daughter took him home to live with her. She had a snug, well-furnished house, and a stout, cheery farmer for a husband, whom Rip recollected for one of the urchins that used to climb upon his back. Rip now resumed his old walks and habits. He soon found many of his former cronies, though all rather the worse for the wear and tear of time, and preferred making friends among the rising generation, with whom he soon grew into great favor.

—*Irving.*

NOTES.—**Rip Van Winkle,** according to Irving's story in "The Sketch Book," was a great drunkard, and was driven from his home in the Catskill Mountains, one night, by his wife. Wandering among the mountains, he fell in with the ghosts of Hendrick Hudson and his crew, with whom he played a game of ninepins. Upon drinking the liquor which they offered him, however, he immediately fell into a deep sleep which lasted for twenty years. The above lesson recounts the events that befell him when he returned to his native village. In the meantime the Revolution of 1776 had taken place.

The **Federals** and the **Democrats** formed the two leading political parties of that time.

Stony Point is a promontory on the Hudson, at the entrance of the Highlands, forty-two miles from New York. It was a fortified post during the Revolution, captured by the British, and again retaken by the Americans under Wayne. **Anthony's Nose** is also a promontory on the Hudson, about fifteen miles above Stony Point.

LXIV. BILL AND JOE.

Oliver Wendell Holmes, 1809-1894, was the son of Abiel Holmes, D.D. He was born in Cambridge, Massachusetts, and graduated at Harvard in 1829, having for classmates several men who have since become distinguished. After graduating, he studied law for about one year, and then turned his attention to medicine. He studied his profession in Paris, and elsewhere in Europe, and took his degree at Cambridge in 1836. In 1838 he was appointed Professor of Anatomy and Physiology in Dartmouth College. He remained here but a short time, and then returned to Boston and entered on the practice of medicine. In 1847 he was appointed professor at Harvard, filling a similar position to the one held at Dartmouth. He discharged the duties of his professorship for more than thirty years, with great success. Literature was never his profession; yet few American authors attained higher success, both as a poet and as a prose writer. His poems are lively and sparkling, abound in wit and humor, but are not wanting in genuine pathos. Many of them were composed for special occasions. His prose writings include works on medicine, essays, and novels; several appeared first as contributions to the "Atlantic Monthly." He gained reputation, also, as a popular lecturer. In person, Dr. Holmes was small and active, with a face expressive of thought and vivacity.

COME, dear old comrade, you and I
Will steal an hour from days gone by—
The shining days when life was new,
And all was bright as morning dew,
The lusty days of long ago,
When you were Bill and I was Joe.

Your name may flaunt a titled trail
Proud as a cockerel's rainbow tail,
And mine as brief appendix wear
As Tam O'Shanter's luckless mare;
To-day, old friend, remember still
That I am Joe and you are Bill.

You've won the great world's envied prize,
And grand you look in people's eyes,
With HON. and LL. D.,
In big, brave letters fair to see,—
Your fist, old fellow! Off they go!—
How are you, Bill? How are you, Joe?

You've worn the judge's ermined robe;
You've taught your name to half the globe;
You've sung mankind a deathless strain;
You've made the dead past live again:
The world may call you what it will,
But you and I are Joe and Bill.

The chaffing young folks stare and say,
"See those old buffers, bent and gray;
They talk like fellows in their teens;
Mad, poor old boys! That's what it means"
And shake their heads; they little know
The throbbing hearts of Bill and Joe—

How Bill forgets his hour of pride,
While Joe sits smiling at his side;
How Joe, in spite of time's disguise,
Finds the old schoolmate in his eyes,—
Those calm, stern eyes, that melt and fill,
As Joe looks fondly up to Bill.

Ah! pensive scholar, what is fame?
A fitful tongue of leaping flame;
A giddy whirlwind's fickle gust,
That lifts a pinch of mortal dust:
A few swift years, and who can show
Which dust was Bill, and which was Joe.

The weary idol takes his stand,
Holds out his bruised and aching hand,
While gaping thousands come and go—
How vain it seems, this empty show!—
Till all at once his pulses thrill:
'T is poor old Joe's, "God bless you, Bill!"

And shall we breathe in happier spheres
The names that pleased our mortal ears;
In some sweet lull of heart and song
For earthborn spirits none too long,
Just whispering of the world below
When this was Bill, and that was Joe?

No matter; while our home is here,
No sounding name is half so dear;
When fades at length our lingering day,
Who cares what pompous tombstones say?
Read on the hearts that love us still,
Hic jacet Joe. *Hic jacet* Bill.

NOTE.—**Hic jacet** (*pro.* hĭc jā′çet) is a Latin phrase, meaning *here lies.* It is frequently used in epitaphs.

LXV. SORROW FOR THE DEAD.

The sorrow for the dead is the only sorrow from which we refuse to be divorced. Every other wound we seek to heal; every other affliction, to forget; but this wound we consider it a duty to keep open. This affliction we cherish, and brood over in solitude. Where is the mother who would willingly forget the infant that has perished like a blossom from her arms, though every recollection is a pang? Where is the child that would willingly forget a tender parent, though to remember be but to lament? Who, even in the hour of agony, would forget the friend over whom he mourns?

No, the love which survives the tomb is one of the noblest attributes of the soul. If it has its woes, it has likewise its delights: and when the overwhelming burst of grief is calmed into the gentle tear of recollection; when the sudden anguish and the convulsive agony over the present ruins of all that we most loved, is softened away into pensive meditation on all that it was in the days of its loveliness, who would root out such a sorrow from the heart? Though it may, sometimes, throw a passing cloud over the bright hour of gayety, or spread a deeper sadness over the hour of gloom; yet, who would exchange it even for the song of pleasure, or the burst of revelry? No, there is a voice from the tomb sweeter than song. There is a remembrance of the dead, to which we turn even from the charms of the living.

Oh, the grave! the grave! It buries every error, covers every defect, extinguishes every resentment! From its peaceful bosom spring none but fond regrets and tender recollections. Who can look down upon the grave even of an enemy, and not feel a compunctious throb, that he should have warred with the poor handful of earth that lies moldering before him? But the grave of those we loved—what a place for meditation! There it is that we call up, in long review, the whole history of virtue and gentleness, and the thousand endearments lavished upon us, almost unheeded in the daily intercourse of intimacy; there it is that we dwell upon the tenderness, the solemn, awful tenderness of the parting scene; the bed of death, with all its stifled griefs, its noiseless attendance, its mute, watchful assiduities! the last testimonies of expiring love! the feeble, fluttering, thrilling,—oh! how thrilling!—pressure of the hand! the last fond look of the glazing eye turning upon us, even from the threshold of existence! the faint, faltering accents, struggling in death to give one more assurance of affection!

Ay, go to the grave of buried love, and meditate! There settle the account with thy conscience for every past benefit unrequited; every past endearment unregarded, of that departed being, who can never—never—never return to be soothed by thy contrition! If thou art a child, and hast ever added a sorrow to the soul, or a furrow to the silvered brow of an affectionate parent; if thou art a husband, and hast ever caused the fond bosom that ventured its whole happiness in thy arms to doubt one moment of thy kindness or thy truth; if thou art a friend, and hast ever wronged, in thought, or word, or deed, the spirit that generously confided in thee; if thou hast given one unmerited pang to that true heart, which now lies cold and still beneath thy feet; then be sure that every unkind look, every ungracious word, every ungentle action, will come thronging back upon thy memory, and knocking dolefully at thy soul; then be sure that thou wilt lie down sorrowing and repentant on the grave, and utter the unheard groan, and pour the unavailing tear; more deep, more bitter, because unheard and unavailing.

Then weave thy chaplet of flowers, and strew the beauties of nature about the grave; console thy broken spirit, if thou canst, with these tender, yet futile, tributes of regret;

but take warning by the bitterness of this, thy contrite affliction over the dead, and henceforth be more faithful and affectionate in the discharge of thy duties to the living.

—*Irving.*

LXVI. THE EAGLE.

James Gates Percival, 1795-1856, was born at Berlin, Connecticut, and graduated at Yale College in 1815, at the head of his class. He was admitted to the practice of medicine in 1820, and went to Charleston, South Carolina. In 1824 he was appointed Professor of Chemistry at West Point, a position which he held but a few months. In 1854 he was appointed State Geologist of Wisconsin, and died at Hazel Green, in that state. Dr. Percival was eminent as a geographer, geologist, and linguist. He began to write poetry at an early age, and his fame rests chiefly upon his writings in this department. In his private life, Percival was always shy, modest, and somewhat given to melancholy. Financially, his life was one of struggle, and he was often greatly straitened for money.

Bird of the broad and sweeping wing!
  Thy home is high in heaven,
Where the wide storms their banners fling,
  And the tempest clouds are driven.
Thy throne is on the mountain top;
  Thy fields, the boundless air;
And hoary peaks, that proudly prop
  The skies, thy dwellings are.

Thou art perched aloft on the beetling crag,
  And the waves are white below,
And on, with a haste that can not lag,
  They rush in an endless flow.
Again thou hast plumed thy wing for flight
  To lands beyond the sea,
And away, like a spirit wreathed in light,
  Thou hurriest, wild and free.

Lord of the boundless realm of air!
  In thy imperial name,
The hearts of the bold and ardent dare
  The dangerous path of fame.
Beneath the shade of thy golden wings,
  The Roman legions bore,
From the river of Egypt's cloudy springs,
  Their pride, to the polar shore.

For thee they fought, for thee they fell,
  And their oath on thee was laid;
To thee the clarions raised their swell,
  And the dying warrior prayed.
Thou wert, through an age of death and fears,
  The image of pride and power,
Till the gathered rage of a thousand years,
  Burst forth in one awful hour.

And then, a deluge of wrath, it came,
  And the nations shook with dread;
And it swept the earth, till its fields were flame,
  And piled with the mingled dead.
Kings were rolled in the wasteful flood,
  With the low and crouching slave;
And together lay, in a shroud of blood,
  The coward and the brave.

Notes.—**Roman legions.** The Roman standard was the image of an eagle. The soldiers swore by it, and the loss of it was considered a disgrace.

One awful hour. Alluding to the destruction of Rome by the northern barbarians.

LXVII. POLITICAL TOLERATION.

Thomas Jefferson, 1743-1826, the third President of the United States, and the author of the Declaration of Independence, was born in Albemarle County, Virginia. He received most of his early education under private tutors, and at the age of seventeen entered William and Mary College, where he remained two years. At college, where he studied industriously, he formed the acquaintance of several distinguished men,—among them was George Wythe, with whom he entered on the study of law. At the age of twenty-four he was admitted to the bar, and soon rose to high standing in his profession. In 1775 he entered the Colonial Congress, having previously served ably in the legislature of his native state. Although one of the youngest men in Congress, he soon took a foremost place in that body. He left Congress in the fall of 1776, and, as a member of the legislature, and later as Governor of Virginia, he was chiefly instrumental in effecting several important reforms in the laws of that state,—the most notable were the abolition of the law of primogeniture, and the passage of a law making all religious denominations equal. From 1785 to 1789 he was Minister to France. On his return to America he was made Secretary of State, in the first Cabinet. While in this office, he became the leader of the Republican or Anti-Federalist party, in opposition to the Federalist party led by Alexander Hamilton. From 1801 to 1809 he was President. On leaving his high office, he retired to his estate at "Monticello," where he passed the closing years of his life, and died on the 4th of July, just fifty years after the passage of his famous Declaration. His compatriot, and sometimes bitter political opponent, John Adams, died on the same day.

Mr. Jefferson, who was never a ready public speaker, was a remarkably clear and forcible writer; his works fill several large volumes. In personal character, he was pure and simple, cheerful, and disposed to look on the bright side. His knowledge of life rendered his conversation highly attractive. The chief enterprise of his later years was the founding of the University of Virginia, at Charlottesville.

During the contest of opinion through which we have passed, the animation of discussions and of exertions has sometimes worn an aspect which might impose on strangers, unused to think freely and to speak and to write what they think; but this being now decided by the voice of the nation, announced according to the rules of the constitution, all will, of course, arrange themselves under the will of the law, and unite in common efforts for the common good.

All, too, will bear in mind this sacred principle, that,

though the will of the majority is, in all cases, to prevail, that will, to be rightful, must be reasonable; that the minority possess their equal rights, which equal laws must protect, and to violate which would be oppression. Let us then, fellow-citizens, unite with one heart and one mind.

Let us restore to social intercourse that harmony and affection, without which liberty, and even life itself, are but dreary things; and let us reflect, that, having banished from our land that religious intolerance under which mankind so long bled and suffered, we have gained little if we countenance a political intolerance as despotic, as wicked, and capable of as bitter and bloody persecutions.

During the throes and convulsions of the ancient world; during the agonizing spasms of infuriated man, seeking, through blood and slaughter, his long-lost liberty; it was not wonderful that the agitation of the billows should reach even this distant and peaceful shore; that this should be more felt and feared by some, and less by others, and should divide opinions as to measures of safety.

But every difference of opinion is not a difference of principle. We have called by different names brethren of the same principle. We are all Republicans; we are all Federalists. If there be any among us who would wish to dissolve this Union, or to change its republican form, let them stand undisturbed as monuments of the safety with which error of opinion may be tolerated when reason is left free to combat it.

I know, indeed, that some honest men fear that a republican government can not be strong; that this government is not strong enough. But would the honest patriot, in the full tide of successful experiment, abandon a government which has so far kept us free and firm, on the theoretic and visionary fear that this government, the world's best hope, may, by possibility, want energy to preserve itself? I trust not; I believe this, on the contrary, the strongest government on earth.

I believe it to be the only one where every man, at the call of the law, would fly to the standard of the law, and would meet invasions of the public order as his own personal concern. Sometimes it is said that man can not be trusted with the government of himself. Can he, then, be trusted with the government of others, or have we found angels, in the form of kings, to govern him? Let history answer this question. Let us, then, with courage and confidence, pursue our own federal and republican principles; our attachment to union and representative government.

NOTE.—At the time of Jefferson's election, party spirit ran very high. He had been defeated by John Adams at the previous presidential election, but the Federal party, to which Adams belonged, became weakened by their management during difficulties with France; and now Jefferson had been elected president over his formerly successful rival. The above selection is from his inaugural address.

LXVIII. WHAT CONSTITUTES A STATE?

Sir William Jones, 1746–1794, was the son of an eminent mathematician; he early distinguished himself by his ability as a student. He graduated at Oxford, became well versed in Oriental literature, studied law, and wrote many able books. In 1783 he was appointed Judge of the Supreme Court of Judicature in Bengal. He was a man of astonishing learning, upright life, and Christian principles.

WHAT constitutes a state?
Not high-raised battlement or labored mound,
Thick wall or moated gate;
Not cities proud with spires and turrets crowned;
Not bays and broad-armed ports,
Where, laughing at the storm, rich navies ride;
Not starred and spangled courts,
Where low-browed baseness wafts perfume to pride.

No:—men, high-minded men,
With powers as far above dull brutes endued
In forest, brake, or den,
As beasts excel cold rocks and brambles rude,—
Men who their duties know,
But know their rights, and, knowing, dare maintain,
Prevent the long-aimed blow,
And crush the tyrant while they rend the chain:
These constitute a state;
And sovereign Law, that state's collected will,
O'er thrones and globes elate,
Sits empress, crowning good, repressing ill.

LXIX. THE BRAVE AT HOME.

Thomas Buchanan Read, 1822-1872, an American poet and painter, was born in Chester County, Pennsylvania. At the age of seventeen he entered a sculptor's studio in Cincinnati. Here he gained reputation as a painter of portraits. From this city he went to New York, Boston, and Philadelphia, and soon after to Florence, Italy. In the later years of his life, he divided his time between Cincinnati, Philadelphia, and Rome. His complete poetical works fill three volumes. Several of his most stirring poems relate to the Revolutionary War, and to the late Civil War in America. Many of his poems are marked by vigor and a ringing power, while smoothness and delicacy distinguish others, no less.

The maid who binds her warrior's sash,
And, smiling, all her pain dissembles,
The while beneath the drooping lash,
One starry tear-drop hangs and trembles;
Though Heaven alone records the tear,
And fame shall never know her story,
Her heart has shed a drop as dear
As ever dewed the field of glory!

The wife who girds her husband's sword,
'Mid little ones who weep and wonder,

And bravely speaks the cheering word,
  What though her heart be rent asunder;—
Doomed nightly in her dreams to hear
  The bolts of war around him rattle,—
Has shed as sacred blood as e'er
  Was poured upon the field of battle!

The mother who conceals her grief,
  While to her breast her son she presses,
Then breathes a few brave words and brief,
  Kissing the patriot brow she blesses;
With no one but her loving God,
  To know the pain that weighs upon her,
Sheds holy blood as e'er the sod
  Received on Freedom's field of honor!

NOTE.—The above selection is from the poem entitled "The Wagoner of the Alleghanies."

LXX. SOUTH CAROLINA.

Robert Young Hayne, 1791–1840, was born in Colleton District, South Carolina, and studied and practiced law at Charleston. He was early elected to the State Legislature, and became Speaker of the House and Attorney-general of the state. He entered the Senate of the United States at the age of thirty-one. He was Governor of South Carolina during the "Nullification" troubles in 1832 and 1833. Mr. Hayne was a clear and àble debater, and a stanch advocate of the extreme doctrine of "State Rights." In the Senate he opposed the Tariff Bill of 1828; and, out of this struggle, grew his famous debate with Daniel Webster in 1830. The following selection is an extract from Mr. Hayne's speech on that memorable occasion.

IF there be one state in the Union, Mr. President, that may challenge comparison with any other, for a uniform, zealous, ardent, and uncalculating devotion to the Union, that state is South Carolina. Sir, from the very commencement of the Revolution, up to this hour, there is no sacrifice, however great, she has not cheerfully made; no service she has ever hesitated to perform.

She has adhered to you in your prosperity; but in your adversity she has clung to you with more than filial affection. No matter what was the condition of her domestic affairs; though deprived of her resources, divided by parties, or surrounded by difficulties, the call of the country has been to her as the voice of God. Domestic discord ceased at the sound; every man became at once reconciled to his brethren, and the sons of Carolina were all seen, crowding to the temple, bringing their gifts to the altar of their common country.

What, sir, was the conduct of the South, during the Revolution? Sir, I honor New England for her conduct in that glorious struggle. But great as is the praise which belongs to her, I think at least equal honor is due to the South. Never were there exhibited, in the history of the world, higher examples of noble daring, dreadful suffering, and heroic endurance, than by the whigs of Carolina, during the Revolution. The whole state, from the mountains to the sea, was overrun by an overwhelming force of the enemy. The fruits of industry perished on the spot where they were produced, or were consumed by the foe.

The plains of Carolina drank up the most precious blood of her citizens. Black, smoking ruins marked the places which had been the habitation of her children. Driven from their homes into the gloomy and almost impenetrable swamps, even there the spirit of liberty survived, and South Carolina, sustained by the example of her Sumters and her Marions, proved, by her conduct, that though her soil might be overrun, the spirit of her people was invincible.

NOTES.—**Thomas Sumter** (b. 1734, d. 1832) was by birth a Virginian, but during the Revolution commanded South Carolina troops. He was one of the most active and able of the

Southern generals, and, after the war, was prominent in politics. He was the last surviving general of the Revolution.

Francis Marion (b. 1732, d. 1795), known as the "Swamp Fox," was a native South Carolinian, of French descent. Marion's brigade became noted during the Revolution for its daring and surprising attacks. See Lesson CXXXV.

LXXI. MASSACHUSETTS AND SOUTH CAROLINA.

Daniel Webster, 1782–1852. This celebrated American statesman and orator was born in Salisbury, New Hampshire. His father, Ebenezer Webster, was a pioneer settler, a soldier in the Old French War and the Revolution, and a man of ability and strict integrity. Daniel attended the common school in his youth, and fitted for college under Rev. Samuel Wood, of Boscawen, graduating at Dartmouth in 1801. He spent a few months of his boyhood at "Phillips Academy," Exeter, where he attained distinction as a student, but was so diffident that he could never give a declamation before his class. During his college course, and later, he taught school several terms in order to increase his slender finances. He was admitted to the bar in Boston in 1805. For the next eleven years, he practiced his profession in his native state. In 1812 he was elected to the United States House of Representatives, and at once took his place as one of the most prominent men of that body. In 1816 he removed to Boston; and in 1827 he was elected to the United States Senate, where he continued for twelve years. In 1841 he was made Secretary of State, and soon after negotiated the famous "Ashburton Treaty" with England, settling the northern boundary of the United States. In 1845 he returned to the Senate; and in 1850 he was reappointed Secretary of State, and continued in office till his death. He died at his country residence in Marshfield, Massachusetts.

Mr. Webster's fame rests chiefly on his state papers and his speeches in Congress; but he took a prominent part in some of the most famous law cases of the present century. Several of his public addresses on occasional themes are well known, also. As a speaker, he was dignified and stately, using clear, straightforward, pure English. He had none of the tricks of oratory. He was large of person, with a massive head, a swarthy complexion, and deep-set, keen, and lustrous eyes. His grand presence added much to his power as a speaker.

THE eulogium pronounced on the character of the State of South Carolina by the honorable gentleman, for her Revolutionary and other merits, meets my hearty concurrence. I shall not acknowledge that the honorable member goes before me, in regard for whatever of distinguished talent or distinguished character South Carolina has produced. I claim part of the honor; I partake in the pride of her great names. I claim them for countrymen, one and all—the Laurenses, the Rutledges, the Pinckneys, the Sumters, the Marions—Americans all—whose fame is no more to be hemmed in by state lines than their talents and patriotism were capable of being circumscribed within the same narrow limits.

In their day and generation, they served and honored the country, and the whole country, and their renown is of the treasures of the whole country. Him whose honored name the gentleman himself bears,—does he suppose me less capable of gratitude for his patriotism, or sympathy for his suffering, than if his eyes had first opened upon the light in Massachusetts, instead of South Carolina? Sir, does he suppose it in his power to exhibit in Carolina a name so bright as to produce envy in my bosom? No, sir,—increased gratification and delight rather. Sir, I thank God that, if I am gifted with little of the spirit which is said to be able to raise mortals to the skies, I have yet none, as I trust, of that other spirit which would drag angels down.

When I shall be found, sir, in my place here in the Senate, or elsewhere, to sneer at public merit because it happened to spring up beyond the little limits of my own state or neighborhood; when I refuse for any such cause, or for any cause, the homage due to American talent, to elevated patriotism, to sincere devotion to liberty and the country; or if I see an uncommon endowment of Heaven; if I see extraordinary capacity or virtue in any son of the South; and if, moved by local prejudice, or gangrened by state jealousy, I get up here to abate a tithe of a hair from his just character and just fame, may my tongue cleave to the roof of my mouth!

Mr. President, I shall enter on no encomium upon Massachusetts. She needs none. There she is; behold her, and judge for yourselves. There is her history; the world knows it by heart. The past, at least, is secure. There is Boston, and Concord, and Lexington, and Bunker Hill; and there they will remain forever. And, sir, where American Liberty raised its first voice, and where its youth was nurtured and sustained, there it still lives, in the strength of its manhood, and full of its original spirit. If discord and disunion shall wound it; if party strife and blind ambition shall hawk at and tear it; if folly and madness, if uneasiness under salutary restraint, shall succeed to separate it from that Union, by which alone its existence is made sure, it will stand, in the end, by the side of that cradle in which its infancy was rocked; it will stretch forth its arm, with whatever of vigor it may still retain, over the friends who gathered around it; and it will fall at last, if fall it must, amid the proudest monuments of its glory and on the very spot of its origin.

NOTES.—The **Laurenses** were of French descent. Henry Laurens was appointed on the commission with Franklin and Jay to negotiate the treaty of peace at Paris at the close of the Revolution. His son, John Laurens, was an aid and secretary of Washington, who was greatly attached to him.

The **Rutledges** were of Irish descent. John Rutledge was a celebrated statesman and lawyer. He was appointed Chief Justice of the United States, but the Senate, for political reasons, refused to confirm his appointment.

Edward Rutledge, brother of the preceding, was Governor of South Carolina during the last two years of his life.

The **Pinckneys** were an old English family who emigrated to Charleston in 1687. Charles Cotesworth Pinckney and his brother Thomas were both active participants in the Revolution. The former was an unsuccessful candidate for the presidency of the United States, in 1800.

Thomas was elected governor of South Carolina in 1789. In the war of 1812 he served as major-general.

Charles Pinckney, a second cousin of the two already mentioned, was four times elected governor of his state.

LXXII. THE CHURCH SCENE FROM EVANGELINE.

Henry Wadsworth Longfellow, 1807-1882, the son of Hon. Stephen Longfellow, an eminent lawyer of Portland, Maine, was born in that city. He graduated, at the age of eighteen, at Bowdoin College. He was soon appointed to the chair of Modern Languages and Literature in that institution, and, to fit himself further for his work, he went abroad and spent four years in Europe. He remained at Bowdoin till 1835, when he was appointed to the chair of Modern Languages and Belles-lettres in Harvard University. On receiving this appointment, he again went to Europe and remained two years. He resigned his professorship in 1854, and after that time resided in Cambridge, pursuing his literary labors and giving to the public, from time to time, the fruits of his pen. In 1868 he made a voyage to England, where he was received with extraordinary marks of honor and esteem. In addition to Mr. Longfellow's original works, both in poetry and in prose, he distinguished himself by several translations; the most famous is that of the works of Dante.

Mr. Longfellow's poetry is always elegant and chaste, showing in every line traces of his careful scholarship. Yet it is not above the popular taste or comprehension, as is shown by the numerous and varied editions of his poems. Many of his poems treat of historical themes; "Evangeline," from which the following selection is taken, is esteemed by many as the most beautiful of all his longer poems; it was first published in 1847.

So passed the morning away. And lo! with a summons sonorous
Sounded the bell from its tower, and over the meadows a drumbeat.
Thronged erelong was the church with men. Without, in the churchyard,
Awaited the women. They stood by the graves, and hung on the headstones
Garlands of autumn leaves and evergreens fresh from the forest.
Then came the guard from the ships, and marching proudly among them
Entered the sacred portal. With loud and dissonant clangor
Echoed the sound of their brazen drums from ceiling and casement,—

Echoed a moment only, and slowly the ponderous portal
Closed, and in silence the crowd awaited the will of the
soldiers.

Then uprose their commander, and spake from the steps of
the altar,
Holding aloft in his hands, with its seals, the royal commission.
'You have convened this day," he said, "by his Majesty's
orders.
Clement and kind has he been; but how you have an-
swered his kindness,
Let your own hearts reply! To my natural make and my
temper
Painful the task is I do, which to you I know must be
grievous.
Yet must I bow and obey, and deliver the will of our
monarch;
Namely, that all your lands, and dwellings, and cattle of
all kinds
Forfeited be to the crown; and that you yourselves from
this province
Be transported to other lands. God grant you may dwell
there
Ever as faithful subjects, a happy and peaceable people!
Prisoners now I declare you; for such is his Majesty's
pleasure!"

As, when the air is serene in the sultry solstice of summer,
Suddenly gathers a storm, and the deadly sling of the
hailstones
Beats down the farmer's corn in the field and shatters his
windows,
Hiding the sun, and strewing the ground with thatch from
the house roofs,

Bellowing fly the herds, and seek to break their inclosure;
So on the hearts of the people descended the words of the speaker.
Silent a moment they stood in speechless wonder, and then rose
Louder and ever louder a wail of sorrow and anger,
And, by one impulse moved, they madly rushed to the doorway.

Vain was the hope of escape; and cries and fierce imprecations
Rang through the house of prayer; and high o'er the heads of the others
Rose, with his arms uplifted, the figure of Basil the blacksmith,
As, on a stormy sea, a spar is tossed by the billows.
Flushed was his face and distorted with passion; and wildly he shouted,—
"Down with the tyrants of England! we never have sworn them allegiance!
Death to these foreign soldiers, who seize on our homes and our harvests!"
More he fain would have said, but the merciless hand of a soldier
Smote him upon the mouth, and dragged him down to the pavement.

In the midst of the strife and tumult of angry contention,
Lo! the door of the chancel opened, and Father Felician
Entered, with serious mien, and ascended the steps of the altar.
Raising his reverend hand, with a gesture he awed into silence
All that clamorous throng; and thus he spake to his people;

Deep were his tones and solemn; in accents measured and mournful
Spake he, as, after the tocsin's alarum, distinctly the clock strikes.

"What is this that ye do, my children? what madness has seized you?
Forty years of my life have I labored among you, and taught you,
Not in word alone, but in deed, to love one another!
Is this the fruit of my toils, of my vigils and prayers and privations?
Have you so soon forgotten all the lessons of love and forgiveness?
This is the house of the Prince of Peace, and would you profane it
Thus with violent deeds and hearts overflowing with hatred?
Lo! where the crucified Christ from his cross is gazing upon you!
See! in those sorrowful eyes what meekness and holy compassion!
Hark! how those lips still repeat the prayer, 'O Father, forgive them!'
Let us repeat that prayer in the hour when the wicked assail us,
Let us repeat it now, and say, 'O Father, forgive them.'"

Few were his words of rebuke, but deep in the hearts of his people
Sank they, and sobs of contrition succeeded the passionate outbreak,
While they repeated his prayer, and said, "O Father, forgive them!"

NOTE.—Nova Scotia was first settled by the French, but, in 1713, was ceded to the English. The inhabitants refusing either to take the oath of allegiance or to bear arms against their fellow-countrymen in the French and Indian War, it was decided to remove the whole people, and distribute them among the other British provinces. This was accordingly done in 1755. The villages were burned to the ground, and the people hurried on board the ships in such a way that but a few families remained undivided.

Longfellow's poem of "Evangeline" is founded on this incident, and the above selection describes the scene where the male inhabitants of Grand-Pré are assembled in the church, and the order for their banishment is first made known to them.

LXXIII. SONG OF THE SHIRT.

Thomas Hood, 1798-1845, the son of a London bookseller, was born in that city. He undertook, after leaving school, to learn the art of an engraver, but soon gave up the business, and turned his attention to literature. His lighter pieces, exhibiting his skill as a wit and punster, soon became well known and popular. In 1821 he became subeditor of the "London Magazine," and formed the acquaintance of the literary men of the metropolis. The last years of his life were clouded by poverty and ill health. Some of his most humorous pieces were written on a sick bed. Hood is best known as a joker—a writer of "whims and oddities"—but he was no mere joker. Some of his pieces are filled with the tenderest pathos; and a gentle spirit, in love with justice and humanity, pervades even his lighter compositions. His "Song of the Shirt" first appeared in the "London Punch."

WITH fingers weary and worn,
With eyelids heavy and red,
A woman sat, in unwomanly rags,
Plying her needle and thread:
Stitch! stitch! stitch!
In poverty, hunger, and dirt,
And still with a voice of dolorous pitch,
She sang the "Song of the Shirt!"

"Work! work! work!
While the cock is crowing aloof!
And work! work! work!
Till the stars shine through the roof!
It is oh to be a slave
Along with the barbarous Turk,
Where woman has never a soul to save,
If this is Christian work!

"Work! work! work!
Till the brain begins to swim;
Work! work! work!
Till the eyes are heavy and dim!
Seam, and gusset, and band,
Band, and gusset, and seam,
Till over the buttons I fall asleep,
And sew them on in a dream!

"O men, with sisters dear!
O men, with mothers and wives!
It is not linen you're wearing out,
But human creatures' lives!
Stitch! stitch! stitch!
In poverty, hunger, and dirt,—
Sewing at once, with a double thread,
A shroud as well as a shirt.

"But why do I talk of Death?
That Phantom of grisly bone,
I hardly fear his terrible shape,
It seems so like my own;
It seems so like my own,
Because of the fasts I keep;
O God! that bread should be so dear,
And flesh and blood so cheap!

"Work! work! work!
My labor never flags;
And what are its wages? A bed of straw,
A crust of bread—and rags,
That shattered roof—and this naked floor—
A table—a broken chair—
And a wall so blank, my shadow I thank
For sometimes falling there.

"Work! work! work!
From weary chime to chime!
Work! work! work!
As prisoners work for crime!
Band, and gusset, and seam,
Seam, and gusset, and band,
Till the heart is sick, and the brain benumbed,
As well as the weary hand.

"Work! work! work!
In the dull December light,
And work! work! work!
When the weather is warm and bright;
While underneath the eaves
The brooding swallows cling,
As if to show me their sunny backs,
And twit me with the spring.

"Oh but to breathe the breath
Of the cowslip and primrose sweet!
With the sky above my head,
And the grass beneath my feet!
For only one short hour
To feel as I used to feel,
Before I knew the woes of want,
And the walk that costs a meal!

"Oh but for one short hour,—
  A respite, however brief!
No blessèd leisure for love or hope,
  But only time for grief!
A little weeping would ease my heart,
  But in their briny bed
My tears must stop, for every drop
  Hinders needle and thread."

With fingers weary and worn,
  With eyelids heavy and red,
A woman sat, in unwomanly rags,
  Plying her needle and thread:
    Stitch! stitch! stitch!
  In poverty, hunger, and dirt,
And still with a voice of dolorous pitch—
Would that its tone could reach the rich!—
    She sang this "Song of the Shirt."

LXXIV. DIAMOND CUT DIAMOND.

Édouard René Lefebvre-Laboulaye, 1811-1883, was a French writer of note. Most of his works involve questions of law and politics, and are considered high authority on the questions discussed. A few works, such as "Abdallah," from which the following extract is adapted, were written as a mere recreation in the midst of law studies; they show great imaginative power. Laboulaye took great interest in the United States, her people, and her literature; and many of his works are devoted to American questions. He translated the works of Dr. William E. Channing into French.

Mansour, the Egyptian merchant, one day repaired to the cadi on account of a suit, the issue of which troubled him but little. A private conversation with the judge had given him hopes of the justice of his cause. The old man asked his son Omar to accompany him in order to accustom him early to deal with the law.

The cadi was seated in the courtyard of the mosque. He was a fat, good-looking man, who never thought, and talked little, which, added to his large turban and his air of perpetual astonishment, gave him a great reputation for justice and gravity.

The spectators were numerous; the principal merchants were seated on the ground on carpets, forming a semicircle around the magistrate. Mansour took his seat a little way from the sheik, and Omar placed himself between the two, his curiosity strongly excited to see how the law was obeyed, and how it was trifled with in case of need.

The first case called was that of a young Banian, as yellow as an orange, with loose-flowing robes and an effeminate air, who had lately landed from India, and who complained of having been cheated by one of Mansour's rivals.

"Having found a casket of diamonds among the effects left by my father," said he, "I set out for Egypt, to live there on the proceeds of their sale. I was obliged by bad weather to put into Jidda, where I soon found myself in want of money. I went to the bazaar, and inquired for a dealer in precious stones. The richest, I was told, was Mansour; the most honest, Ali, the jeweler. I applied to Ali.

"He welcomed me as a son, as soon as he learned that I had diamonds to sell, and carried me home with him. He gained my confidence by every kind of attention, and advanced me all the money I needed. One day, after dinner, at which wine was not wanting, he examined the diamonds, one by one, and said, 'My child, these diamonds are of little value; my coffers are full of such stones. The rocks of the desert furnish them by thousands.'

"To prove the truth of what he said, he opened a box, and, taking therefrom a diamond thrice as large as any of mine, gave it to the slave that was with me. 'What will become of me?' I cried; 'I thought myself rich, and here I am, poor, and a stranger.'

"'My child,' replied Ali, 'Leave this casket with me, and I will give you a price for it such as no one else would offer. Choose whatever you wish in Jidda, and in two hours I will give you an equal weight of what you have chosen in exchange for your Indian stones.'

"On returning home, night brought reflection. I learned that Ali had been deceiving me. What he had given to the slave was nothing but a bit of crystal. I demanded my casket. Ali refused to restore it. Venerable magistrate, my sole hope is in your justice."

It was now Ali's turn to speak. "Illustrious cadi," said he, "It is true that we made a bargain, which I am ready to keep. The rest of the young man's story is false. What matters it what I gave the slave? Did I force the stranger to leave the casket in my hands? Why does he accuse me of treachery? Have I broken my word, and has he kept his?"

"Young man," said the cadi to the Banian, "have you witnesses to prove that Ali deceived you? If not, I shall put the accused on his oath, as the law decrees." A Koran was brought. Ali placed his hand on it, and swore three times that he had not deceived the stranger. "Wretch," said the Banian, "thou art among those whose feet go down to destruction. Thou hast thrown away thy soul."

Omar smiled, and while Ali was enjoying the success of his ruse, he approached the stranger, and asked, "Do you wish me to help you gain the suit?" "Yes," was the reply; "but you are only a child—you can do nothing."

"Have confidence in me a few moments," said Omar; "accept Ali's bargain; let me choose in your stead, and fear nothing."

The stranger bowed his head, and murmured, "What can I fear after having lost all?" Then, turning to the cadi, and bowing respectfully, "Let the bargain be consummated," said he, "since the law decrees it, and let this young man choose in my stead what I shall receive in payment."

A profound silence ensued. Omar rose, and, bowing to the cadi, "Ali," said he to the jeweler, "you have doubtless brought the casket, and can tell us the weight thereof."

"Here it is," said Ali; "it weighs twenty pounds. Choose what you will; if the thing asked for is in Jidda, you shall have it within two hours, otherwise the bargain is null and void."

"What we desire," said Omar, raising his voice, "is ants' wings, half male and half female. You have two hours in which to furnish the twenty pounds you have promised us." "This is absurd," cried the jeweler; "it is impossible. I should need half a score of persons and six months labor to satisfy so foolish a demand."

"Are there any winged ants in Jidda?" asked the cadi. "Of course," answered the merchants, laughing; "they are one of the plagues of Egypt. Our houses are full of them, and it would be doing us a great service to rid us of them."

"Then Ali must keep his promise or give back the casket," said the cadi. "This young man was mad to sell his diamonds weight for weight; he is mad to exact such payment. So much the better for Ali the first time: so much the worse for him the second. Justice has not two weights and measures. Every bargain holds good before the law. Either furnish twenty pounds of ants' wings, or restore the casket to the Banian." "A righteous judgment," shouted the spectators, wonder-struck at such equity.

The stranger, beside himself with joy, took from the casket three diamonds of the finest water; he forced them on Omar, who put them in his girdle, and seated himself by his father, his gravity unmoved by the gaze of the assembly. "Well done," said Mansour; but it is my turn

Kappes

now; mark me well, and profit by the lesson I shall give you. Stop, young man!" he cried to the Banian, "we have an account to settle."

"The day before yesterday," continued he, "this young man entered my shop, and, bursting into tears, kissed my hand and entreated me to sell him a necklace which I had already sold to the Pasha of Egypt, saying that his life and that of a lady depended upon it. 'Ask of me what you will, my father,' said he, 'but I must have these gems or die.'

"I have a weakness for young men, and, though I knew the danger of disappointing my master the pasha, I was unable to resist his supplications. 'Take the necklace,' said I to him, 'but promise to give whatever I may ask in exchange.' 'My head itself, if you will,' he replied, 'for you have saved my life.' We were without witnesses, but," added Mansour, turning to the Banian, "is not my story true?"

"Yes," said the young man, "and I beg your pardon for not having satisfied you sooner: you know the cause. Ask of me what you desire."

"What I desire," said Mansour, "is the casket with all its contents. Illustrious magistrate, you have declared that all bargains hold good before the law; this young man has promised to give me what I please; now I declare that nothing pleases me but these diamonds."

The cadi raised his head and looked about the assembly, as if to interrogate the faces, then stroked his beard, and relapsed into his meditations.

"Ali is defeated," said the sheik to Omar, with a smile. "The fox is not yet born more cunning than the worthy Mansour."

"I am lost!" cried the Banian. "O Omar, have you saved me only to cast me down from the highest pinnacle of joy to the depths of despair? Persuade your father to spare me, that I may owe my life to you a second time."

"Well, my son," said Mansour, "doubtless you are shrewd, but this will teach you that your father knows rather more than you do. The cadi is about to decide: try whether you can dictate his decree."

"It is mere child's play," answered Omar, shrugging his shoulders; "but since you desire it, my father, you shall lose your suit." He rose, and taking a piaster from his girdle, put it into the hand of the Banian, who laid it before the judge.

"Illustrious cadi," said Omar, "this young man is ready to fulfill his engagement. This is what he offers Mansour—a piaster. In itself this coin is of little value; but examine it closely, and you will see that it is stamped with the likeness of the sultan, our glorious master. May God destroy and confound all who disobey his highness!

"It is this precious likeness that we offer you," added he, turning to Mansour; "if it pleases you, you are paid; to say that it displeases you is an insult to the pasha, a crime punishable by death; and I am sure that our worthy cadi will not become your accomplice—he who has always been and always will be the faithful servant of all the sultans."

When Omar had finished speaking, all eyes turned toward the cadi, who, more impenetrable than ever, stroked his face and waited for the old man to come to his aid. Mansour was agitated and embarrassed. The silence of the cadi and the assembly terrified him, and he cast a supplicating glance toward his son.

"My father," said Omar, "permit this young man to thank you for the lesson of prudence which you have given him by frightening him a little. He knows well that it was you who sent me to his aid, and that all this is a farce. No one is deceived by hearing the son oppose the father, and who has ever doubted Mansour's experience and generosity?"

"No one," interrupted the cadi, starting up like a man

suddenly awakened from a dream, "and I least of all; and this is why I have permitted you to speak, my young Solomon. I wished to honor in you the wisdom of your father; but another time avoid meddling with his highness's name; it is not safe to sport with the lion's paws. The matter is settled. The necklace is worth a hundred thousand piasters, is it not, Mansour? This madcap, shall give you, therefore, a hundred thousand piasters, and all parties will be satisfied."

NOTES.—A **cadi** in the Mohammedan countries, corresponds to our magistrate.

A **sheik** among the Arabs and Moors, may mean simply an old man, or, as in this case, a man of eminence.

A **Banian** is a Hindoo merchant, particularly one who visits foreign countries on business.

Jidda is a city in Arabia, on the Red Sea.

A **pasha** is the governor of a Turkish province.

The Turkish **piaster** was formerly worth twenty-five cents: it is now worth only about eight cents.

LXXV. THANATOPSIS.

To him who in the love of Nature holds
Communion with her visible forms, she speaks
A various language: for his gayer hours
She has a voice of gladness, and a smile
And eloquence of beauty; and she glides
Into his darker musings, with a mild
And healing sympathy, that steals away
Their sharpness, ere he is aware.

When thoughts
Of the last bitter hour come like a blight
Over thy spirit, and sad images

Of the stern agony, and shroud, and pall,
And breathless darkness, and the narrow house,
Make thee to shudder, and grow sick at heart;—
Go forth, under the open sky, and list
To Nature's teachings, while from all around—
Earth and her waters, and the depths of air—
Comes a still voice,—

Yet a few days, and thee
The all-beholding sun shall see no more
In all his course; nor yet in the cold ground,
Where thy pale form was laid, with many tears,
Nor in the embrace of ocean, shall exist
Thy image. Earth, that nourished thee, shall claim
Thy growth, to be resolved to earth again;
And, lost each human trace, surrendering up
Thine individual being, shalt thou go
To mix forever with the elements;
To be a brother to the insensible rock
And to the sluggish clod, which the rude swain
Turns with his share, and treads upon. The oak
Shall send his roots abroad, and pierce thy mold.

Yet not to thine eternal resting place
Shalt thou retire alone, nor couldst thou wish
Couch more magnificent. Thou shalt lie down
With patriarchs of the infant world,—with kings,
The powerful of the earth,—the wise, the good,
Fair forms, and hoary seers of ages past,—
All in one mighty sepulcher.

The hills,
Rock-ribbed, and ancient as the sun; the vales
Stretching in pensive quietness between;
The venerable woods; rivers that move

In majesty, and the complaining brooks,
That make the meadows green; and, poured round all,
Old Ocean's gray and melancholy waste,—
Are but the solemn decorations all
Of the great tomb of man. The golden sun,
The planets, all the infinite host of heaven,
Are shining on the sad abodes of death,
Through the still lapse of ages.

All that tread
The globe are but a handful to the tribes
That slumber in its bosom. Take the wings
Of morning, pierce the Barcan wilderness,
Or lose thyself in the continuous woods
Where rolls the Oregon, and hears no sound
Save his own dashings,—yet the dead are there:
And millions in those solitudes, since first
The flight of years began, have laid them down
In their last sleep,—the dead reign there alone.

So shalt thou rest; and what if thou withdraw
In silence from the living, and no friend
Take note of thy departure? All that breathe
Will share thy destiny. The gay will laugh
When thou art gone, the solemn brood of care
Plod on, and each one as before will chase
His favorite phantom; yet all these shall leave
Their mirth and their employments, and shall come
And make their bed with thee. As the long train
Of ages glide away, the sons of men—
The youth in life's green spring, and he who goes
In the full strength of years, matron and maid,
The speechless babe, and the gray-headed man—
Shall one by one be gathered to thy side
By those who in their turn shall follow them.

So live, that when thy summons comes to join
The innumerable caravan, which moves
To that mysterious realm, where each shall take
His chamber in the silent halls of death,
Thou go not, like the quarry slave at night,
Scourged to his dungeon, but, sustained and soothed
By an unfaltering trust, approach thy grave,
Like one who wraps the drapery of his couch
About him, and lies down to pleasant dreams.

—*Bryant.*

NOTES.—**Thanatopsis** is composed of two Greek words, θάνᾰτος (thanatos), meaning *death*, and ὄψις (opsis), a *view.* The word, therefore, signifies *a view of death*, or *reflections on death.*

Barca is in the northeastern part of Africa: the southern and eastern portions of the country are a barren desert.

The **Oregon** (or Columbia) River is the most important river of the United States emptying into the Pacific. The Lewis and Clark Expedition (1803–1806) had first explored the country through which it flows only five years before the poem was written.

LXXVI. INDIAN JUGGLERS.

William Hazlitt, 1778–1830, was born in Maidstone, England. His father was a Unitarian clergyman, and he was sent to a college of that denomination to be educated for the ministry; but having a greater taste for art than theology, he resolved, on leaving school, to devote himself to painting. He succeeded so well in his efforts as to meet the warmest commendation of his friends, but did not succeed in satisfying his own fastidious taste. On this account he threw away his pencil and took up his pen. His works, though numerous, are, with the exception of a life of Napoleon, chiefly criticisms on literature and art.

Hazlitt is thought to have treated his contemporaries with an unjust severity; but his genial appreciation of the English classics, and the thorough and loving manner in which he discusses their merits, make his essays the delight of every lover of those perpetual wellsprings of intellectual pleasure. His "Table Talk," "Characters of Shakespeare's Plays," "Lectures on the English Poets," and "Lectures on the Literature of the Elizabethan Age," are the works that exhibit his style and general merits in their most favorable light.

Coming forward and seating himself on the ground, in his white dress and tightened turban, the chief of the Indian jugglers begins with tossing up two brass balls, which is what any of us could do, and concludes by keeping up four at the same time, which is what none of us could do to save our lives, not if we were to take our whole lives to do it in.

Is it then a trifling power we see at work, or is it not something next to miraculous? It is the utmost stretch of human ingenuity, which nothing but the bending the faculties of body and mind to it from the tenderest infancy with incessant, ever-anxious application up to manhood, can accomplish or make even a slight approach to. Man, thou art a wonderful animal, and thy ways past finding out! Thou canst do strange things, but thou turnest them to small account!

To conceive of this extraordinary dexterity, distracts the imagination and makes admiration breathless. Yet it costs nothing to the performer, any more than if it were a mere mechanical deception with which he had nothing to do, but to watch and laugh at the astonishment of the spectators. A single error of a hair's breadth, of the smallest conceivable portion of time, would be fatal; the precision of the movements must be like a mathematical truth; their rapidity is like lightning.

To catch four balls in succession, in less than a second of time, and deliver them back so as to return with seeming consciousness to the hand again; to make them revolve around him at certain intervals, like the planets in their spheres; to make them chase each other like sparkles of fire, or shoot up like flowers or meteors; to throw them behind his back, and twine them round his neck like ribbons, or like serpents; to do what appears an impossibility, and to do it with all the ease, the grace, the carelessness imaginable; to laugh at, to play with the glittering mockeries, to follow them with his eye as if he could fascinate them with its lambent fire, or as if he had only to see that they kept time with the music on the stage—there is something in all this which he who does not admire may be quite sure he never really admired anything in the whole course of his life. It is skill surmounting difficulty, and beauty triumphing over skill. It seems as if the difficulty, once mastered, naturally resolved itself into ease and grace, and as if, to be overcome at all, it must be overcome without an effort. The smallest awkwardness or want of pliancy or self-possession would stop the whole process. It is the work of witchcraft, and yet sport for children.

Some of the other feats are quite as curious and wonderful—such as the balancing the artificial tree, and shooting a bird from each branch through a quill—though none of them have the elegance or facility of the keeping up of the brass balls. You are in pain for the result, and glad when the experiment is over; they are not accompanied with the same unmixed, unchecked delight as the former; and I would not give much to be merely astonished without being pleased at the same time. As to the swallowing of the sword, the police ought to interfere to prevent it.

When I saw the Indian juggler do the same things before, his feet were bare, and he had large rings on his toes, which he kept turning round all the time of the performance, as if they moved of themselves.

The hearing a speech in Parliament drawled or stammered out by the honorable member or the noble lord, the ringing the changes on their commonplaces, which any one could repeat after them as well as they, stirs me not a jot,—shakes not my good opinion of myself. I ask what there is that I can do as well as this. Nothing. What have I been doing all my life? Have I been idle, or have I nothing to show for all my labor and pains? Or have I passed my time in pouring words like water into empty sieves, rolling a stone up a hill and then down again, trying to prove an argument in the teeth of facts, and looking for causes in the dark, and not finding them? Is there no one thing in which I can challenge competition, that I can bring as an instance of exact perfection, in which others can not find a flaw?

The utmost I can pretend to is to write a description of what this fellow can do. I can write a book: so can many others who have not even learned to spell. What abortions are these essays! What errors, what ill-pieced transitions, what crooked reasons, what lame conclusions! How little is made out, and that little how ill! Yet they are the best I can do.

I endeavor to recollect all I have ever heard or thought upon a subject, and to express it as neatly as I can. Instead of writing on four subjects at a time, it is as much as I can manage, to keep the thread of one discourse clear and unentangled. I have also time on my hands to correct my opinions and polish my periods; but the one I can not, and the other I will not, do. I am fond of arguing; yet, with a good deal of pains and practice, it is often much as I can do to beat my man, though he may be a very indifferent hand. A common fencer would disarm his adversary in the twinkling of an eye, unless he were a professor like himself. A stroke of wit will sometimes produce this effect, but there is no such power or superiority in sense or reasoning. There is no complete mastery of execution to be shown there; and you hardly know the professor from the impudent pretender or the mere clown.

LXXVII. ANTONY OVER CÆSAR'S DEAD BODY.

FRIENDS, Romans, countrymen, lend me your ears:
I come to bury Cæsar, not to praise him.
The evil that men do lives after them;
The good is oft interrèd with their bones;

So let it be with Cæsar. The noble Brutus
Hath told you Cæsar was ambitious:
If it were so, it was a grievous fault,
And grievously hath Cæsar answered it.
Here, under leave of Brutus and the rest—
For Brutus is an honorable man;
So are they all, all honorable men—
Come I to speak in Cæsar's funeral.

He was my friend, faithful and just to me:
But Brutus says he was ambitious;
And Brutus is an honorable man.
He hath brought many captives home to Rome,
Whose ransoms did the general coffers fill:
Did this in Cæsar seem ambitious?
When that the poor have cried, Cæsar hath wept:
Ambition should be made of sterner stuff:
Yet Brutus says he was ambitious;
And Brutus is an honorable man.

You all did see, that on the Lupercal,
I thrice presented him a kingly crown,
Which he did thrice refuse. Was this ambition?
Yet Brutus says he was ambitious;
And, sure, he is an honorable man.
I speak not to disprove what Brutus spoke,
But here I am to speak what I do know.
You all did love him once, not without cause;
What cause withholds you, then, to mourn for him?
O judgment! thou art fled to brutish beasts,
And men have lost their reason. Bear with me;
My heart is in the coffin there with Cæsar,
And I must pause till it come back to me.

But yesterday the word of Cæsar might
Have stood against the world; now lies he there,

And none so poor to do him reverence.
O masters! if I were disposed to stir
Your hearts and minds to mutiny and rage,
I should do Brutus wrong, and Cassius wrong,
Who, you all know, are honorable men:
I will not do them wrong; I rather choose
To wrong the dead, to wrong myself and you,
Than I will wrong such honorable men.

But here's a parchment with the seal of Cæsar;
I found it in his closet; 't is his will:
Let but the commons hear this testament—
Which, pardon me, I do not mean to read—
And they would go and kiss dead Cæsar's wounds,
And dip their napkins in his sacred blood;
Yea, beg a hair of him for memory,
And, dying, mention it within their wills,
Bequeathing it as a rich legacy
Unto their issue.

Citizen. We'll hear the will: read it, Mark Antony.
All. The will, the will; we will hear Cæsar's will.
Ant. Have patience, gentle friends, I must not read it;
It is not meet you know how Cæsar loved you.
You are not wood, you are not stones, but men;
And, being men, hearing the will of Cæsar,
It will inflame you, it will make you mad;
'T is good you know not that you are his heirs;
For, if you should, Oh what would come of it!
Cit. Read the will; we'll hear it, Antony;
You shall read the will, Cæsar's will.
Ant. Will you be patient? Will you stay awhile?
I have o'ershot myself to tell you of it:
I fear I wrong the honorable men
Whose daggers have stabbed Cæsar. I do fear it.
Cit. They were traitors: honorable men!
All. The will! the testament!

Ant. You will compel me, then, to read the will?
Then make a ring about the corpse of Cæsar,
And let me show you him that made the will.

(*He comes down from the pulpit.*)

If you have tears, prepare to shed them now.
You all do know this mantle: I remember
The first time ever Cæsar put it on;
'Twas on a summer's evening, in his tent,
That day he overcame the Nervii:
Look! in this place, ran Cassius' dagger through:
See what a rent the envious Casca made:
Through this, the well belovèd Brutus stabbed;
And, as he plucked his cursèd steel away,
Mark how the blood of Cæsar followed it,
As rushing out of doors, to be resolved
If Brutus so unkindly knocked, or no;
For Brutus, as you know, was Cæsar's angel:
Judge, O you gods, how dearly Cæsar loved him!

This was the most unkindest cut of all;
For, when the noble Cæsar saw him stab,
Ingratitude, more strong than traitors' arms,
Quite vanquished him: then burst his mighty heart;
And, in his mantle muffling up his face,
Even at the base of Pompey's statua,
Which all the while ran blood, great Cæsar fell.

Oh, what a fall was there, my countrymen!
Then I, and you, and all of us fell down,
Whilst bloody treason flourished over us.
Oh, now you weep; and, I perceive, you feel
The dint of pity: these are gracious drops.
Kind souls, what, weep you when you but behold

Our Cæsar's vesture wounded? Look you here,
Here is himself, marred, as you see, with traitors.

1st Cit. O piteous spectacle!

2d Cit. O noble Cæsar!

3d Cit. We will be revenged!

All. Revenge! About! Seek! Burn! Fire!
Kill! Slay! Let not a traitor live.

Ant. Stay, countrymen.

1st Cit. Peace there! hear the noble Antony.

2d Cit. We'll hear him, we'll follow him, we'll die with him.

Ant. Good friends, sweet friends, let me not stir you up
To such a sudden flood of mutiny.
They that have done this deed are honorable:
What private griefs they have, alas, I know not,
That made them do it; they are wise and honorable,
And will, no doubt, with reasons answer you.

I come not, friends, to steal away your hearts:
I am no orator, as Brutus is;
But, as you know me all, a plain, blunt man,
That love my friend; and that they know full well
That gave me public leave to speak of him:
For I have neither wit, nor words, nor worth,
Action, nor utterance, nor the power of speech,
To stir men's blood: I only speak right on:
I tell you that which you yourselves do know;
Show you sweet Cæsar's wounds, poor, poor, dumb mouths,
And bid them speak for me: but were I Brutus,
And Brutus Antony, there were an Antony
Would ruffle up your spirits, and put a tongue
In every wound of Cæsar, that should move
The stones of Rome to rise and mutiny.

Shakespeare.—Julius Cæsar, Act iii, Scene ii.

NOTES.—**Gaius Julius Cæsar** (b. 102, d. 44 B. C.) was the most remarkable genius of the ancient world. Cæsar ruled Rome as imperator five years and a half, and, in the intervals of seven campaigns during that time, spent only fifteen months in Rome. Under his rule Rome was probably at her best, and his murder at once produced a state of anarchy.

The conspirators against Cæsar—among whom were **Brutus, Cassius** and **Casca**—professed to be moved by honest zeal for the good of Rome; but their own ambition was no doubt the true motive, except with Brutus.

Mark Antony was a strong friend of Julius Cæsar. Upon the latter's death, Antony, by his funeral oration, incited the people and drove the conspirators from Rome.

The **Lupercal** was a festival of purification and expiation held in Rome on the 15th of February. Antony was officiating as priest at this festival when he offered the crown to Cæsar.

In his **will** Cæsar left to every citizen of Rome a sum of money, and bequeathed his private gardens to the public.

The **Nervii** were one of the most warlike tribes of Celtic Gaul. Cæsar almost annihilated them in 57 B. C.

Pompey, once associated with Cæsar in the government of Rome, was afterwards at war with him. He was murdered by those who thought to propitiate Cæsar, but the latter wept when Pompey's head was sent to him, and had the murderers put to death.

Statua is the Latin form of *statue*, in common use in Shakespeare's time; this form is required here by the meter.

LXXVIII. THE ENGLISH CHARACTER.

William Hickling Prescott, 1796-1859, the historian, was the son of William Prescott, an eminent jurist, and the grandson of Col. William Prescott, who commanded the Americans at the battle of Bunker Hill. He was born in Salem, Massachusetts, graduated at Harvard University in 1814, and died in Boston. Just as he was completing his college course, the careless sport of a fellow-student injured one of his eyes so seriously that he never recovered from it. He had intended to adopt law as his profession; but, from his defective eyesight, he was obliged to choose work in which he could regulate his hours of labor, and could employ the aid of a secretary. He chose to be a historian; and followed his

choice with wonderful system, perseverance, and success till the close of **his** life. His works are: "The Reign of Ferdinand and Isabella," "The Conquest of Mexico," "The Conquest of Peru," "The Reign of Philip II.," and a volume of "Miscellanies." He had not completed the history of Philip at the time of his death. As a writer of history, Mr. Prescott ranks with the first for accuracy, precision, clearness, and beauty of style. As a man, he was genial, kind-hearted and even-tempered.

On the whole, what I have seen raises my preconceived estimate of the English character. It is full of generous, true, and manly qualities; and I doubt if there ever was so high a standard of morality in an aristocracy which has such means for self-indulgence at its command, and which occupies a position that secures it so much deference. In general, they do not seem to abuse their great advantages. The respect for religion — at least for the forms of it — is universal, and there are few, I imagine, of the great proprietors who are not more or less occupied with improving their estates, and with providing for the comfort of their tenantry, while many take a leading part in the great political movements of the time. There never was an aristocracy which combined so much practical knowledge and industry with the advantages of exalted rank.

The Englishman is seen to most advantage in his country home. For he is constitutionally both domestic and rural in his habits. His fireside and his farm — these are the places in which one sees his simple and warm-hearted nature more freely unfolded. There is a shyness in an Englishman, — a natural reserve, which makes him cold to strangers, and difficult to approach. But once corner him in his own house, a frank and full expansion will be given to his feelings that we should look for in vain in the colder Yankee, and a depth not to be found in the light and superficial Frenchman, — speaking of nationalities, not of individualities.

The Englishman is the most truly rural in his tastes and habits of any people in the world. I am speaking of the higher classes. The aristocracy of other countries affect the camp and the city. But the English love their old castles and country seats with a patriotic love. They are fond of country sports. Every man shoots or hunts. No man is too old to be in the saddle some part of the day, and men of seventy years and more follow the hounds, and take a five-barred gate at a leap. The women are good whips, are fond of horses and dogs, and other animals. Duchesses have their cows, their poultry, their pigs,—all watched over and provided with accommodations of Dutch-like neatness. All this is characteristic of the people. It may be thought to detract something from the feminine graces which in other lands make a woman so amiably dependent as to be nearly imbecile. But it produces a healthy and blooming race of women to match the hardy Englishman,—the finest development of the physical and moral nature which the world has witnessed. For we are not to look on the English gentleman as a mere Nimrod. With all his relish for field sports and country usages, he has his house filled with collections of art and with extensive libraries. The tables of the drawing-rooms are covered with the latest works, sent down by the London publisher. Every guest is provided with an apparatus for writing, and often a little library of books for his own amusement. The English country gentleman of the present day is anything but a Squire Western, though he does retain all his relish for field sports.

The character of an Englishman, under its most refined aspect, has some disagreeable points which jar unpleasantly on the foreigner not accustomed to them. The consciousness of national superiority, combined with natural feelings of independence, gives him an air of arrogance, though it must be owned that this is never betrayed in his own house,—I may almost say in his own country. But abroad, when he seems to institute a comparison between himself and the people he is thrown with, it becomes so obvious that he is the most unpopular, not to say odious, person in the world. Even the open hand with which he dispenses his bounty will not atone for the violence he offers to national vanity.

There are other defects, which are visible even in his most favored circumstances. Such is his bigotry, surpassing everything in a quiet passive form, that has been witnessed since the more active bigotry of the times of the Spanish Philips. Such, too, is the exclusive, limited range of his knowledge and conceptions of all political and social topics and relations. The Englishman, the cultivated Englishman, has no standard of excellence borrowed from mankind. His speculation never travels beyond his own little—great little—island. That is the world to him. True, he travels, shoots lions among the Hottentots, chases the grizzly bear over the Rocky Mountains, kills elephants in India and salmon on the coast of Labrador, comes home, and very likely makes a book. But the scope of his ideas does not seem to be enlarged by all this. The body travels, not the mind. And, however he may abuse his own land, he returns home as hearty a John Bull, with all his prejudices and national tastes as rooted, as before. The English—the men of fortune—all travel. Yet how little sympathy they show for other people or institutions, and how slight is the interest they take in them! They are islanders, cut off from the great world. But their island is, indeed, a world of its own. With all their faults, never has the sun shone—if one may use the expression in reference to England—on a more noble race, or one that has done more for the great interests of humanity.

NOTES.—**Nimrod** is spoken of in Genesis (x. 9) as "a mighty hunter." Thus the name came to be applied to any one devoted to hunting.

Squire Western is a character in Fielding's "Tom Jones." He is represented as an ignorant, prejudiced, irascible, but, withal, a jolly, good-humored English country gentleman.

LXXIX. THE SONG OF THE POTTER.

TURN, turn, my wheel! Turn round and round,
Without a pause, without a sound:
So spins the flying world away!
This clay, well mixed with marl and sand,
Follows the motion of my hand;
For some must follow, and some command,
Though all are made of clay!

Turn, turn, my wheel! All things must change
To something new, to something strange;
Nothing that is can pause or stay;
The moon will wax, the moon will wane,
The mist and cloud will turn to rain,
The rain to mist and cloud again,
To-morrow be to-day.

Turn, turn, my wheel! All life is brief;
What now is bud will soon be leaf,
What now is leaf will soon decay;
The wind blows east, the wind blows west;
The blue eggs in the robin's nest
Will soon have wings and beak and breast,
And flutter and fly away.

Turn, turn, my wheel! This earthen jar
A touch can make, a touch can mar;
And shall it to the Potter say,
What makest thou? Thou hast no hand?
As men who think to understand
A world by their Creator planned,
Who wiser is than they.

Turn, turn, my wheel! 'Tis nature's plan
The child should grow into the man,
  The man grow wrinkled, old, and gray;
In youth the heart exults and sings,
The pulses leap, the feet have wings;
In age the cricket chirps, and brings
  The harvest home of day.

Turn, turn, my wheel! The human race,
Of every tongue, of every place,
  Caucasian, Coptic, or Malay,
All that inhabit this great earth,
Whatever be their rank or worth,
Are kindred and allied by birth,
  And made of the same clay.

Turn, turn, my wheel! What is begun
At daybreak must at dark be done,
  To-morrow will be another day;
To-morrow the hot furnace flame
Will search the heart and try the frame,
And stamp with honor or with shame
  These vessels made of clay.

Stop, stop, my wheel! Too soon, too soon
The noon will be the afternoon,
  Too soon to-day be yesterday;
Behind us in our path we cast
The broken potsherds of the past,
And all are ground to dust at last,
  And trodden into clay.

—*Longfellow.*

NOTE.—**Coptic** was formerly the language of Egypt, and is preserved in the inscriptions of the ancient monuments found there; it has now given place entirely to Arabic.

LXXX. A HOT DAY IN NEW YORK.

William Dean Howells, 1837——, was born in Belmont County, Ohio. In boyhood he learned the printer's trade, at which he worked for several years. He published a volume of poems in 1860, in connection with John J. Piatt. From 1861 to 1865 he was United States Consul at Venice. On his return he resided for a time in New York City, and was one of the editors of the "Nation." In 1871 he was appointed editor in chief of the "Atlantic Monthly." He held the position ten years, and then retired in order to devote himself to his own writings. Since then, he has been connected with other literary magazines.

Mr. Howells has written several books: novels and sketches: his writings are marked by an artistic finish, and a keen but subtile humor. The following selection is an extract from "Their Wedding Journey."

When they alighted, they took their way up through one of the streets of the great wholesale businesses, to Broadway. On this street was a throng of trucks and wagons, lading and unlading; bales and boxes rose and sank by pulleys overhead; the footway was a labyrinth of packages of every shape and size; there was no flagging of the pitiless energy that moved all forward, no sign of how heavy a weight lay on it, save in the reeking faces of its helpless instruments.

It was four o'clock, the deadliest hour of the deadly summer day. The spiritless air seemed to have a quality of blackness in it, as if filled with the gloom of low-hovering wings. One half the street lay in shadow, and one half in sun; but the sunshine itself was dim, as if a heat greater than its own had smitten it with languor. Little gusts of sick, warm wind blew across the great avenue at the corners of the intersecting streets. In the upward distance, at which the journeyers looked, the loftier roofs and steeples lifted themselves dim out of the livid atmosphere, and far up and down the length of the street swept a stream of tormented life.

All sorts of wheeled things thronged it, conspicuous among which rolled and jarred the gaudily painted stages, with quivering horses driven each by a man who sat in the shade of a branching, white umbrella, and suffered with a moody truculence of aspect, and as if he harbored the bitterness of death in his heart for the crowding passengers within, when one of them pulled the strap about his legs, and summoned him to halt.

Most of the foot passengers kept to the shady side, and to the unaccustomed eyes of the strangers they were not less in number than at any other time, though there were fewer women among them. Indomitably resolute of soul, they held their course with the swift pace of custom, and only here and there they showed the effect of the heat.

One man, collarless, with waistcoat unbuttoned, and hat set far back from his forehead, waved a fan before his death-white, flabby face, and set down one foot after the other with the heaviness of a somnambulist. Another, as they passed him, was saying huskily to the friend at his side, "I can't stand this much longer. My hands tingle as if they had gone to sleep; my heart—" But still the multitude hurried on, passing, repassing, encountering, evading, vanishing into shop doors, and emerging from them, dispersing down the side streets, and swarming out of them.

It was a scene that possessed the beholder with singular fascination, and in its effect of universal lunacy, it might well have seemed the last phase of a world presently to be destroyed. They who were in it, but not of it, as they fancied—though there was no reason for this—looked on it amazed, and at last their own errands being accomplished, and themselves so far cured of the madness of purpose, they cried with one voice that it was a hideous sight, and strove to take refuge from it in the nearest place where the soda fountain sparkled.

It was a vain desire. At the front door of the apothecary's hung a thermometer, and as they entered they heard the next comer cry out with a maniacal pride in the affliction laid upon mankind, "Ninety-seven degrees!" Behind them, at the door, there poured in a ceaseless stream of people, each pausing at the shrine of heat, before he tossed off the hissing draught that two pale, close-clipped boys served them from either side of the fountain. Then, in the order of their coming, they issued through another door upon the side street, each, as he disappeared, turning his face half round, and casting a casual glance upon a little group near another counter.

The group was of a very patient, half-frightened, half-puzzled looking gentleman who sat perfectly still on a stool, and of a lady who stood beside him, rubbing all over his head a handkerchief full of pounded ice, and easing one hand with the other when the first became tired. Basil drank his soda, and paused to look upon this group, which he felt would commend itself to realistic sculpture as eminently characteristic of the local life, and, as "The Sunstroke," would sell enormously in the hot season.

"Better take a little more of that," the apothecary said, looking up from his prescription, and, as the organized sympathy of the seemingly indifferent crowd, smiling very kindly at his patient, who thereupon tasted something in the glass he held.

"Do you still feel like fainting?" asked the humane authority. "Slightly, now and then," answered the other, "but I'm hanging on hard to the bottom curve of that icicled S on your soda fountain, and I feel that I'm all right as long as I can see that. The people get rather hazy occasionally, and have no features to speak of. But I don't know that I look very impressive myself," he added in the jesting mood which seems the natural condition of Americans in the face of all embarrassments.

"Oh, you'll do!" the apothecary answered, with a laugh; but he said, in an answer to an anxious question from the lady, "He mustn't be moved for an hour yet," and gayly pestled away at a prescription, while she resumed her office of grinding the pounded ice round and round upon her husband's skull. Isabel offered her the commiseration of friendly words, and of looks kinder yet, and then, seeing that they could do nothing, she and Basil fell into the endless procession, and passed out of the side door.

"What a shocking thing," she whispered. "Did you see how all the people looked, one after another, so indifferently at that couple, and evidently forgot them the next instant? It was dreadful. I should n't like to have you sun-struck in New York."

"That's very considerate of you; but place for place, if any accident must happen to me among strangers, I think I should prefer to have it in New York. The biggest place is always the kindest as well as the cruelest place. Amongst the thousands of spectators the good Samaritan as well as the Levite would be sure to be. As for a sunstroke, it requires peculiar gifts. But if you compel me to a choice in the matter, then I say give me the busiest part of Broadway for a sunstroke. There is such experience of calamity there that you could hardly fall the first victim to any misfortune."

LXXXI. DISCONTENT.—AN ALLEGORY.

Joseph Addison, 1672-1719, the brilliant essayist and poet, has long occupied an exalted place in English literature. He was the son of an English clergyman, was born in Wiltshire, and educated at Oxford; he died at "Holland House" (the property of his wife, to whom he had been married but about two years), and was buried in Westminster Abbey. Several years of his life were spent in the political affairs of his time, he held several public offices, and was, for ten years, a member of Parliament. His fame as an author rests chiefly upon his "Hymns," his tragedy of "Cato," and his "Essays" contributed principally to the "Tatler" and the "Spectator." The excellent style of his essays, their genial wit and sprightly humor, made them conspicuous in an age when coarseness, bitterness, and exaggeration deformed the writings of the most eminent; and these characteristics have given them an unquestioned place among the classics of our language.

Mr. Addison was shy and diffident, but genial and lovable; his moral character was above reproach, excepting that he is said to have been too fond of wine.

It is a celebrated thought of Socrates, that if all the misfortunes of mankind were cast into a public stock, in order to be equally distributed among the whole species, those who now think themselves the most unhappy, would prefer the share they are already possessed of before that which would fall to them by such a division. Horace has carried this thought a great deal farther, and supposes that the hardships or misfortunes we lie under, are more easy to us than those of any other person would be, in case we could change conditions with him.

As I was ruminating on these two remarks, and seated in my elbowchair, I insensibly fell asleep; when, on a sudden, methought there was a proclamation made by Jupiter, that every mortal should bring in his griefs and calamities, and throw them together in a heap. There was a large plain appointed for this purpose. I took my stand in the center of it, and saw, with a great deal of pleasure, the whole human species marching one after another, and throwing down their several loads, which immediately grew up into a prodigious mountain, that seemed to rise above the clouds.

There was a certain lady of a thin, airy shape, who was very active in this solemnity. She carried a magnifying glass in one of her hands, and was clothed in a loose, flowing robe, embroidered with several figures of fiends and specters, that discovered themselves in a thousand chimerical shapes as her garment hovered in the wind. There was something wild and distracted in her looks. Her name was Fancy. She led up every mortal to the appointed place, after having officiously assisted him in making up his pack, and laying it upon his shoulders. My heart melted within me to see my fellow-creatures groaning under their respective burdens, and to consider that prodigious bulk of human calamities which lay before me.

There were, however, several persons who gave me great diversion upon this occasion. I observed one bringing in a fardel, very carefully concealed under an old embroidered cloak, which, upon his throwing it into the heap, I discovered to be poverty. Another, after a great deal of puffing, threw down his luggage, which, upon examining, I found to be his wife.

There were multitudes of lovers saddled with very whimsical burdens, composed of darts and flames; but, what was very odd, though they sighed as if their hearts would break under these bundles of calamities, they could not persuade themselves to cast them into the heap, when they came up to it; but, after a few faint efforts, shook their heads, and marched away as heavy loaden as they came.

I saw multitudes of old women throw down their wrinkles, and several young ones who stripped themselves of a tawny skin. There were very great heaps of red noses, large lips, and rusty teeth. The truth of it is, I was surprised to see the greatest part of the mountain made up of bodily deformities. Observing one advancing toward the heap with a larger cargo than ordinary upon his back, I found, upon his near approach, that it was only a natural hump, which he disposed of with great joy of heart among this collection of human miseries.

There were, likewise, distempers of all sorts, though I could not but observe that there were many more imaginary than real. One little packet I could not but take notice of, which was a complication of all the diseases incident to human nature, and was in the hand of a great many fine people. This was called the spleen. But what most of all surprised me was, that there was not a single vice or folly thrown into the whole heap: at which I was very much astonished, having concluded within myself that everyone would take this opportunity of getting rid of his passions, prejudices, and frailties.

I took notice in particular of a very profligate fellow, who, I did not question, came loaden with his crimes, but upon searching into his bundle, I found that instead of throwing his guilt from him, he had only laid down his memory. He was followed by another worthless rogue, who flung away his modesty instead of his ignorance.

When the whole race of mankind had thus cast their burdens, the phantom which had been so busy on this occasion, seeing me an idle spectator of what passed, approached toward me. I grew uneasy at her presence, when, of a sudden, she held her magnifying glass full before my eyes. I no sooner saw my face in it, but was startled at the shortness of it, which now appeared to me in its utmost aggravation. The immoderate breadth of the features made me very much out of humor with my own countenance, upon which I threw it from me like a mask. It happened very luckily that one who stood by me had just before thrown down his visage, which, it seems, was too long for him. It was, indeed, extended to a most shameful length; I believe the very chin was, modestly speaking, as long as my whole face. We had both of us an opportunity of mending ourselves; and all the contributions being now brought in, every man was at liberty to exchange his misfortunes for those of another person.

As we stood round the heap, and surveyed the several materials of which it was composed, there was scarcely a mortal in this vast multitude who did not discover what he thought pleasures and blessings of life, and wondered how the owners of them ever came to look upon them as burthens and grievances. As we were regarding very attentively this confusion of miseries, this chaos of calamity, Jupiter issued out a second proclamation, that everyone was now at liberty to exchange his affliction, and to return to his habitation with any such other bundle as should be delivered to him. Upon this, Fancy began again to bestir herself, and, parceling out the whole heap with incredible activity, recommended to everyone his particular packet. The hurry and confusion at this time was not to be expressed. Some observations, which I made upon the occasion, I shall communicate to the public.

A venerable, gray-headed man, who had laid down the colic, and who, I found, wanted an heir to his estate, snatched up an undutiful son that had been thrown into the heap by an angry father. The graceless youth, in less than a quarter of an hour, pulled the old gentleman by the beard, and had liked to have knocked his brains out; so that meeting the true father, who came toward him with a fit of the gripes, he begged him to take his son again, and give him back his colic; but they were incapable, either of them, to recede from the choice they had made. A poor galley slave, who had thrown down his chains, took up the gout in their stead, but made such wry faces that one might easily perceive he was no great gainer by the bargain.

The female world were very busy among themselves in bartering for features; one was trucking a lock of gray hairs for a carbuncle; and another was making over a short waist for a pair of round shoulders; but on all these occasions there was not one of them who did not think the new blemish, as soon as she had got it into her possession, much more disagreeable than the old one.

I must not omit my own particular adventure. My friend with the long visage had no sooner taken upon him my short face, but he made such a grotesque figure in it, that as I looked upon him, I could not forbear laughing at myself, insomuch that I put my own face out of countenance. The poor gentleman was so sensible of the ridicule, that I found he was ashamed of what he had done. On the other side, I found that I myself had no great reason to triumph, for as I went to touch my forehead, I missed the place, and clapped my finger upon my upper lip. Besides, as my nose was exceedingly prominent, I gave it two or three unlucky knocks as I was playing my hand about my face, and aiming at some other part of it.

I saw two other gentlemen by me who were in the same ridiculous circumstances. These had made a foolish swap between a couple of thick bandy legs and two long trapsticks that had no calves to them. One of these looked like a man walking upon stilts, and was so lifted up into the air, above his ordinary height, that his head turned round with it, while the other made such awkward circles, as he attempted to walk, that he scarcely knew how to move forward upon his new supporters. Observing him to be a pleasant kind of a fellow, I stuck my cane in the ground, and told him I would lay him a bottle of wine that he did not march up to it on a line that I drew for him, in a quarter of an hour.

The heap was at last distributed among the two sexes, who made a most piteous sight, as they wandered up and down under the pressure of their several burthens. The whole plain was filled with murmurs and complaints, groans and lamentations. Jupiter, at length taking compassion on the poor mortals, ordered them a second time to lay down their loads, with a design to give everyone his own again. They discharged themselves with a great deal of pleasure; after which, the phantom who had led them into such gross delusions, was commanded to disappear. There was sent in her stead a goddess of a quite different figure: her motions were steady and composed, and her aspect serious but cheerful. She every now and then cast her eyes toward heaven, and fixed them upon Jupiter. Her name was Patience. She had no sooner placed herself by the Mount of Sorrows, but, what I thought very remarkable, the whole heap sunk to such a degree that it did not appear a third part so big as it was before. She afterward returned every man his own proper calamity, and, teaching him how to bear it in the most commodious manner, he marched off with it contentedly, being very well pleased that he had not been left to his own choice as to the kind of evil which fell to his lot.

Beside the several pieces of morality to be drawn out of this vision, I learnt from it never to repine at my own misfortunes, or to envy the happiness of another, since it is impossible for any man to form a right judgment of his neighbor's sufferings; for which reason, also, I have determined never to think too lightly of another's complaints, but to regard the sorrows of my fellow-creatures with sentiments of humanity and compassion.

Notes.—**Horace** (b. 65, d. 8 B. C.) was a celebrated Roman poet.

Jupiter, according to mythology, was the greatest of the Greek and Roman gods; he was thought to be the supreme ruler of both mortals and immortals.

LXXXII. JUPITER AND TEN.

James T. Fields, 1817–1881, was born at Portsmouth, New Hampshire. For many years he was partner in the well-known firm of Ticknor & Fields (later Fields, Osgood & Co.), the leading publishers of standard American literature. For eight years, he was chief editor of the "Atlantic Monthly;" and, after he left that position, he often enriched its pages by the productions of his pen. During his latter years Mr. Fields gained some reputation as a lecturer. His literary abilities were of no mean order: but he did not do so much in producing literature himself, as in aiding others in its production.

Mrs. Chub was rich and portly,
  Mrs. Chub was very grand,
Mrs. Chub was always reckoned
  A lady in the land.

You shall see her marble mansion
  In a very stately square,—
Mr. C. knows what it cost him,
  But that's neither here nor there.

Mrs. Chub was so sagacious,
  Such a patron of the arts,
And she gave such foreign orders
  That she won all foreign hearts.

Mrs. Chub was always talking,
  When she went away from home,
Of a most prodigious painting
  Which had just arrived from Rome.

"Such a treasure," she insisted,
  "One might never see again!"
"What's the subject?" we inquired.
  "It is Jupiter and Ten!"

"Ten what?" we blandly asked her
  For the knowledge we did lack,
"Ah! that I can not tell you,
  But the name is on the back.

"There it stands in printed letters,—
  Come to-morrow, gentlemen,—
Come and see our splendid painting,
  Our fine Jupiter and Ten!"

When Mrs. Chub departed,
  Our brains began to rack,—
She could not be mistaken
  For the name was on the back.

So we begged a great Professor
  To lay aside his pen,
And give some information
  Touching "Jupiter and Ten."

And we pondered well the subject,
  And our Lemprière we turned,
To find out who the Ten were;
  But we could not, though we burned.

But when we saw the picture,—
  O Mrs. Chub! Oh, fie! O!
We perused the printed label,
  And 't was JUPITER AND IO!

NOTES.—**John Lemprière,** an Englishman, was the author of a "Classical Dictionary" which until the middle of the present century was the chief book of reference on ancient mythology.

Io is a mythical heroine of Greece, with whom Jupiter was enamored.

LXXXIII. SCENE FROM "THE POOR GENTLEMAN."

George Colman, 1762-1836, was the son of George Colman, a writer of dramas, who in 1777 purchased the "Haymarket Theater," in London. Owing to the illness of the father, Colman the younger assumed the management of the theater in 1785, which post he held for a long time. He was highly distinguished as a dramatic author and wit. "The Poor Gentleman," from which the following selection is adapted, is perhaps the best known of his works.

SIR ROBERT BRAMBLE *and* HUMPHREY DOBBINS.

Sir R. I'LL tell you what, Humphrey Dobbins, there is not a syllable of sense in all you have been saying. But I suppose you will maintain there is.

Hum. Yes.

Sir R. Yes! Is that the way you talk to me, you old boor? What's my name?

Hum. Robert Bramble.

Sir R. An't I a baronet? Sir Robert Bramble, of Blackberry Hall, in the county of Kent? 'Tis time you should know it, for you have been my clumsy, two-fisted valet these thirty years: can you deny that?

Hum. Hem!

Sir R. Hem? What do you mean by hem? Open that rusty door of your mouth, and make your ugly voice walk out of it. Why don't you answer my question?

Hum. Because, if I contradict you, I shall tell you a lie, and whenever I agree with you, you are sure to fall out.

Sir R. Humphrey Dobbins, I have been so long endeavoring to beat a few brains into your pate that all your hair has tumbled off before my point is carried.

Hum. What then? Our parson says my head is an emblem of both our honors.

Sir R. Ay; because honors, like your head, are apt to be empty.

Hum. No; but if a servant has grown bald under his master's nose, it looks as if there was honesty on one side, and regard for it on the other.

Sir R. Why, to be sure, old Humphrey, you are as honest as a—pshaw! the parson means to palaver us; but, to return to my position, I tell you I don't like your flat contradiction.

Hum. Yes, you do.

Sir R. I tell you I don't. I only love to hear men's arguments. I hate their flummery.

Hum. What do you call flummery?

Sir R. Flattery, blockhead! a dish too often served up by paltry poor men to paltry rich ones.

Hum. I never serve it up to you.

Sir R. No, you give me a dish of a different description.

Hum. Hem! what is it?

Sir R. **Sauerkraut, you old crab.**

Hum. I have held you a stout tug at argument this many a year.

Sir R. And yet I could never teach you a syllogism. Now mind, when a poor man assents to what a rich man says, I suspect he means to flatter him: now I am rich, and hate flattery. *Ergo*—when a poor man subscribes to my opinion, I hate him.

Hum. That's wrong.

Sir R. Very well; *negatur;* now prove it.

Hum. Put the case then, I am a poor man.

Sir R. You an't, you scoundrel. You know you shall never want while I have a shilling.

Hum. Bless you!

Sir R. Pshaw! Proceed.

Hum. Well, then, I am a poor—I must be a poor man now, or I never shall get on.

Sir R. Well, get on, be a poor man.

Hum. I am a poor man, and I argue with you, and convince you, you are wrong; then you call yourself a blockhead, and I am of your opinion: now, that's no flattery.

Sir R. Why, no; but when a man's of the same opinion with me, he puts an end to the argument, and that puts an end to the conversation, and so I hate him for that. But where's my nephew Frederic?

Hum. Been out these two hours.

Sir R. An undutiful cub! Only arrived from Russia last night, and though I told him to stay at home till I rose, he's scampering over the fields like a Calmuck Tartar.

Hum. He's a fine fellow.

Sir R. He has a touch of our family. Don't you think he is a little like me, Humphrey?

Hum. No, not a bit; you are as ugly an old man as ever I clapped my eyes on.

Sir R. Now that's plaguy impudent, but there's no

flattery in it, and it keeps up the independence of argument. His father, my brother Job, is of as tame a spirit —Humphrey, you remember my brother Job?

Hum. Yes, you drove him to Russia five and twenty years ago.

Sir R. I did not drive him.

Hum. Yes, you did. You would never let him be at peace in the way of argument.

Sir R. At peace! Zounds, he would never go to war.

Hum. He had the merit to be calm.

Sir R. So has a duck pond. He was a bit of still life; a chip; weak water gruel; a tame rabbit, boiled to rags, without sauce or salt. He received my arguments with his mouth open, like a poorbox gaping for half-pence, and, good or bad, he swallowed them all without any resistance. We could n't disagree, and so we parted.

Hum. And the poor, meek gentleman went to Russia for a quiet life.

Sir R. A quiet life! Why, he married the moment he got there, tacked himself to the shrew relict of a Russian merchant, and continued a speculation with her in furs, flax, potashes, tallow, linen, and leather; what 's the consequence? Thirteen months ago he broke.

Hum. Poor soul, his wife should have followed the business for him.

Sir R. I fancy she did follow it, for she died just as he broke, and now this madcap, Frederic, is sent over to me for protection. Poor Job, now he is in distress, I must not neglect his son.

Hum. Here comes his son; that's Mr. Frederic.

Enter FREDERIC.

Fred. Oh, my dear uncle, good morning! Your park is nothing but beauty.

Sir R. Who bid you caper over my beauty? I told you to stay in doors till I got up.

Fred. So you did, but I entirely forgot it.

Sir R. And pray, what made you forget it?

Fred. The sun.

Sir R. The sun! he's mad! you mean the moon, I believe.

Fred. Oh, my dear uncle, you don't know the effect of a fine spring morning upon a fellow just arrived from Russia. The day looked bright, trees budding, birds singing, the park was so gay that I took a leap out of your old balcony, made your deer fly before me like the wind, and chased them all around the park to get an appetite for breakfast, while you were snoring in bed, uncle.

Sir R. Oh, oh! So the effect of English sunshine upon a Russian, is to make him jump out of a balcony, and worry my deer.

Fred. I confess it had that influence upon me.

Sir R. You had better be influenced by a rich old uncle, unless you think the sun likely to leave you a fat legacy.

Fred. I hate legacies.

Sir R. Sir, that's mighty singular. They are pretty solid tokens, at least.

Fred. Very melancholy tokens, uncle; they are the posthumous dispatches Affection sends to Gratitude, to inform us we have lost a gracious friend.

Sir R. How charmingly the dog argues!

Fred. But I own my spirits ran away with me this morning. I will obey you better in future; for they tell me you are a very worthy, good sort of old gentleman.

Sir R. Now who had the familiar impudence to tell you that?

Fred. Old rusty, there.

Sir R. Why Humphrey, you didn't?

Hum. Yes, but I did though.

Fred. Yes, he did, and on that score I shall be anxious to show you obedience, for 'tis as meritorious to attempt sharing a good man's heart, as it is paltry to have designs upon a rich man's money. A noble nature aims its attentions full breast high, uncle; a mean mind levels its dirty assiduities at the pocket.

Sir R. (*Shaking him by the hand.*) Jump out of every window I have in my house; hunt my deer into high fevers, my fine fellow! Ay, that's right. This is spunk, and plain speaking. Give me a man who is always flinging his dissent to my doctrines smack in my teeth.

Fred. I disagree with you there, uncle.

Hum. And so do I.

Fred. You! you forward puppy! If you were not so old, I'd knock you down.

Sir R. I'll knock you down, if you do. I won't have my servants thumped into dumb flattery.

Hum. Come, you are ruffled. Let us go to the business of the morning.

Sir R. I hate the business of the morning. Don't you see we are engaged in discussion. I tell you, I hate the business of the morning.

Hum. No you don't.

Sir R. Don't I? Why not?

Hum. Because 'tis charity.

Sir R. Pshaw! Well, we must not neglect the business, if there be any distress in the parish. Read the list, Humphrey.

Hum. (*Taking out a paper and reading.*) "Jonathan Huggins, of Muck Mead, is put in prison for debt."

Sir R. Why, it was only last week that Gripe, the attorney, recovered two cottages for him by law, worth sixty pounds.

Hum. Yes, and charged a hundred for his trouble; so seized the cottages for part of his bill, and threw Jonathan into jail for the remainder.

Sir R. A harpy! I must relieve the poor fellow's distress.

Fred. And I must kick his attorney.

Hum. (*Reading.*) "The curate's horse is dead."

Sir R. Pshaw! There's no distress in that.

Hum. Yes, there is, to a man that must go twenty miles every Sunday to preach three sermons, for thirty pounds a year.

Sir R. Why won't the vicar give him another nag?

Hum. Because 't is cheaper to get another curate ready mounted.

Sir R. Well, send him the black pad which I purchased last Tuesday, and tell him to work him as long as he lives. What else have we upon the list?

Hum. Something out of the common; there's one Lieutenant Worthington, a disabled officer and a widower, come to lodge at Farmer Harrowby's, in the village; he is, it seems, very poor, and more proud than poor, and more honest than proud.

Sir R. And so he sends to me for assistance?

Hum. He'd see you hanged first! No, he'd sooner die than ask you or any man for a shilling! There's his daughter, and his wife's aunt, and an old corporal that served in the wars with him, he keeps them all upon his half pay.

Sir R. Starves them all, I'm afraid, Humphrey.

Fred. (*Going.*) Good morning, uncle.

Sir R. You rogue, where are you running now?

Fred. To talk with Lieutenant Worthington.

Sir R. And what may you be going to say to him?

Fred. I can't tell till I encounter him; and then, uncle, when I have an old gentleman by the hand, who has been disabled in his country's service, and is struggling to support his motherless child, a poor relation, and a faithful servant, in honorable indigence, impulse will supply me with words to express my sentiments.

Sir R. Stop, you rogue; I must be before you in this business.

Fred. That depends on who can run the fastest; so, start fair, uncle, and here goes.—(*Runs out.*)

Sir R. Stop, stop; why, Frederic—a jackanapes—to take my department out of my hands! I'll disinherit the dog for his assurance.

Hum. No, you won't.

Sir R. Won't I? Hang me if I—but we'll argue that point as we go. So, come along Humphrey.

NOTES.—**Ergo** (pro. ẽr′ḡo) is a Latin word meaning *therefore.* **Negatur** (pro. ne-ḡā′tur) is a Latin verb, and means *it is denied.*

The **Tartars** are a branch of the Mongolian race, embracing among other tribes the **Calmucks.** The latter are a fierce, nomadic people inhabiting parts of the Russian and Chinese empires.

LXXXIV. MY MOTHER'S PICTURE.

William Cowper, 1731-1800, was the son of an English clergyman; both his parents were descended from noble families. He was always of a gentle, timid disposition; and the roughness of his schoolfellows increased his weakness in this respect. He studied law, and was admitted to the bar, but never practiced his profession. When he was about thirty years of age, he was appointed to a clerkship in the House of Lords, but could not summon courage to enter upon the discharge of its duties. He was so disturbed by this affair that he became insane, sought to destroy himself, and had to be consigned to a private asylum. Soon after his recovery, he found a congenial home in the family of the Rev. Mr. Unwin. On the death of this gentleman, a few years later, he continued to reside with his widow till her death, a short time before that of Cowper. Most of this time their home was at Olney. His first writings were published in 1782. He wrote several beautiful hymns, "The Task," and some minor poems. These, with his translations of Homer and his correspondence, make up his published works. His life was always pure and gentle; he took great pleasure in simple, natural objects, and in playing with animals. His insanity returned from time to time, and darkened his life at its close. When six years of age, he lost his mother; and the following selection is part of a touching tribute to her memory, written many years later.

Oh that those lips had language! Life has passed
With me but roughly since I heard them last.
My mother, when I learned that thou wast dead,
Say, wast thou conscious of the tears I shed?
Hovered thy spirit o'er thy sorrowing son,
Wretch even then, life's journey just begun?
Perhaps thou gavest me, though unfelt, a kiss,
Perhaps a tear, if souls can weep in bliss.
Ah, that maternal smile! it answers — Yes!

I heard the bell tolled on thy burial day;
I saw the hearse that bore thee slow away;
And, turning from my nursery window, drew
A long, long sigh, and wept a last adieu!
But was it such? It was. Where thou art gone,
Adieus and farewells are a sound unknown.
May I but meet thee on that peaceful shore,
The parting word shall pass my lips no more.

Thy maidens, grieved themselves at my concern,
Oft gave me promise of thy quick return;
What ardently I wished, I long believed;
And, disappointed still, was still deceived;
By expectation, every day beguiled,
Dupe of to-morrow, even when a child.
Thus many a sad to-morrow came and went,
Till, all my stock of infant sorrows spent,
I learned at last submission to my lot;
But, though I less deplored thee, ne'er forgot.

My boast is not that I deduce my birth
From loins enthroned, and rulers of the earth;
But higher far my proud pretensions rise,—
The son of parents passed into the skies.

And now, farewell! Time, unrevoked, has run
His wonted course, yet what I wished is done.

By Contemplation's help, not sought in vain,
I seem to have lived my childhood o'er again;
To have renewed the joys that once were mine,
Without the sin of violating thine;
And, while the wings of Fancy still are free,
And I can view this mimic show of thee,
Time has but half succeeded in his theft,—
Thyself removed, thy power to soothe me left.

LXXXV. DEATH OF SAMSON.

John Milton, 1608–1674, was born in London—eight years before the greatest English poet, Shakespeare, died. His father followed the profession of a scrivener, in which he acquired a competence. As a boy, Milton was exceedingly studious, continuing his studies till midnight. He graduated at Christ's College, Cambridge, where his singular beauty, his slight figure, and his fastidious morality caused his companions to nickname him "the lady of Christ's." On leaving college he spent five years more in study, and produced his lighter poems. He then traveled on the continent, returning about the time the civil war broke out. For a time he taught a private school, but soon threw himself with all the power of his able and tried pen into the political struggle. He was the champion of Parliament and of Cromwell for about twenty years. On the accession of Charles II., he concealed himself for a time, but was soon allowed to live quietly in London. His eyesight had totally failed in 1654; but now, in blindness, age, family affliction, and comparative poverty, he produced his great work "Paradise Lost." In 1667 he sold the poem for £5 in cash, with a promise of £10 more on certain contingencies; the sum total received by himself and family for the immortal poem, was £23. Later, he produced "Paradise Regained" and "Samson Agonistes," from the latter of which the following extract is taken. Milton is a wonderful example of a man, who, by the greatness of his own mind, triumphed over trials, afflictions, hardships, and the evil influence of bitter political controversy.

OCCASIONS drew me early to this city;
And, as the gates I entered with sunrise,
The morning trumpets festival proclaimed

Through each high street: little I had dispatched,
When all abroad was rumored that this day
Samson should be brought forth, to show the people
Proof of his mighty strength in feats and games.
I sorrowed at his captive state,
But minded not to be absent at that spectacle.

The building was a spacious theater
Half-round, on two main pillars vaulted high,
With seats where all the lords, and each degree
Of sort, might sit in order to behold;
The other side was open, where the throng
On banks and scaffolds under sky might stand:
I among these aloof obscurely stood.
The feast and noon grew high, and sacrifice
Had filled their hearts with mirth, high cheer, and wine,
When to their sports they turned. Immediately
Was Samson as a public servant brought,
In their state livery clad: before him pipes
And timbrels; on each side went armèd guards;
Both horse and foot before him and behind,
Archers and slingers, cataphracts, and spears.
At sight of him the people with a shout
Rifted the air, clamoring their god with praise,
Who had made their dreadful enemy their thrall.

He, patient, but undaunted, where they led him,
Came to the place; and what was set before him,
Which without help of eye might be essayed,
To heave, pull, draw, or break, he still performed
All with incredible, stupendous force,
None daring to appear antagonist.

At length for intermission sake, they led him
Between the pillars; he his guide requested,

As overtired, to let him lean awhile
With both his arms on those two massy pillars,
That to the archèd roof gave main support.

He unsuspicious led him; which when Samson
Felt in his arms, with head awhile inclined,
And eyes fast fixed, he stood, as one who prayed,
Or some great matter in his mind revolved:
At last, with head erect, thus cried aloud:—
"Hitherto, lords, what your commands imposed
I have performed, as reason was, obeying,
Not without wonder or delight beheld;
Now, of my own accord, such other trial
I mean to show you of my strength yet greater,
As with amaze shall strike all who behold."

This uttered, straining all his nerves, he bowed;
As with the force of winds and waters pent
When mountains tremble, those two massy pillars
With horrible convulsion to and fro
He tugged, he shook, till down they came, and drew
The whole roof after them with burst of thunder
Upon the heads of all who sat beneath,—
Lords, ladies, captains, counselors, or priests,
Their choice nobility and flower, not only
Of this, but each Philistian city round,
Met from all parts to solemnize this feast.
Samson, with these immixed, inevitably
Pulled down the same destruction on himself;
The vulgar only 'scaped who stood without.

NOTE.—The person supposed to be speaking is a Hebrew who chanced to be present at Gaza when the incidents related took place. After the catastrophe he rushes to Manoah, the father of Samson, to whom and his assembled friends he relates what he saw. (Cf. Bible, Judges xvi, 23.)

LXXXVI. AN EVENING ADVENTURE.

Not long since, a gentleman was traveling in one of the counties of Virginia, and about the close of the day stopped at a public house to obtain refreshment and spend the night. He had been there but a short time, before an old man alighted from his gig, with the apparent intention of becoming his fellow guest at the same house.

As the old man drove up, he observed that both the shafts of his gig were broken, and that they were held together by withes, formed from the bark of a hickory sapling. Our traveler observed further that he was plainly clad, that his knee buckles were loosened, and that something like negligence pervaded his dress. Conceiving him to be one of the honest yeomanry of our land, the courtesies of strangers passed between them, and they entered the tavern. It was about the same time, that an addition of three or four young gentlemen was made to their number; most, if not all of them, of the legal profession.

As soon as they became conveniently accommodated, the conversation was turned, by one of the latter, upon the eloquent harangue which had that day been displayed at the bar. It was replied by the other that he had witnessed, the same day, a degree of eloquence no doubt equal, but it was from the pulpit. Something like a sarcastic rejoinder was made as to the eloquence of the pulpit, and a warm and able altercation ensued, in which the merits of the Christian religion became the subject of discussion. From six o'clock until eleven, the young champions wielded the sword of argument, adducing with ingenuity and ability everything that could be said *pro* and *con*.

During this protracted period, the old gentleman listened with the meekness and modesty of a child, as if he were adding new information to the stores of his own mind; or perhaps he was observing, with a philosophic eye, the faculties of the youthful mind, and how new energies are evolved by repeated action; or perhaps, with patriotic emotion, he was reflecting upon the future destinies of his country, and on the rising generation, upon whom those future destinies must devolve; or, most probably, with a sentiment of moral and religious feeling, he was collecting an argument which no art would be "able to elude, and no force to resist." Our traveler remained a spectator, and took no part in what was said.

At last one of the young men, remarking that it was impossible to combat with long and established prejudices, wheeled around, and with some familiarity exclaimed, "Well, my old gentleman, what think you of these things?" "If," said the traveler, "a streak of vivid lightning had at that moment crossed the room, their amazement could not have been greater than it was from what followed." The most eloquent and unanswerable appeal that he had ever heard or read, was made for nearly an hour by the old gentleman. So perfect was his recollection, that every argument urged against the Christian religion was met in the order in which it was advanced. Hume's sophistry on the subject of miracles, was, if possible, more perfectly answered than it had already been done by Campbell. And in the whole lecture there was so much simplicity and energy, pathos and sublimity, that not another word was uttered.

"An attempt to describe it," said the traveler, "would be an attempt to paint the sunbeams." It was now a matter of curiosity and inquiry who the old gentleman was. The traveler concluded that it was the preacher from whom the pulpit eloquence was heard; but no, it was John Marshall, the Chief Justice of the United States.

NOTES.—**David Hume** (b. 1711, d. 1776) was a celebrated Scotch historian and essayist. His most important work is

"The History of England." He was a skeptic in matters of religion, and was a peculiarly subtle writer.

George Campbell (b. 1719, d. 1796) was a distinguished Scotch minister. He wrote "A Dissertation on Miracles," ably answering Hume's "Essay on Miracles."

John Marshall (b. 1755, d. 1835) was Chief Justice of the United States from 1801 until his death. He was an eminent jurist, and wrote a "Life of Washington," which made him famous as an author.

LXXXVII. THE BAREFOOT BOY.

John Greenleaf Whittier, 1807-1892, was born in Haverhill, Mass., and, with short intervals of absence, he always resided in that vicinity. His parents were Friends or "Quakers," and he always held to the same faith. He spent his boyhood on a farm, occasionally writing verses for the papers even then. Two years of study in the academy seem to have given him all the special opportunity for education that he ever enjoyed. In 1829 he edited a newspaper in Boston, and the next year assumed a similar position in Hartford. For two years he was a member of the Massachusetts legislature. In 1836 he edited an anti-slavery paper in Philadelphia, and was secretary of the American Anti-Slavery Society.

Mr. Whittier wrote extensively both in prose and verse. During the later years of his life he published several volumes of poems, and contributed frequently to the pages of the "Atlantic Monthly." An earnest opponent of slavery, some of his poems bearing on that subject are fiery and even bitter; but, in general, their sentiment is gentle, and often pathetic. As a poet, he took rank among those most highly esteemed by his countrymen. "Snow-Bound," published in 1865, is one of the longest and best of his poems. Several of his shorter pieces are marked by much smoothness and sweetness.

Blessings on thee, little man,
Barefoot boy, with cheek of tan!
With thy turned-up pantaloons,
And thy merry whistled tunes;
With thy red lip, redder still
Kissed by strawberries on the hill;
With the sunshine on thy face,
Through thy torn brim's jaunty grace;

From my heart I give thee joy,—
I was once a barefoot boy!
Prince thou art,—the grown-up man
Only is republican.
Let the million-dollared ride!
Barefoot, trudging at his side,
Thou hast more than he can buy
In the reach of ear and eye,—
Outward sunshine, inward joy:
Blessings on thee, barefoot boy!

Oh for boyhood's painless play,
Sleep that wakes in laughing day,
Health that mocks the doctor's rules,
Knowledge never learned of schools,
Of the wild bee's morning chase,
Of the wild flower's time and place,
Flight of fowl and habitude
Of the tenants of the wood;
How the tortoise bears his shell,
How the woodchuck digs his cell,
And the ground mole sinks his well
How the robin feeds her young,
How the oriole's nest is hung;
Where the whitest lilies blow,
Where the freshest berries grow,
Where the groundnut trails its vine,
Where the wood grape's clusters shine;
Of the black wasp's cunning way,
Mason of his walls of clay,
And the architectural plans
Of gray hornet artisans!—
For, eschewing books and tasks,
Nature answers all he asks;
Hand in hand with her he walks,

Face to face with her he talks,
Part and parcel of her joy,—
Blessings on thee, barefoot boy!

Oh for boyhood's time of June,
Crowding years in one brief moon,
When all things I heard or saw
Me, their master, waited for.
I was rich in flowers and trees,
Humming birds and honeybees;
For my sport the squirrel played,
Plied the snouted mole his spade;
For my taste the blackberry cone
Purpled over hedge and stone;
Laughed the brook for my delight
Through the day and through the night,
Whispering at the garden wall,
Talked with me from fall to fall;
Mine the sand-rimmed pickerel pond,
Mine the walnut slopes beyond,
Mine, on bending orchard trees,
Apples of Hesperides!
Still, as my horizon grew,
Larger grew my riches too;
All the world I saw or knew
Seemed a complex Chinese toy,
Fashioned for a barefoot boy!

Oh for festal dainties spread,
Like my bowl of milk and bread,—
Pewter spoon and bowl of wood,
On the doorstone, gray and rude!
O'er me, like a regal tent,
Cloudy-ribbed, the sunset bent,
Purple-curtained, fringed with gold,

Looped in many a wind-swung fold;
While for music came the play
Of the pied frog's orchestra;
And to light the noisy choir,
Lit the fly his lamp of fire.
I was monarch: pomp and joy
Waited on the barefoot boy!

Cheerily, then, my little man,
Live and laugh, as boyhood can!
Though the flinty slopes be hard,
Stubble-speared the new-mown sward,
Every morn shall lead thee through
Fresh baptisms of the dew;
Every evening from thy feet
Shall the cool wind kiss the heat:
All too soon these feet must hide
In the prison cells of pride,
Lose the freedom of the sod,
Like a colt's for work be shod,
Made to tread the mills of toil,
Up and down in ceaseless moil:
Happy if their track be found
Never on forbidden ground;
Happy if they sink not in
Quick and treacherous sands of sin.
Ah! that thou shouldst know thy joy
Ere it passes, barefoot boy!

NOTE.—The **Hesperides**, in Grecian mythology, were four sisters (some traditions say three, and others, seven) who guarded the golden apples given to Juno as a wedding present. The locality of the garden of the Hesperides is a disputed point with mythologists.

LXXXVIII. THE GLOVE AND THE LIONS.

James Henry Leigh Hunt, 1784-1859. Leigh Hunt, as he is commonly called, was prominent before the public for fifty years as "a writer of essays, poems, plays, novels, and criticisms." He was born at Southgate, Middlesex, England. His mother was an American lady. He began to write for the public at a very early age. In 1808, in connection with his brother, he established "The Examiner," a newspaper advocating liberal opinions in politics. For certain articles offensive to the government, the brothers were fined £500 each and condemned to two years' imprisonment. Leigh fitted up his prison like a boudoir, received his friends here, and wrote several works during his confinement. Mr. Hunt was intimate with Byron, Shelley, Moore, and Keats, and was associated with Byron and Shelley in the publication of a political and literary journal. His last years were peacefully devoted to literature, and in 1847 he received a pension from the government.

KING Francis was a hearty king, and loved a royal sport,
And one day, as his lions fought, sat looking on the court;
The nobles filled the benches round, the ladies by their side,
And 'mongst them sat the Count de Lorge, with one for whom he sighed:
And truly 't was a gallant thing to see that crowning show,
Valor and love, and a king above, and the royal beasts below.

Ramped and roared the lions, with horrid laughing jaws;
They bit, they glared, gave blows like beams, a wind went with their paws;
With wallowing might and stifled roar, they rolled on one another:
Till all the pit, with sand and mane, was in a thunderous smother;
The bloody foam above the bars came whizzing through the air:
Said Francis, then, "Faith, gentlemen, we're better here than there."

De Lorge's love o'erheard the king,—a beauteous, lively dame,
With smiling lips, and sharp, bright eyes, which always seemed the same;
She thought, "The Count, my lover, is brave as brave can be,
He surely would do wondrous things to show his love for me;
King, ladies, lovers, all look on; the occasion is divine;
I'll drop my glove to prove his love; great glory will be mine."

She dropped her glove to prove his love, then looked at him and smiled;
He bowed, and in a moment leaped among the lions wild;
The leap was quick, return was quick, he soon regained his place,
Then threw the glove, but not with love, right in the lady's face.
"In faith," cried Francis, "rightly done!" and he rose from where he sat;
"No love," quoth he, "but vanity, sets love a task like that."

NOTE.—**King Francis.** This is supposed to have been Francis I. of France (b. 1494, d. 1547). He was devoted to sports of this nature.

LXXXIX. THE FOLLY OF INTOXICATION.

Iago. WHAT, are you hurt, lieutenant?
Cassio. Ay, past all surgery.
Iago. Marry, heaven forbid!
Cas. Reputation, reputation, reputation! Oh, I have

lost my reputation! I have lost the immortal part of myself, and what remains is bestial. My reputation! Iago, my reputation!

Iago. As I am an honest man, I thought you had received some bodily wound; there is more sense in that than in reputation. Reputation is an idle and most false imposition: oft got without merit, and lost without deserving: you have lost no reputation at all, unless you repute yourself such a loser. What, man! there are ways to recover the general again. Sue to him again, and he 's yours.

Cas. I will rather sue to be despised than to deceive so good a commander with so slight, so drunken, and so indiscreet an officer. Drunk? and speak parrot? and squabble? swagger? swear? and discourse fustian with one's own shadow? O thou invisible spirit of wine, if thou hast no name to be known by, let us call thee devil!

Iago. What was he that you followed with your sword? What had he done to you?

Cas. I know not.

Iago. Is 't possible?

Cas. I remember a mass of things, but nothing distinctly; a quarrel, but nothing wherefore. Oh that men should put an enemy in their mouths to steal away their brains! that we should, with joy, revel, pleasure, and applause, transform ourselves into beasts!

Iago. Why, but you are now well enough: how came you thus recovered?

Cas. It hath pleased the devil, Drunkenness, to give place to the devil, Wrath; one unperfectness shows me another, to make me frankly despise myself.

Iago. Come, you are too severe a moraler. As the time, the place, and the condition of this country stands, I could heartily wish this had not befallen; but since it is as it is, mend it for your own good.

Cas. I will ask him for my place again: he shall tell

me I am a drunkard! Had I as many mouths as Hydra, such an answer would stop them all. To be now a sensible man, by and by a fool, and presently a beast! Oh strange!—Every inordinate cup is unblessed, and the ingredient is a devil!

Iago. Come, come; good wine is a good familiar creature, if it be well used; exclaim no more against it. And, good lieutenant, I think you think I love you.

Cas. I have well approved it, sir,—I, drunk!

Iago. You or any man living may be drunk at a time, man. I'll tell you what you shall do. Our general's wife is now the general. Confess yourself freely to her; importune her help to put you in your place again. She is of so free, so kind, so apt, so blessed a disposition, she holds it a vice in her goodness not to do more than she is requested. This broken joint between you and her husband, entreat her to splinter; and, my fortunes against any lay worth naming, this crack of your love shall grow stronger than it was before.

Cas. You advise me well.

Iago. I protest in the sincerity of love and honest kindness.

Cas. I think it freely, and betimes in the morning, I will beseech the virtuous Desdemona to undertake for me: I am desperate of my fortunes if they check me here.

Iago. You are in the right. Good night, lieutenant, I must to the watch.

Cas. Good night, honest Iago.

Shakespeare.—Othello, Act ii, Scene iii.

NOTES.—**Iago** is represented as a crafty, unscrupulous villain. He applies for the position of lieutenant under Othello, but the latter has already appointed **Cassio**—who is honest, but of a weak character—to that position; he, however, makes Iago his ensign. Then Iago, to revenge himself for this and other fancied wrongs, enters upon a systematic course of villainy, part of which is to bring about the intoxication of Cassio, and his consequent discharge from the lieutenancy.

The **Hydra** was a fabled monster of Grecian mythology, having nine heads, one of which was immortal.

Desdemona was the wife of Othello.

XC. STARVED ROCK.

Francis Parkman, 1823–1893, the son of a clergyman of the same name, was born in Boston, and graduated at Harvard University in 1844. He spent more than twenty years in a careful study of the early French explorations and settlements in America; and he published the fruits of his labor in twelve large volumes. Although troubled with an affection of the eyes, which sometimes wholly prevented reading or writing, his work was most carefully and successfully done. His narratives are written in a clear and animated style, and his volumes are a rich contribution to American history.

The cliff called "Starved Rock," now pointed out to travelers as the chief natural curiosity of the region, rises, steep on three sides as a castle wall, to the height of a hundred and twenty-five feet above the river. In front, it overhangs the water that washes its base; its western brow looks down on the tops of the forest trees below; and on the east lies a wide gorge, or ravine, choked with the mingled foliage of oaks, walnuts, and elms; while in its rocky depths a little brook creeps down to mingle with the river.

From the rugged trunk of the stunted cedar that leans forward from the brink, you may drop a plummet into the river below, where the catfish and the turtles may plainly be seen gliding over the wrinkled sands of the clear and shallow current. The cliff is accessible only from the south, where a man may climb up, not without difficulty, by a steep and narrow passage. The top is about an acre in extent.

Here, in the month of December, 1682, La Salle and Tonty began to entrench themselves. They cut away the forest that crowned the rock, built storehouses and dwellings of its remains, dragged timber up the rugged pathway, and encircled the summit with a palisade. Thus the winter was passed, and meanwhile the work of negotiation went prosperously on. The minds of the Indians had been already prepared. In La Salle they saw their champion against the Iroquois, the standing terror of all this region. They gathered around his stronghold like the timorous peasantry of the Middle Ages around the rock-built castle of their feudal lord.

From the wooden ramparts of St. Louis,—for so he named his fort,—high and inaccessible as an eagle's nest, a strange scene lay before his eye. The broad, flat valley of the Illinois was spread beneath him like a map, bounded in the distance by its low wall of wooded hills. The river wound at his feet in devious channels among islands bordered with lofty trees; then, far on the left, flowed calmly westward through the vast meadows, till its glimmering blue ribbon was lost in hazy distance.

There had been a time, and that not remote, when these fair meadows were a waste of death and desolation, scathed with fire, and strewn with the ghastly relics of an Iroquois victory. Now, all was changed. La Salle looked down from his rock on a concourse of wild human life. Lodges of bark and rushes, or cabins of logs, were clustered on the open plain, or along the edges of the bordering forests. Squaws labored, warriors lounged in the sun, naked children whooped and gamboled on the grass.

Beyond the river, a mile and a half on the left, the banks were studded once more with the lodges of the Illinois, who, to the number of six thousand, had returned, since their defeat, to this their favorite dwelling place. Scattered along the valley, among the adjacent hills, or over the neighboring prairie, were the cantonments of a half score of other tribes, and fragments of tribes, gathered under the protecting ægis of the French.

NOTES.—The curious elevation called **Starved Rock** is on the south side of Illinois River, between La Salle and Ottawa. There is a legend according to which it is said that over one hundred years ago, a party of Illinois Indians took refuge here from the Pottawatomies; their besiegers, however, confined them so closely that the whole party perished of starvation, or, as some say, of thirst. From this circumstance the rock takes its name.

La Salle (b. 1643, d. 1687) was a celebrated French explorer and fur trader. He established many forts throughout the Mississippi Valley,—among them, Fort St. Louis, in 1683.

Tonty was an Italian, who formerly served in both the French army and navy, and afterwards joined La Salle in his explorations.

XCI. PRINCE HENRY AND FALSTAFF.

PRINCE HENRY *and* POINS, *in a back room, in a tavern.* *Enter* FALSTAFF, GADSHILL, BARDOLPH, *and* PETO.

Poins. WELCOME, Jack. Where hast thou been?

Falstaff. A plague of all cowards, I say, and a vengeance too! marry, and amen! Give me a cup of sack, boy. Ere I lead this life long, I'll sew nether stocks, and mend them, and foot them, too. A plague of all cowards! Give me a cup of sack, rogue. Is there no virtue extant? (*He drinks, and then continues.*) You rogue, here's lime in this sack, too; there is nothing but roguery to be found in villainous man: yet a coward is worse than a cup of sack with lime in it. A villainous coward! Go thy ways, old Jack; die when thou wilt: if manhood, good manhood, be not forgot upon the face of the earth, then am I a shotten herring. There live not three good men unhanged, in England; and one of them is fat and grows old; a bad world, I say! I would I were a weaver; I could sing psalms, or anything. A plague of all cowards, I say still.

Prince Henry. How now, woolsack? What mutter you?

Fal. A king's son! If I do not beat thee out of thy kingdom with a dagger of lath, and drive all thy subjects afore thee like a flock of wild geese, I'll never wear hair on my face more. You, Prince of Wales!

P. Henry. Why, you baseborn dog! What's the matter?

Fal. Are you not a coward? Answer me to that; and Poins there?

Poins. Ye fat braggart, an ye call me coward, I'll stab thee.

Fal. I call thee coward? I'll see thee gibbeted ere I call thee coward: but I would give a thousand pounds I could run as fast as thou canst. You are straight enough in the shoulders, you care not who sees your back: call you that backing of your friends? A plague upon such backing! Give me them that will face me. Give me a cup of sack. I am a rogue, if I have drunk to-day.

P. Henry. O villain! thy lips are scarce wiped since thou drunkest last.

Fal. All's one for that. A plague of all cowards, still say I. (*He drinks.*)

P. Henry. What's the matter?

Fal. What's the matter! There be four of us here have ta'en a thousand pounds this morning.

P. Henry. Where is it, Jack? where is it?

Fal. Where is it? Taken from us it is; a hundred upon poor four of us.

P. Henry. What! a hundred, man?

Fal. I am a rogue, if I were not at half-sword with a

dozen of them two hours together. I have 'scaped by miracle. I am eight times thrust through the doublet; four, through the hose; my buckler cut through and through; my sword hacked like a handsaw; look here! (*shows his sword.*) I never dealt better since I was a man; all would not do. A plague of all cowards! Let them speak (*pointing to* GADSHILL, BARDOLPH, *and* PETO); if they speak more or less than truth, they are villains and the sons of darkness.

P. Henry. Speak, sirs; how was it?

Gadshill. We four set upon some dozen—

Fal. Sixteen, at least, my lord.

Gad. And bound them.

Peto. No, no, they were not bound.

Fal. You rogue, they were bound, every man of them; or I am a Jew, else—an Ebrew Jew.

Gad. As we were sharing, some six or seven fresh men set upon us—

Fal. And unbound the rest; and then come in the other.

P. Henry. What! fought ye with them all?

Fal. All? I know not what ye call all; but if I fought not with fifty of them, I am a bunch of radish: if there were not two or three and fifty upon poor old Jack, then I am no two-legged creature.

P. Henry. Pray heaven, you have not murdered some of them.

Fal. Nay, that's past praying for; for I have peppered two of them; two I am sure I have paid; two rogues in buckram suits. I tell thee what, Hal, if I tell thee a lie, spit in my face, and call me a horse. Thou knowest my old ward; (*he draws his sword and stands if about to fight*) here I lay, and thus I bore my point. Four rogues in buckram let drive at me—

P. Henry. What! four? Thou saidst but two even now.

Fal. Four, Hal; I told thee four.

Poins. Ay, ay, he said four.

Fal. These four came all afront, and mainly thrust at me. I made no more ado, but took all their seven points in my target, thus.

P. Henry. Seven? Why, there were but four, even now.

Fal. In buckram?

Poins. Ay, four, in buckram suits.

Fal. Seven, by these hilts, or I am a villain else.

P. Henry. Prithee, let him alone; we shall have more anon.

Fal. Dost thou hear me, Hal?

P. Henry. Ay, and mark thee, too, Jack.

Fal. Do so, for it is worth the listening to. These nine in buckram, that I told thee of—

P. Henry. So, two more already.

Fal. Their points being broken, began to give me ground; but I followed me close, came in foot and hand; and, with a thought, seven of the eleven I paid.

P. Henry. O, monstrous! eleven buckram men grown out of two!

Fal. But three knaves, in Kendal green, came at my back, and let drive at me; for it was so dark, Hal, that thou couldst not see thy hand.

P. Henry. These lies are like the father of them; gross as a mountain, open, palpable. Why, thou clay-brained, nott-pated fool; thou greasy tallow keech—

Fal. What! Art thou mad! Art thou mad? Is not the truth the truth?

P. Henry. Why, how couldst thou know these men in Kendal green, when it was so dark thou couldst not see thy hand? Come, tell us your reason; what sayest thou to this?

Poins. Come, your reason, Jack, your reason.

Fal. What, upon compulsion? No, were I at the strappado, or all the racks in the world, I would not tell you on compulsion. Give you a reason on compulsion! If reasons were as plentiful as blackberries, I would give no man a reason on compulsion, I.

P. Henry. I'll be no longer guilty of this sin: this sanguine coward, this horseback breaker, this huge hill of flesh—

Fal. Away! you starveling, you eel skin, you dried neat's tongue, you stockfish! Oh for breath to utter what is like thee!—you tailor's yard, you sheath, you bow case, you—

P. Henry. Well, breathe awhile, and then to it again; and when thou hast tired thyself in base comparisons, hear me speak but this.

Poins. Mark, Jack.

P. Henry. We two saw you four set on four; you bound them, and were masters of their wealth. Mark now, how a plain tale shall put you down. Then did we two set on you four, and with a word outfaced you from your prize, and have it; yea, and can show it you here in the house.—And, Falstaff, you carried yourself away as nimbly, with as quick dexterity, and roared for mercy, and still ran and roared, as ever I heard a calf. What a slave art thou, to hack thy sword as thou hast done, and then say it was in fight! What trick, what device, what starting hole, canst thou now find out to hide thee from this open and apparent shame?

Poins. Come, let's hear, Jack. What trick hast thou now?

Fal. Why, I knew ye as well as he that made ye. Why, hear ye, my masters: was it for me to kill the heir apparent? Should I turn upon the true prince? Why, thou knowest I am as valiant as Hercules; but beware instinct; the lion will not touch the true prince; instinct is a great matter; I was a coward on instinct. I shall think the better of myself and thee during my life; I for a valiant lion, and thou for a true prince. But, lads, I am glad you have the money. Hostess, clap to the doors. Watch to-night, pray to-morrow. Gallants, lads, boys, hearts of gold; all the titles of good-fellowship come to you! What! shall we be merry? Shall we have a play *extempore?*

P. Henry. Content; and the argument shall be thy running away.

Fal. Ah, no more of that, Hal, an thou lovest me!

Shakespeare.—Henry IV, Part I, Act ii, Scene iv.

NOTES.—The **lime** is a fruit allied to the lemon, but smaller, and more intensely sour.

The **strappado** was an instrument of torture by which the victim's limbs were wrenched out of joint and broken.

Hercules is a hero of fabulous history, remarkable for his great strength and wonderful achievements.

XCII. STUDIES.

Sir Francis Bacon, 1561-1626. This eminent man was the youngest son of Sir Nicholas Bacon, lord keeper of the seal in the early part of Elizabeth's reign, and Anne Bacon, one of the most learned women of the time, daughter of Sir Anthony Cooke. He was born in London, and educated at Cambridge. He was a laborious and successful student, but even in his boyhood conceived a great distrust of the methods of study pursued at the seats of learning,—methods which he exerted his great powers to correct in his maturer years. Much of his life was spent in the practice of law, in the discharge of the duties of high office, and as a member of Parliament; but, to the end of life, he busied himself with philosophical pursuits, and he will be known to posterity chiefly for his deep and clear writings on these subjects. His constant direction in philosophy is to break away from assumption and tradition, and to be led only by sound induction based on a knowledge of observed phenomena. His "Novum Organum" and "Advancement of Learning" embody his ideas on philosophy and the true methods of seeking knowledge.

Bacon rose to no very great distinction during the reign of Elizabeth; but, under James I, he was promoted to positions of great honor and influence. In 1618 he was made Baron of Verulam; and, three years later, he was made Viscount of St. Albans. During much of his life, Bacon was in pecuniary straits, which was doubtless one reason of his downfall; for, in 1621, he was accused of taking bribes, a charge to which he pleaded guilty. His disgrace followed, and he passed the last years of his life in retirement. Among the distinguished names in English literature, none stands higher in his department than that of Francis Bacon.

STUDIES serve for delight, for ornament, and for ability. Their chief use for delight is in privateness, and retiring; for ornament, is in discourse; and for ability, is in the judgment and disposition of business; for expert men can execute, and perhaps judge of the particulars, one by one; but the general counsels, and the plots and marshaling of affairs, come best from those that are learned.

To spend too much time in studies, is sloth; to use them too much for ornament, is affectation; to make judgment wholly by their rules, is the humor of a scholar; they perfect nature and are perfected by experience — for natural abilities are like natural plants, that need pruning by study; and studies themselves do give forth directions too much at large, except they be bounded in by experience. Crafty men contemn studies, simple men admire them, and wise men use them, for they teach not their own use; but that is a wisdom without them, and above them, won by observation.

Read not to contradict and confute, nor to believe and take for granted, nor to find talk and discourse, but to weigh and consider. Some books are to be tasted, others to be swallowed, and some few to be chewed and digested; that is, some books are to be read only in parts; others to be read, but not curiously; and some few to be read wholly, and with diligence and attention. Some books also may be read by deputy, and extracts made of them by others; but that would be only in the less important arguments, and the meaner sort of books; else distilled books are like common distilled waters, flashy things.

Reading maketh a full man, conference a ready man, and writing an exact man; and, therefore, if a man write little, he had need have a great memory; if he confer little, he had need have a present wit; and if he read little, he had need have much cunning, to seem to know that he doth not. Histories make men wise; poets, witty; the mathematics, subtle; natural philosophy, deep; moral philosophy, grave; logic and rhetoric, able to contend.

XCIII. SURRENDER OF GRANADA.

Sir Edward George Bulwer-Lytton, 1806–1873, was born in Norfolk County, England. His father died when he was young; his mother was a woman of strong literary tastes, and did much to form her son's mind. In 1844, by royal license, he took the surname of Lytton from his mother's family. Bulwer graduated at Cambridge. He began to publish in 1826, and his novels and plays followed rapidly. "Pelham," "The Caxtons," "My Novel," "What will he do with it?" and "Kenelm Chillingly" are among the best known of his numerous novels; and "The Lady of Lyons" and "Richelieu" are his most successful plays. His novels are extensively read on the continent, and have been translated into most of the languages spoken there. "Leila, or the Siege of Granada," from which this selection is adapted, was published in 1840.

Day dawned upon Granada, and the beams of the winter sun, smiling away the clouds of the past night, played cheerily on the murmuring waves of the Xenil and the Darro. Alone, upon a balcony commanding a view of the beautiful landscape, stood Boabdil, the last of the Moorish kings. He had sought to bring to his aid all the lessons of the philosophy he had cultivated.

"What are we," thought the musing prince, "that we should fill the world with ourselves—we kings? Earth resounds with the crash of my falling throne; on the ear of races unborn the echo will live prolonged. But what have I lost? Nothing that was necessary to my happiness, my repose: nothing save the source of all my wretchedness, the Marah of my life! Shall I less enjoy heaven and earth, or thought or action, or man's more material luxuries of food or sleep—the common and the cheap desires of all? Arouse thee, then, O heart within me! Many and deep emotions of sorrow or of joy are yet left to break the monotony of existence. . . . But it is time to depart." So saying, he descended to the court, flung himself on his barb, and, with a small and saddened train, passed through the gate which we yet survey, by a blackened and crumbling tower, overgrown with vines and ivy; thence, amidst gardens now appertaining to the convent of the victor faith, he took his mournful and unwitnessed way.

When he came to the middle of the hill that rises above those gardens, the steel of the Spanish armor gleamed upon him, as the detachment sent to occupy the palace marched over the summit in steady order and profound silence. At the head of this vanguard, rode, upon a snow-white palfrey, the Bishop of Avila, followed by a long train of barefooted monks. They halted as Boabdil approached, and the grave bishop saluted him with the air of one who addresses an infidel and inferior. With the quick sense of dignity common to the great, and yet more to the fallen, Boabdil felt, but resented not, the pride of the ecclesiastic. "Go, Christian," said he, mildly, "the gates of the Alhambra are open, and Allah has bestowed the palace and the city upon your king; may his virtues atone the faults of Boabdil!" So saying, and waiting no answer, he rode on without looking to the right or the left. The Spaniards also pursued their way.

The sun had fairly risen above the mountains, when Boabdil and his train beheld, from the eminence on which they were, the whole armament of Spain; and at the same moment, louder than the tramp of horse or the clash of arms, was heard distinctly the solemn chant of *Te Deum*, which preceded the blaze of the unfurled and lofty standards. Boabdil, himself still silent, heard the groans and exclamations of his train; he turned to cheer or chide them, and then saw, from his own watchtower, with the sun shining full upon its pure and dazzling surface, the silver cross of Spain. His Alhambra was already in the hands of the foe; while beside that badge of the holy war waved the gay and flaunting flag of St. Iago, the canonized Mars of the chivalry of Spain. At that sight the King's voice died within him; he gave the rein to his barb, impatient to close the fatal ceremonial, and did not slacken his speed till almost within bowshot of the first ranks of the army.

Never had Christian war assumed a more splendid and imposing aspect. Far as the eye could reach, extended the glittering and gorgeous lines of that goodly power, bristling with sunlit spears and blazoned banners; while beside, murmured, and glowed, and danced, the silver and laughing Xenil, careless what lord should possess, for his little day, the banks that bloomed by its everlasting course. By a small mosque halted the flower of the army. Surrounded by the archpriests of that mighty hierarchy, the peers and princes of a court that rivaled the Rolands of Charlemagne, was seen the kingly form of Ferdinand himself, with Isabel at his right hand, and the highborn dames of Spain, relieving, with their gay colors and sparkling gems, the sterner splendor of the crested helmet and polished mail. Within sight of the royal group, Boabdil halted, composed his aspect so as best to conceal his soul, and, a little in advance of his scanty train, but never in mien and majesty more a king, the son of Abdallah met his haughty conqueror.

At the sight of his princely countenance and golden hair, his comely and commanding beauty, made more touching by youth, a thrill of compassionate admiration ran through that assembly of the brave and fair. Ferdinand and Isabel slowly advanced to meet their late rival,—their new subject; and, as Boabdil would have dismounted, the Spanish king placed his hand upon his shoulder. "Brother and prince," said he, "forget thy sorrows; and may our friendship hereafter console thee for reverses, against which thou hast contended as a hero and a king — resisting man, but resigned at length to God."

Boabdil did not affect to return this bitter but unintentional mockery of compliment. He bowed his head, and remained a moment silent; then motioning to his train, four of his officers approached, and, kneeling beside Ferdinand, proffered to him, upon a silver buckler, the keys of the city. "O king!" then said Boabdil, "accept the keys of the last hold which has resisted the arms of Spain! The empire of the Moslem is no more. Thine are the city and the people of Granada; yielding to thy prowess, they yet confide in thy mercy." "They do well," said the king; "our promises shall not be broken. But since we know the gallantry of Moorish cavaliers, not to us, but to gentler hands, shall the keys of Granada be surrendered."

Thus saying, Ferdinand gave the keys to Isabel, who would have addressed some soothing flatteries to Boabdil, but the emotion and excitement were too much for her compassionate heart, heroine and queen though she was; and when she lifted her eyes upon the calm and pale features of the fallen monarch, the tears gushed from them irresistibly, and her voice died in murmurs. A faint flush overspread the features of Boabdil, and there was a momentary pause of embarrassment, which the Moor was the first to break.

"Fair queen," said he, with mournful and pathetic dignity, "thou canst read the heart that thy generous sympathy touches and subdues; this is thy last, nor least glorious conquest. But I detain ye; let not my aspect cloud your triumph. Suffer me to say farewell." "Farewell, my brother," replied Ferdinand, "and may fair fortune go with you! Forget the past!" Boabdil smiled bitterly, saluted the royal pair with profound and silent reverence, and rode slowly on, leaving the army below as he ascended the path that led to his new principality beyond the Alpuxarras. As the trees snatched the Moorish cavalcade from the view of the king, Ferdinand ordered the army to recommence its march; and trumpet and cymbal presently sent their music to the ear of the Moslems.

Boabdil spurred on at full speed, till his panting charger halted at the little village where his mother, his slaves, and his faithful wife, Amine—sent on before—awaited him. Joining these, he proceeded without delay upon his melancholy path. They ascended that eminence which is the pass into the Alpuxarras. From its height, the vale, the rivers, the spires, and the towers of Granada broke gloriously upon the view of the little band. They halted mechanically and abruptly; every eye was turned to the beloved scene. The proud shame of baffled warriors, the tender memories of home, of childhood, of fatherland, swelled every heart, and gushed from every eye.

Suddenly the distant boom of artillery broke from the citadel, and rolled along the sunlit valley and crystal river. A universal wail burst from the exiles; it smote,—it overpowered the heart of the ill-starred king, in vain seeking to wrap himself in Eastern pride or stoical philosophy. The tears gushed from his eyes, and he covered his face with his hands. The band wound slowly on through the solitary defiles; and that place where the king wept is still called The Last Sigh of the Moor.

NOTES.—**Granada** was the capital of an ancient Moorish kingdom of the same name, in the southeastern part of Spain. The **Darro** River flows through it, emptying into the **Xenil** (or Jenil) just outside the city walls. King Ferdinand of Spain drove out the Moors, and captured the city in 1492.

Marah. See Exodus xv. 23.

Avila is an episcopal city in Spain, capital of a province of the same name.

The **Te Deum** is an ancient Christian hymn, composed by St. Ambrose; it is so called from the first Latin words, "Te Deum laudamus," *We praise thee, O God.*

Mars, in mythology, the god of war.

The **Alhambra** is the ancient palace of the Moorish kings, at Granada.

Allah is the Mohammedan name for the Supreme Being.

Roland was a nephew of Charlemagne, or Charles the Great, emperor of the West and king of France. He was one of the most famous knights of the chivalric romances.

The **Alpuxarras** is a mountainous region in the old province of Granada, where the Moors were allowed to remain some time after their subjugation by Ferdinand.

XCIV. HAMLET'S SOLILOQUY.

To be, or not to be; that is the question:—
Whether 't is nobler in the mind to suffer
The slings and arrows of outrageous fortune,
Or to take arms against a sea of troubles,
And by opposing end them? To die,—to sleep,—
No more: and by a sleep to say we end
The heartache and the thousand natural shocks
That flesh is heir to,—'t is a consummation
Devoutly to be wished. To die,—to sleep:—
To sleep! perchance to dream:—ay, there's the rub;
For in that sleep of death what dreams may come
When we have shuffled off this mortal coil,
Must give us pause. There's the respect
That makes calamity of so long life;
For who would bear the whips and scorns of time,
The oppressor's wrong, the proud man's contumely,
The pangs of despised love, the law's delay,
The insolence of office, and the spurns
That patient merit of the unworthy takes,

When he himself might his quietus make
With a bare bodkin? Who would fardels bear,
To grunt and sweat under a weary life,
But that the dread of something after death,—
The undiscovered country from whose bourn
No traveler returns,—puzzles the will
And makes us rather bear those ills we have
Than fly to others that we know not of?
Thus conscience doth make cowards of us all;
And thus the native hue of resolution
Is sicklied o'er with the pale cast of thought,
And enterprises of great pith and moment
With this regard their currents turn awry,
And lose the name of action.

Shakespeare.—Hamlet, Act iii, Scene i.

XCV. GINEVRA.

Samuel Rogers, 1763-1855, was the son of a London banker, and, in company with his father, followed the banking business for some years. He began to write at an early age, and published his "Pleasures of Memory," perhaps his most famous work, in 1792. The next year his father died, leaving him an ample fortune. He now retired from business and established himself in an elegant house in St. James's Place. This house was a place of resort for literary men during fifty years. In 1822 he published his longest poem, "Italy," after which he wrote but little. He wrote with care, spending, as he said, nine years on the "Pleasures of Memory," and sixteen on "Italy." "His writings are remarkable for elegance of diction, purity of taste, and beauty of sentiment." It is said that he was very agreeable in conversation and manners, and benevolent in his disposition; but he was addicted to ill-nature and satire in some of his criticisms.

IF thou shouldst ever come by choice or chance
To Modena,—where still religiously
Among her ancient trophies, is preserved
Bologna's bucket (in its chain it hangs

Within that reverend tower, the Guirlandine),—
Stop at a palace near the Reggio gate,
Dwelt in of old by one of the Orsini.
Its noble gardens, terrace above terrace,
And rich in fountains, statues, cypresses,
Will long detain thee; through their archèd walks,
Dim at noonday, discovering many a glimpse
Of knights and dames such as in old romance,
And lovers such as in heroic song,—
Perhaps the two, for groves were their delight,
That in the springtime, as alone they sate,
Venturing together on a tale of love.
Read only part that day.—A summer sun
Sets ere one half is seen; but, ere thou go,
Enter the house—prithee, forget it not—
And look awhile upon a picture there.

'T is of a lady in her earliest youth,
The very last of that illustrious race,
Done by Zampieri—but by whom I care not.
He who observes it, ere he passes on,
Gazes his fill, and comes and comes again,
That he may call it up when far away.

She sits, inclining forward as to speak,
Her lips half-open, and her finger up,
As though she said, "Beware!" her vest of gold,
Broidered with flowers, and clasped from head to foot,
An emerald stone in every golden clasp;
And on her brow, fairer than alabaster,
A coronet of pearls. But then her face,
So lovely, yet so arch, so full of mirth,
The overflowings of an innocent heart,—
It haunts me still, though many a year has fled,
Like some wild melody!

Alone it hangs
Over a moldering heirloom, its companion,
An oaken chest, half-eaten by the worm,
But richly carved by Antony of Trent
With scripture stories from the life of Christ;
A chest that came from Venice, and had held
The ducal robes of some old ancestors—
That, by the way, it may be true or false—
But don't forget the picture; and thou wilt not,
When thou hast heard the tale they told me there.

She was an only child; from infancy
The joy, the pride, of an indulgent sire;
The young Ginevra was his all in life,
Still as she grew, forever in his sight;
And in her fifteenth year became a bride,
Marrying an only son, Francesco Doria,
Her playmate from her birth, and her first love.

Just as she looks there in her bridal dress,
She was all gentleness, all gayety,
Her pranks the favorite theme of every tongue.
But now the day was come, the day, the hour;
Now, frowning, smiling, for the hundredth time,
The nurse, that ancient lady, preached decorum:
And, in the luster of her youth, she gave
Her hand, with her heart in it, to Francesco.

Great was the joy; but at the bridal feast,
When all sate down, the bride was wanting there.
Nor was she to be found! Her father cried,
"'Tis but to make a trial of our love!"
And filled his glass to all; but his hand shook,
And soon from guest to guest the panic spread.
'Twas but that instant she had left Francesco,

Laughing and looking back and flying still,
Her ivory tooth imprinted on his finger.
But now, alas! she was not to be found;
Nor from that hour could anything be guessed,
But that she was not! —Weary of his life,
Francesco flew to Venice, and forthwith
Flung it away in battle with the Turk.
Orsini lived; and long was to be seen
An old man wandering as in quest of something,
Something he could not find—he knew not what.
When he was gone, the house remained a while
Silent and tenantless—then went to strangers.

Full fifty years were past, and all forgot,
When on an idle day, a day of search
'Mid the old lumber in the gallery,
That moldering chest was noticed; and 't was said
By one as young, as thoughtless as Ginevra,
"Why not remove it from its lurking place?"
'T was done as soon as said; but on the way
It burst, it fell; and lo! a skeleton,
With here and there a pearl, an emerald stone,
A golden clasp, clasping a shred of gold.
All else had perished, save a nuptial ring,
And a small seal, her mother's legacy,
Engraven with a name, the name of both,
"Ginevra."——There then had she found a grave!
Within that chest had she concealed herself,
Fluttering with joy, the happiest of the happy;
When a spring lock, that lay in ambush there,
Fastened her down forever!

NOTES.—The above selection is part of the poem, "Italy." Of the story Rogers says, "This story is, I believe, founded on fact; though the time and place are uncertain. Many old houses in England lay claim to it."

Modena is the capital of a province of the same name in northern Italy.

Bologna's bucket. This is affirmed to be the very bucket which Tassoni, an Italian poet, has celebrated in his mock heroics as the cause of a war between Bologna and Modena.

Reggio is a city about sixteen miles northwest of Modena.

The **Orsini.** A famous Italian family in the Middle Ages.

Zampieri, Domenichino (b. 1581, d. 1641), was one of the most celebrated of the Italian painters.

XCVI. INVENTIONS AND DISCOVERIES.

John Caldwell Calhoun, 1782-1850. This great statesman, and champion of southern rights and opinions, was born in Abbeville District, South Carolina. In the line of both parents, he was of Irish Presbyterian descent. In youth he was very studious, and made the best use of such opportunities for education as the frontier settlement afforded. He graduated at Yale College in 1804, and studied law at Litchfield, Connecticut. In 1808 he was elected to the Legislature of South Carolina; and, three years later, he was chosen to the National House of Representatives. During the six years that he remained in the House, he took an active and prominent part in the stirring events of the time. In 1817 he was appointed Secretary of War, and held the office seven years. From 1825 to 1832 he was Vice President of the United States. He then resigned this office, and took his seat as senator from South Carolina. In 1844 President Tyler called him to his Cabinet as Secretary of State; and, in 1845, he returned to the Senate, where he remained till his death. During all his public life Mr. Calhoun was active and outspoken. His earnestness and logical force commanded the respect of those who differed most widely from him in opinion. He took the most advanced ground in favor of "State Rights," and defended slavery as neither morally nor politically wrong. His foes generally conceded his honesty, and respected his ability; while his friends regarded him as little less than an oracle.

In private life Mr. Calhoun was highly esteemed and respected. His home was at "Fort Hill," in the northwestern district of South Carolina; and here he spent all the time he could spare from his public duties, in the enjoyments of domestic life and in cultivating his plantation. In his home he was remarkable for kindness, cheerfulness, and sociability.

To comprehend more fully the force and bearing of public opinion, and to form a just estimate of the changes to which, aided by the press, it will probably lead, politically and socially, it will be necessary to consider it in connection with the causes that have given it an influence so great as to entitle it to be regarded as a new political element. They will, upon investigation, be found in the many discoveries and inventions made in the last few centuries.

All these have led to important results. Through the invention of the mariner's compass, the globe has been circumnavigated and explored; and all who inhabit it, with but few exceptions, are brought within the sphere of an all-pervading commerce, which is daily diffusing over its surface the light and blessings of civilization.

Through that of the art of printing, the fruits of observation and reflection, of discoveries and inventions, with all the accumulated stores of previously acquired knowledge, are preserved and widely diffused. The application of gunpowder to the art of war has forever settled the long conflict for ascendency between civilization and barbarism, in favor of the former, and thereby guaranteed that, whatever knowledge is now accumulated, or may hereafter be added, shall never again be lost.

The numerous discoveries and inventions, chemical and mechanical, and the application of steam to machinery, have increased many fold the productive powers of labor and capital, and have thereby greatly increased the number who may devote themselves to study and improvement, and the amount of means necessary for commercial exchanges, especially between the more and the less advanced and civilized portions of the globe, to the great advantage of both, but particularly of the latter.

The application of steam to the purposes of travel and transportation, by land and water, has vastly increased the facility, cheapness, and rapidity of both: diffusing, with them, information and intelligence almost as quickly and as freely as if borne by the winds; while the electrical wires outstrip them in velocity, rivaling in rapidity even thought itself.

The joint effect of all this has been a great increase and diffusion of knowledge; and, with this, an impulse to progress and civilization heretofore unexampled in the history of the world, accompanied by a mental energy and activity unprecedented.

To all these causes, public opinion, and its organ, the press, owe their origin and great influence. Already they have attained a force in the more civilized portions of the globe sufficient to be felt by all governments, even the most absolute and despotic. But, as great as they now are, they have, as yet, attained nothing like their maximum force. It is probable that not one of the causes which have contributed to their formation and influence, has yet produced its full effect; while several of the most powerful have just begun to operate; and many others, probably of equal or even greater force, yet remain to be brought to light.

When the causes now in operation have produced their full effect, and inventions and discoveries shall have been exhausted—if that may ever be—they will give a force to public opinion, and cause changes, political and social, difficult to be anticipated. What will be their final bearing, time only can decide with any certainty.

That they will, however, greatly improve the condition of man ultimately, it would be impious to doubt; it would be to suppose that the all-wise and beneficent Being, the Creator of all, had so constituted man as that the employment of the high intellectual faculties with which He has been pleased to endow him, in order that he might develop the laws that control the great agents of the material world, and make them subservient to his use, would prove to him the cause of permanent evil, and not of permanent good.

NOTE.—This selection is an extract from "A Disquisition on Government." Mr. Calhoun expected to revise his manuscript before it was printed, but death interrupted his plans.

XCVII. ENOCH ARDEN AT THE WINDOW.

Alfred Tennyson, 1809-1892, was born in Somerby, Lincolnshire, England; his father was a clergyman noted for his energy and physical stature. Alfred, with his two older brothers, graduated at Trinity College, Cambridge. His first volume of poems appeared in 1830; it made little impression, and was severely treated by the critics. On the publication of his third series, in 1842, his poetic genius began to receive general recognition. On the death of Wordsworth he was made poet laureate, and he was then regarded as the foremost living poet of England. "In Memoriam," written in memory of his friend Arthur Hallam, appeared in 1850; the "Idyls of the King," in 1858; and "Enoch Arden," a touching story in verse, from which the following selection is taken, was published in 1864. In 1883 he accepted a peerage as Baron **Tennyson** of Aldworth, Sussex, and of Freshwater, Isle of Wight.

But Enoch yearned to see her face again;
"If I might look on her sweet face again
And know that she is happy." So the thought
Haunted and harassed him, and drove him forth,
At evening when the dull November day
Was growing duller twilight, to the hill.
There he sat down gazing on all below;
There did a thousand memories roll upon him,
Unspeakable for sadness. By and by
The ruddy square of comfortable light,
Far-blazing from the rear of Philip's house,
Allured him, as the beacon blaze allures
The bird of passage, till he madly strikes
Against it, and beats out his weary life.

For Philip's dwelling fronted on the street,
The latest house to landward; but behind,
With one small gate that opened on the waste,
Flourished a little garden, square and walled:
And in it throve an ancient evergreen,
A yew tree, and all round it ran a walk
Of shingle, and a walk divided it:
But Enoch shunned the middle walk, and stole

Up by the wall, behind the yew; and thence
That which he better might have shunned, if griefs
Like his have worse or better, Enoch saw.

For cups and silver on the burnished board
Sparkled and shone; so genial was the hearth:
And on the right hand of the hearth he saw
Philip, the slighted suitor of old times,
Stout, rosy, with his babe across his knees;
And o'er her second father stooped a girl,
A later but a loftier Annie Lee,
Fair-haired and tall, and from her lifted hand
Dangled a length of ribbon and a ring
To tempt the babe, who reared his creasy arms,
Caught at and ever missed it, and they laughed:
And on the left hand of the hearth he saw
The mother glancing often toward her babe,
But turning now and then to speak with him,
Her son, who stood beside her tall and strong,
And saying that which pleased him, for he smiled.

Now when the dead man come to life beheld
His wife, his wife no more, and saw the babe,
Hers, yet not his, upon the father's knee,
And all the warmth, the peace, the happiness.
And his own children tall and beautiful,
And him, that other, reigning in his place,
Lord of his rights and of his children's love,
Then he, tho' Miriam Lane had told him all,
Because things seen are mightier than things heard,
Staggered and shook, holding the branch, and feared
To send abroad a shrill and terrible cry,
Which in one moment, like the blast of doom,
Would shatter all the happiness of the hearth.

He, therefore, turning softly like a thief,
Lest the harsh shingle should grate underfoot,
And feeling all along the garden wall,
Lest he should swoon and tumble and be found,
Crept to the gate, and opened it, and closed,
As lightly as a sick man's chamber door,
Behind him, and came out upon the waste.
And there he would have knelt but that his knees
Were feeble, so that falling prone he dug
His fingers into the wet earth, and prayed.

"Too hard to bear! why did they take me thence?
O God Almighty, blessed Savior, Thou
That did'st uphold me on my lonely isle,
Uphold me, Father, in my loneliness
A little longer! aid me, give me strength
Not to tell her, never to let her know.
Help me not to break in upon her peace.
My children too! must I not speak to these?
They know me not. I should betray myself.
Never!—no father's kiss for me!—the girl
So like her mother, and the boy, my son!"

There speech and thought and nature failed a little,
And he lay tranced; but when he rose and paced
Back toward his solitary home again,
All down the long and narrow street he went
Beating it in upon his weary brain,
As tho' it were the burden of a song,
"Not to tell her, never to let her know."

NOTE.—**Enoch Arden** had been wrecked on an uninhabited island, and was supposed to be dead. After many years he was rescued, and returned home, where he found his wife happily married a second time. For her happiness, he kept his existence a secret, but soon died of a broken heart.

XCVIII. LOCHINVAR.

Oh, young Lochinvar is come out of the west,
Through all the wide Border his steed was the best;
And save his good broadsword, he weapon had none,
He rode all unarmed, and he rode all alone!
So faithful in love, and so dauntless in war,
There never was knight like the young Lochinvar!

He stayed not for brake, and he stopped not for stone,
He swam the Eske River where ford there was none;
But ere he alighted at Netherby gate,
The bride had consented, the gallant came late:
For a laggard in love, and a dastard in war,
Was to wed the fair Ellen of brave Lochinvar!

So boldly he entered the Netherby hall,
Among bridesmen, and kinsmen, and brothers, and all:
Then spoke the bride's father, his hand on his sword—
For the poor craven bridegroom said never a word—
"Oh, come ye in peace here, or come ye in war,
Or to dance at our bridal, young Lord Lochinvar?"

"I long wooed your daughter, my suit you denied;—
Love swells like the Solway, but ebbs like its tide—
And now am I come, with this lost love of mine,
To lead but one measure, drink one cup of wine.
There are maidens in Scotland more lovely by far,
That would gladly be bride to the young Lochinvar."

The bride kissed the goblet; the knight took it up,
He quaffed off the wine, and he threw down the cup.
She looked down to blush, and she looked up to sigh,

With a smile on her lips, and a tear in her eye.
He took her soft hand, ere her mother could bar,
"Now tread we a measure!" said young Lochinvar.

So stately his form, and so lovely her face,
That never a hall such a galliard did grace;
While her mother did fret, and her father did fume,
And the bridegroom stood dangling his bonnet and plume;
And the bridemaidens whispered, "'Twere better by far
To have matched our fair cousin with young Lochinvar."

One touch to her hand, and one word in her ear,
When they reached the hall door, and the charger stood near,
So light to the croup the fair lady he swung,
So light to the saddle before her he sprung!
"She is won! we are gone, over bank, bush, and scaur:
They'll have fleet steeds that follow," quoth young Lochinvar.

There was mounting 'mong Græmes of the Netherby clan;
Forsters, Fenwicks, and Musgraves, they rode and they ran;
There was racing and chasing on Cannobie Lee,
But the lost bride of Netherby ne'er did they see.
So daring in love, and so dauntless in war,
Have ye e'er heard of gallant like young Lochinvar?

— *Walter Scott.*

NOTES.—The above selection is a song taken from Scott's poem of "Marmion." It is in a slight degree founded on a ballad called "Katharine Janfarie," to be found in the "Minstrelsy of the Scottish Border."

The **Solway** Frith, on the southwest coast of Scotland, is remarkable for its high spring tides.

Bonnet is the ordinary name in Scotland for a man's cap.

XCIX. SPEECH ON THE TRIAL OF A MURDERER.

Against the prisoner at the bar, as an individual, I can not have the slightest prejudice. I would not do him the smallest injury or injustice. But I do not affect to be indifferent to the discovery and the punishment of this deep guilt. I cheerfully share in the opprobrium, how much soever it may be, which is cast on those who feel and manifest an anxious concern that all who had a part in planning, or a hand in executing this deed of midnight assassination, may be brought to answer for their enormous crime at the bar of public justice.

This is a most extraordinary case. In some respects it has hardly a precedent anywhere; certainly none in our New England history. This bloody drama exhibited no suddenly excited, ungovernable rage. The actors in it were not surprised by any lionlike temptation springing upon their virtue, and overcoming it before resistance could begin. Nor did they do the deed to glut savage vengeance, or satiate long-settled and deadly hate. It was a cool, calculating, money-making murder. It was all "hire and salary, not revenge." It was the weighing of money against life; the counting out of so many pieces of silver against so many ounces of blood.

An aged man, without an enemy in the world, in his own house, and in his own bed, is made the victim of a butcherly murder for mere pay. Truly, here is a new lesson for painters and poets. Whoever shall hereafter draw the portrait of murder, if he will show it as it has been exhibited in an example, where such example was last to have been looked for, in the very bosom of our New England society, let him not give it the grim visage of Moloch, the brow knitted by revenge, the face black with settled hate, and the bloodshot eye emitting livid fires of malice. Let him draw, rather, a decorous, smooth-faced, bloodless demon; a picture in repose, rather than in action; not so much an example of human nature in its depravity, and in its paroxysms of crime, as an infernal nature, a fiend in the ordinary display and development of his character.

The deed was executed with a degree of self-possession and steadiness equal to the wickedness with which it was planned. The circumstances, now clearly in evidence, spread out the whole scene before us. Deep sleep had fallen on the destined victim, and on all beneath his roof. A healthful old man, to whom sleep was sweet,—the first sound slumbers of the night held him in their soft but strong embrace. The assassin enters through the window, already prepared, into an unoccupied apartment. With noiseless foot he paces the lonely hall, half-lighted by the moon; he winds up the ascent of the stairs, and reaches the door of the chamber. Of this, he moves the lock by soft and continued pressure till it turns on its hinges without noise; and he enters, and beholds his victim before him. The room was uncommonly open to the admission of light. The face of the innocent sleeper was turned from the murderer, and the beams of the moon, resting on the gray locks of his aged temple, showed him where to strike. The fatal blow is given! and the victim passes, without a struggle or a motion, from the repose of sleep to the repose of death!

It is the assassin's purpose to make sure work; and he yet plies the dagger, though it was obvious that life had been destroyed by the blow of the bludgeon. He even raises the aged arm, that he may not fail in his aim at the heart; and replaces it again over the wounds of the poniard! To finish the picture, he explores the wrist for the pulse! He feels for it, and ascertains that it beats no longer! It is accomplished. The deed is done. He retreats, retraces his steps to the window, passes out through it as he came in, and escapes. He has done the murder;

no eye has seen him, no ear has heard him. The secret is his own, and it is safe!

Ah! gentlemen, that was a dreadful mistake. Such a secret can be safe nowhere. The whole creation of God has neither nook nor corner where the guilty can bestow it, and say it is safe. Not to speak of that eye which glances through all disguises, and beholds everything as in the splendor of noon; such secrets of guilt are never safe from detection, even by men. True it is, generally speaking, that "murder will out." True it is that Providence hath so ordained, and doth so govern things, that those who break the great law of Heaven by shedding man's blood, seldom succeed in avoiding discovery. Especially, in a case exciting so much attention as this, discovery must come, and will come, sooner or later. A thousand eyes turn at once to explore every man, everything, every circumstance connected with the time and place; a thousand ears catch every whisper; a thousand excited minds intensely dwell on the scene, shedding all their light, and ready to kindle the slightest circumstance into a blaze of discovery.

Meantime, the guilty soul can not keep its own secret. It is false to itself, or rather it feels an irresistible impulse of conscience to be true to itself. It labors under its guilty possession, and knows not what to do with it. The human heart was not made for the residence of such an inhabitant. It finds itself preyed on by a torment, which it dares not acknowledge to God nor man. A vulture is devouring it, and it can ask no sympathy or assistance either from heaven or earth. The secret which the murderer possesses soon comes to possess him; and, like the evil spirits of which we read, it overcomes him, and leads him whithersoever it will. He feels it beating at his heart, rising to his throat, and demanding disclosure. He thinks the whole world sees it in his face, reads it in his eyes, and almost hears its workings in the very silence of his thoughts. It has become his master. It betrays his discretion, it breaks down his courage, it conquers his prudence. When suspicions from without begin to embarrass him, and the net of circumstance to entangle him, the fatal secret struggles with still greater violence to burst forth. It must be confessed, it will be confessed; there is no refuge from confession but suicide, and suicide is confession.

—*Daniel Webster.*

NOTE.—The above extract is from Daniel Webster's argument in the trial of John F. Knapp for the murder of Mr. White, a very wealthy and respectable citizen of Salem, Mass. Four persons were arrested as being concerned in the conspiracy; one confessed the plot and all the details of the crime, implicating the others, but he afterwards refused to testify in court. The man who, by this confession, was the actual murderer, committed suicide, and Mr. Webster's assistance was obtained in prosecuting the others. John F. Knapp was convicted as principal, and the other two as accessaries in the murder.

C. THE CLOSING YEAR.

George Denison Prentice, 1802-1870, widely known as a political writer, a poet, and a wit, was born in Preston, Connecticut, and graduated at Brown University in 1823. He studied law, but never practiced his profession. He edited a paper in Hartford for two years; and, in 1831, he became editor of the "Louisville Journal," which position he held for nearly forty years. As an editor, Mr. Prentice was an able, and sometimes bitter, political partisan, abounding in wit and satire; as a poet, he not only wrote gracefully himself, but he did much by his kindness and sympathy to develop the poetical talents of others. Some who have since taken high rank, first became known to the world through the columns of the "Louisville Journal."

'T IS midnight's holy hour, and silence now
Is brooding like a gentle spirit o'er
The still and pulseless world. Hark! on the winds,
The bell's deep notes are swelling; 't is the knell
Of the departed year.

No funeral train
Is sweeping past; yet, on the stream and wood,
With melancholy light, the moonbeams rest
Like a pale, spotless shroud; the air is stirred
As by a mourner's sigh; and, on yon cloud,
That floats so still and placidly through heaven,
The spirits of the Seasons seem to stand—
Young Spring, bright Summer, Autumn's solemn form,
And Winter, with his aged locks—and breathe
In mournful cadences, that come abroad
Like the far wind harp's wild and touching wail,
A melancholy dirge o'er the dead year,
Gone from the earth forever.

'T is a time
For memory and for tears. Within the deep,
Still chambers of the heart, a specter dim,
Whose tones are like the wizard voice of Time,
Heard from the tomb of ages, points its cold
And solemn finger to the beautiful
And holy visions, that have passed away,
And left no shadow of their loveliness
On the dead waste of life. That specter lifts
The coffin lid of Hope, and Joy, and Love,
And, bending mournfully above the pale,
Sweet forms that slumber there, scatters dead flowers
O'er what has passed to nothingness.

The year
Has gone, and, with it, many a glorious throng
Of happy dreams. Its mark is on each brow,
Its shadow in each heart. In its swift course
It waved its scepter o'er the beautiful,
And they are not. It laid its pallid hand
Upon the strong man; and the haughty form

Is fallen, and the flashing eye is dim.
It trod the hall of revelry, where thronged
The bright and joyous; and the tearful wail
Of stricken ones is heard, where erst the song
And reckless shout resounded. It passed o'er
The battle plain, where sword, and spear, and shield
Flashed in the light of midday; and the strength
Of serried hosts is shivered, and the grass,
Green from the soil of carnage, waves above
The crushed and moldering skeleton. It came,
And faded like a wreath of mist at eve;
Yet, ere it melted in the viewless air,
It heralded its millions to their home
In the dim land of dreams.

Remorseless Time!—
Fierce spirit of the glass and scythe!—what power
Can stay him in his silent course, or melt
His iron heart to pity! On, still on
He presses, and forever. The proud bird,
The condor of the Andes, that can soar
Through heaven's unfathomable depths, or brave
The fury of the northern hurricane,
And bathe his plumage in the thunder's home,
Furls his broad wings at nightfall, and sinks down
To rest upon his mountain crag; but Time
Knows not the weight of sleep or weariness;
And Night's deep darkness has no chain to bind
His rushing pinion.

Revolutions sweep
O'er earth, like troubled visions o'er the breast
Of dreaming sorrow; cities rise and sink
Like bubbles on the water; fiery isles
Spring blazing from the ocean, and go back

To their mysterious caverns; mountains rear
To heaven their bald and blackened cliffs, and bow
Their tall heads to the plain; new empires rise,
Gathering the strength of hoary centuries,
And rush down, like the Alpine avalanche,
Startling the nations; and the very stars,
Yon bright and burning blazonry of God,
Glitter awhile in their eternal depths,
And, like the Pleiad, loveliest of their train,
Shoot from their glorious spheres, and pass away,
To darkle in the trackless void; yet Time,
Time the tomb builder, holds his fierce career,
Dark, stern, all pitiless, and pauses not
Amid the mighty wrecks that strew his path,
To sit and muse, like other conquerors,
Upon the fearful ruin he has wrought.

CI. A NEW CITY IN COLORADO.

Helen Hunt Jackson, 1830-1885, was the daughter of the late Professor Nathan W. Fiske, of Amherst College. She was born in Amherst, and educated at Ipswich, Massachusetts, and at New York. Mrs. Jackson was twice married. In the latter years of her life, she became deeply interested in the Indians, and wrote two books, "Ramona," a novel, and "A Century of Dishonor," setting forth vividly the wrongs to which the red race has been subjected. She had previously published several books of prose and poetry, less important but charming in their way. The following selection is adapted from "Bits of Travel at Home."

Garland City is six miles from Fort Garland. The road to it from the fort lies for the last three miles on the top of a sage-grown plateau. It is straight as an arrow, looks in the distance like a brown furrow on the pale gray plain, and seems to pierce the mountains beyond. Up to within an eighth of a mile of Garland City, there is no trace of human habitation. Knowing that the city must be near, you look in all directions for a glimpse of it; the hills ahead of you rise sharply across your way. Where is the city? At your very feet, but you do not suspect it.

The sunset light was fading when we reached the edge of the ravine in which the city lies. It was like looking unawares over the edge of a precipice; the gulch opened beneath us as suddenly as if the earth had that moment parted and made it. With brakes set firm, we drove cautiously down the steep road; the ravine twinkled with lights, and almost seemed to flutter with white tents and wagon tops. At the farther end it widened, opening out on an inlet of the San Luis Park; and, in its center, near this widening mouth, lay the twelve-days-old city. A strange din arose from it.

"What is going on?" we exclaimed. "The building of the city," was the reply. "Twelve days ago there was not a house here. To-day there are one hundred and five, and in a week more there will be two hundred; each man is building his own home, and working day and night to get it done ahead of his neighbor. There are four sawmills going constantly, but they can't turn out lumber half fast enough. Everybody has to be content with a board at a time. If it were not for that, there would have been twice as many houses done as there are."

We drove on down the ravine. A little creek on our right was half hid in willow thickets. Hundreds of white tents gleamed among them: tents with poles; tents made by spreading sailcloth over the tops of bushes; round tents; square tents; big tents; little tents; and for every tent a camp fire; hundreds of white-topped wagons, also, at rest for the night, their great poles propped up by sticks, and their mules and drivers lying and standing in picturesque groups around them.

It was a scene not to be forgotten. Louder and louder sounded the chorus of the hammers as we drew near the center of the "city;" more and more the bustle thickened; great ox teams swaying unwieldily about, drawing logs and planks, backing up steep places; all sorts of vehicles driving at reckless speed up and down; men carrying doors; men walking along inside of window sashes, — the easiest way to carry them; men shoveling; men wheeling wheelbarrows; not a man standing still; not a man with empty hands; every man picking up something, and running to put it down somewhere else, as in a play; and, all the while, "Clink! clink! clink!" ringing above the other sounds, — the strokes of hundreds of hammers, like the "Anvil Chorus."

"Where is Perry's Hotel?" we asked. One of the least busy of the throng spared time to point to it with his thumb, as he passed us. In some bewilderment we drew up in front of a large unfinished house, through the many uncased apertures of which we could see only scaffoldings, rough boards, carpenters' benches, and heaps of shavings. Streams of men were passing in and out through these openings, which might be either doors or windows; no steps led to any of them.

"Oh, yes! oh, yes! can accommodate you all!" was the landlord's reply to our hesitating inquiries. He stood in the doorway of his dining-room; the streams of men we had seen going in and out were the fed and the unfed guests of the house. It was supper time; we also were hungry. We peered into the dining room: three tables full of men; a huge pile of beds on the floor, covered with hats and coats; a singular wall, made entirely of doors propped upright; a triangular space walled off by sailcloth, — this is what we saw. We stood outside, waiting among the scaffolding and benches. A black man was lighting the candles in a candelabrum made of two narrow bars of wood nailed across each other at right angles, and perforated with holes. The candles sputtered, and the hot fat fell on the shavings below.

"Dangerous way of lighting a room full of shavings," some one said. The landlord looked up at the swinging candelabra and laughed. "Tried it pretty often," he said. "Never burned a house down yet."

I observed one peculiarity in the speech at Garland City. Personal pronouns, as a rule, were omitted; there was no time for a superfluous word.

"Took down this house at Wagon Creek," he continued, "just one week ago; took it down one morning while the people were eating breakfast; took it down over their heads; putting it up again over their heads now."

This was literally true. The last part of it we ourselves were seeing while he spoke, and a friend at our elbow had seen the Wagon Creek crisis.

"Waiting for that round table for you," said the landlord; "'ll bring the chairs out here's fast's they quit 'em. That's the only way to get the table."

So, watching his chances, as fast as a seat was vacated, he sprang into the room, seized the chair and brought it out to us; and we sat there in our "reserved seats," biding the time when there should be room enough vacant at the table for us to take our places.

What an indescribable scene it was! The strange-looking wall of propped doors which we had seen, was the *impromptu* wall separating the bedrooms from the dining-room. Bedrooms? Yes, five of them; that is, five bedsteads in a row, with just space enough between them to hang up a sheet, and with just room enough between them and the propped doors for a moderate-sized person to stand upright if he faced either the doors or the bed. Chairs? Oh, no! What do you want of a chair in a bedroom which has a bed in it? Washstands? One tin basin out in the unfinished room. Towels? Uncertain.

The little triangular space walled off by the sailcloth was a sixth bedroom, quite private and exclusive; and the big pile of beds on the dining-room floor was to be made up into seven bedrooms more between the tables, after everybody had finished supper.

Luckily for us we found a friend here,—a man who has been from the beginning one of Colorado's chief pioneers; and who is never, even in the wildest wilderness, without resources of comfort.

"You can't sleep here," he said. "I can do better for you than this."

"Better!"

He offered us luxury. How movable a thing is one's standard of comfort! A two-roomed pine shanty, board walls, board floors, board ceilings, board partitions not reaching to the roof, looked to us that night like a palace. To have been entertained at Windsor Castle would not have made us half so grateful.

It was late before the "city" grew quiet; and, long after most of the lights were out, and most of the sounds had ceased, I heard one solitary hammer in the distance, clink, clink, clink. I fell asleep listening to it.

CII. IMPORTANCE OF THE UNION.

Mr. President: I am conscious of having detained you and the Senate much too long. I was drawn into the debate with no previous deliberation, such as is suited to the discussion of so grave and important a subject. But it is a subject of which my heart is full, and I have not been willing to suppress the utterance of its spontaneous sentiments. I can not, even now, persuade myself to relinquish it, without expressing once more my deep conviction, that, since it respects nothing less than the union of the states, it is of most vital and essential importance to the public happiness.

I profess, sir, in my career hitherto, to have kept steadily in view the prosperity and honor of the whole country, and the preservation of our federal Union. It is to that Union we owe our safety at home, and our consideration and dignity abroad. It is to that Union that we are chiefly indebted for whatever makes us most proud of our country. That Union we reached only by the discipline of our virtues, in the severe school of adversity. It had its origin in the necessities of disordered finance, prostrate commerce, and ruined credit. Under its benign influences, these great interests immediately awoke, as from the dead, and sprang forth with newness of life. Every year of its duration has teemed with fresh proofs of its utility and its blessings; and, although our territory has stretched out wider and wider, and our population spread farther and farther, they have not outrun its protection or its benefits. It has been to us all a copious fountain of national, social, and personal happiness.

I have not allowed myself, sir, to look beyond the Union, to see what might lie hidden in the dark recess behind. I have not coolly weighed the chances of preserving liberty, when the bonds that unite us together shall be broken asunder. I have not accustomed myself to hang over the precipice of disunion, to see whether, with my short sight, I can fathom the depth of the abyss below; nor could I regard him as a safe counselor in the affairs of this government, whose thoughts should be mainly bent on considering, not how the Union should be best preserved, but how tolerable might be the condition of the people when it shall be broken up and destroyed.

While the Union lasts, we have high, exciting, gratifying prospects spread out before us, for us and our children. Beyond that, I seek not to penetrate the veil. God grant that in my day, at least, that curtain may not rise. God grant that on my vision never may be opened what lies behind. When my eyes shall be turned to behold, for the last time, the sun in heaven, may I not see him shining on the broken and dishonored fragments of a once glorious Union; on States dissevered, discordant, belligerent; on a land rent with civil feuds, or drenched, it may be, in fraternal blood.

Let their last feeble and lingering glance rather behold the gorgeous ensign of the Republic, now known and honored throughout the earth, still full high advanced, its arms and trophies streaming in their original luster, not a stripe erased or polluted, not a single star obscured—bearing for its motto no such miserable interrogatory as, What is all this worth? nor those other words of delusion and folly, Liberty first, and Union afterwards—but everywhere, spread all over in characters of living light, blazing on all its ample folds, as they float over the sea and over the land, and in every wind under the whole heavens, that other sentiment, dear to every true American heart—Liberty and Union, now and forever, one and inseparable!

—*Daniel Webster.*

NOTE.—This selection is the peroration of Mr. Webster's speech in reply to Mr. Hayne during the debate in the Senate on Mr. Foot's Resolution in regard to the Public Lands.

CIII. THE INFLUENCES OF THE SUN.

John Tyndall, 1820-1893, one of the most celebrated modern scientists, was an Irishman by birth. He was a pupil of the distinguished Faraday. In 1853 he was appointed Professor of Natural Philosophy in the Royal Institution of London. He is known chiefly for his brilliant experiments and clear writing respecting heat, light, and sound. He also wrote one or two interesting books concerning the Alps and their glaciers. He visited America, and delighted the most intelligent audiences by his scientific lectures and his brilliant experiments. The scientific world is indebted to him for several remarkable discoveries.

As surely as the force which moves a clock's hands is derived from the arm which winds up the clock, so surely is all terrestrial power drawn from the sun. Leaving out of account the eruptions of volcanoes, and the ebb and flow of the tides, every mechanical action on the earth's surface, every manifestation of power, organic and inorganic, vital and physical, is produced by the sun. His warmth keeps the sea liquid, and the atmosphere a gas, and all the storms which agitate both are blown by the mechanical force of the sun. He lifts the rivers and the glaciers up to the mountains; and thus the cataract and the avalanche shoot with an energy derived immediately from him.

Thunder and lightning are also his transmitted strength. Every fire that burns and every flame that glows, dispenses light and heat which originally belonged to the sun. In these days, unhappily, the news of battle is familiar to us, but every shock and every charge is an application or misapplication of the mechanical force of the sun. He blows the trumpet, he urges the projectile, he bursts the bomb. And, remember, this is not poetry, but rigid mechanical truth.

He rears, as I have said, the whole vegetable world, and through it the animal; the lilies of the field are his workmanship, the verdure of the meadows, and the cattle upon a thousand hills. He forms the muscles, he urges the blood, he builds the brain. His fleetness is in the lion's foot; he springs in the panther, he soars in the eagle, he slides in the snake. He builds the forest and hews it down, the power which raised the tree, and which wields the ax, being one and the same. The clover sprouts and blossoms, and the scythe of the mower swings, by the operation of the same force.

The sun digs the ore from our mines, he rolls the iron; he rivets the plates, he boils the water; he draws the train. He not only grows the cotton, but he spins the fiber and weaves the web. There is not a hammer raised, a wheel turned, or a shuttle thrown, that is not raised, and turned, and thrown by the sun.

His energy is poured freely into space, but our world is a halting place where this energy is conditioned. Here the

Proteus works his spells; the selfsame essence takes a million shapes and hues, and finally dissolves into its primitive and almost formless form. The sun comes to us as heat; he quits us as heat; and between his entrance and departure the multiform powers of our globe appear. They are all special forms of solar power—the molds into which his strength is temporarily poured in passing from its source through infinitude.

NOTE.—**Proteus** (pro. Prō′te-ŭs) was a mythological divinity. His distinguishing characteristic was the power of assuming different shapes.

CIV. COLLOQUIAL POWERS OF FRANKLIN.

William Wirt, 1772-1834, an American lawyer and author, was born at Bladensburg, Maryland. Left an orphan at an early age, he was placed in care of his uncle. He improved his opportunities for education so well that he became a private tutor at fifteen. In 1792 he was admitted to the bar, and began the practice of law in Virginia; he removed to Richmond in 1799. From 1817 to 1829 he was Attorney-general of the United States. His last years were spent in Baltimore. Mr. Wirt was the author of several books; his "Letters of a British Spy," published in 1803, and "Life of Patrick Henry," published in 1817, are the best known of his writings.

NEVER have I known such a fireside companion. Great as he was both as a statesman and philosopher, he never shone in a light more winning than when he was seen in a domestic circle. It was once my good fortune to pass two or three weeks with him, at the house of a private gentleman, in the back part of Pennsylvania, and we were confined to the house during the whole of that time by the unintermitting constancy and depth of the snows. But confinement never could be felt where Franklin was an inmate. His cheerfulness and his colloquial powers spread around him a perpetual spring.

When I speak, however, of his colloquial powers, I do

not mean to awaken any notion analogous to that which Boswell has given us of Johnson. The conversation of the latter continually reminds one of the "pomp and circumstance of glorious war." It was, indeed, a perpetual contest for victory, or an arbitrary or despotic exaction of homage to his superior talents. It was strong, acute, prompt, splendid, and vociferous; as loud, stormy, and sublime as those winds which he represents as shaking the Hebrides, and rocking the old castle which frowned on the dark-rolling sea beneath.

But one gets tired of storms, however sublime they may be, and longs for the more orderly current of nature. Of Franklin, no one ever became tired. There was no ambition of eloquence, no effort to shine in anything which came from him. There was nothing which made any demand upon either your allegiance or your admiration. His manner was as unaffected as infancy. It was nature's self. He talked like an old patriarch; and his plainness and simplicity put you at once at your ease, and gave you the full and free possession and use of your faculties. His thoughts were of a character to shine by their own light, without any adventitious aid. They only required a medium of vision like his pure and simple style, to exhibit to the highest advantage their native radiance and beauty.

His cheerfulness was unremitting. It seemed to be as much the effect of a systematic and salutary exercise of the mind, as of its superior organization. His wit was of the first order. It did not show itself merely in occasional coruscations; but, without any effort or force on his part, it shed a constant stream of the purest light over the whole of his discourse. Whether in the company of commons or nobles, he was always the same plain man; always most perfectly at his ease, with his faculties in full play, and the full orbit of his genius forever clear and unclouded.

And then, the stores of his mind were inexhaustible. He had commenced life with an attention so vigilant that nothing had escaped his observation; and a judgment so solid that every incident was turned to advantage. His youth had not been wasted in idleness, nor overcast by intemperance. He had been, all his life, a close and deep reader, as well as thinker; and by the force of his own powers, had wrought up the raw materials which he had gathered from books, with such exquisite skill and felicity, that he has added a hundred fold to their original value, and justly made them his own.

NOTES.—**Benjamin Franklin** (b. 1706, d. 1790) was one of the most prominent men in the struggle of the American colonies for liberty. He was renowned as a statesman, and, although not an author by profession, was a very prolific writer. His "Autobiography," which was first printed in France, is now a household volume in America. See page 431.

Boswell, James, (b. 1740, d. 1795,) was a Scotch lawyer, and is chiefly known as the biographer of Dr. Johnson, of whom he was the intimate friend and companion.

Johnson, Samuel. See biographical notice, page 78.

CV. THE DREAM OF CLARENCE.

SCENE—*Room in the Tower of London. Enter* CLARENCE *and* BRAKENBURY.

Brak. WHY looks your grace so heavily to-day?
Clar. O, I have passed a miserable night,
So full of ugly sights, of ghastly dreams,
That, as I am a Christian, faithful man,
I would not spend another such a night,
Though 't were to buy a world of happy days,
So full of dismal terror was the time!
Brak. What was your dream? I long to hear you tell it.
Clar. Methoughts that I had broken from the Tower,
And was embarked to cross to Burgundy;

And, in my company, my brother Gloster;
Who, from my cabin, tempted me to walk
Upon the hatches; thence we looked toward England,
And cited up a thousand fearful times,
During the wars of York and Lancaster,
That had befallen us. As we paced along
Upon the giddy footing of the hatches,
Methought that Gloster stumbled; and, in falling,
Struck me, that thought to stay him, overboard,
Into the tumbling billows of the main.
Oh, then, methought, what pain it was to drown!
What dreadful noise of waters in mine ears!
What ugly sights of death within mine eyes!
Methought I saw a thousand fearful wrecks;
Ten thousand men that fishes gnawed upon;
Wedges of gold, great anchors, heaps of pearl,
Inestimable stones, unvalued jewels,
All scattered in the bottom of the sea.
Some lay in dead men's skulls; and, in those holes
Where eyes did once inhabit, there were crept,
As 't were in scorn of eyes, reflecting gems,
Which wooed the slimy bottom of the deep,
And mocked the dead bones that lay scattered by.

Brak. Had you such leisure in the time of death,
To gaze upon the secrets of the deep?

Clar. Methought I had; and often did I strive
To yield the ghost: but still the envious flood
Kept in my soul, and would not let it forth
To seek the empty, vast, and wandering air;
But smothered it within my panting bulk,
Which almost burst to belch it in the sea.

Brak. Awaked you not with this sore agony?

Clar. Oh, no; my dream was lengthened after life;
Oh, then began the tempest to my soul,
Who passed, methought, the melancholy flood,
With that grim ferryman which poets write of,

Unto the kingdom of perpetual night.
The first that there did greet my stranger soul,
Was my great father-in-law, renownèd Warwick;
Who cried aloud, "What scourge for perjury
Can this dark monarchy afford false Clarence?"
And so he vanished. Then came wandering by
A shadow like an angel, with bright hair
Dabbled in blood; and he shrieked out aloud:
"Clarence is come! false, fleeting, perjured Clarence!
That stabbed me in the field by Tewksbury:
Seize on him, Furies, take him to your torments!"
With that, methoughts, a legion of foul fiends
Environed me, and howlèd in mine ears
Such hideous cries, that, with the very noise,
I, trembling, waked, and, for a season after,
Could not believe but that I was in hell;
Such terrible impression made the dream.

Brak. No marvel, lord, though it affrighted you;
I am afraid, methinks, to hear you tell it.

Clar. O Brakenbury, I have done those things,
Which now bear evidence against my soul,
For Edward's sake; and see how he requites me!
O God! if my deep prayers can not appease thee,
But thou wilt be avenged on my misdeeds,
Yet execute thy wrath in me alone:
Oh, spare my guiltless wife and my poor children!
——I pray thee, gentle keeper, stay by me;
My soul is heavy, and I fain would sleep.

Brak. I will, my lord: God give your grace good rest!

CLARENCE *reposes himself on a chair.*

Sorrow breaks seasons and reposing hours,
Makes the night morning, and the noontide night.

Shakespeare.—Richard III, Act i, Scene iv.

Notes.—The houses of **York** and **Lancaster** were at war for the possession of the English throne. The Duke of **Clarence** and the Duke of **Gloster** were brothers of King **Edward** IV., who was head of the house of York. Clarence married the daughter of the Earl of **Warwick**, and joined the latter in several insurrections against the king. They finally plotted with Queen Margaret of the Lancaster party for the restoration of the latter house to the English throne, but Clarence betrayed Warwick and the Queen, and killed the latter's son at the battle of **Tewksbury**. Through the plots of Gloster, Clarence was imprisoned in the Tower of London, and there murdered.

Brakenbury was lieutenant of the Tower.

The **ferryman** referred to is Charon, of Greek mythology, who was supposed to ferry the souls of the dead over the river Acheron to the infernal regions.

CVI. HOMEWARD BOUND.

Richard H. Dana, Jr., 1815-1882, was the son of Richard H. Dana, the poet. He was born in Cambridge, Mass. In his boyhood he had a strong desire to be a sailor, but by his father's advice chose a student's life, and entered Harvard University. At the age of nineteen an affection of the eyes compelled him to suspend his studies. He now made a voyage to California as a common sailor, and was gone two years. On his return, he resumed his studies and graduated in 1837. He afterwards studied law, and entered upon an active and successful practice. Most of his life was spent in law and politics, although he won distinction in literature.

The following extract is from his "Two Years before the Mast," a book published in 1840, giving an account of his voyage to California. This book details, in a most clear and entertaining manner, the every-day life of a common sailor on shipboard, and is the best known of all Mr. Dana's works.

It is usual, in voyages round the Cape from the Pacific, to keep to the eastward of the Falkland Islands; but, as there had now set in a strong, steady, and clear south-wester, with every prospect of its lasting, and we had had enough of high latitudes, the captain determined to stand immediately to the northward, running inside the Falkland

Islands. Accordingly, when the wheel was relieved at eight o'clock, the order was given to keep her due north, and all hands were turned up to square away the yards and make sail.

In a moment the news ran through the ship that the captain was keeping her off, with her nose straight for Boston, and Cape Horn over her taffrail. It was a moment of enthusiasm. Everyone was on the alert, and even the two sick men turned out to lend a hand at the halyards. The wind was now due southwest, and blowing a gale to which a vessel close-hauled could have shown no more than a single close-reefed sail; but as we were going before it, we could carry on. Accordingly, hands were sent aloft and a reef shaken out of the topsails, and the reefed foresail set. When we came to masthead the topsail yards, with all hands at the halyards, we struck up, "Cheerly, men," with a chorus which might have been heard halfway to Staten Island.

Under her increased sail, the ship drove on through the water. Yet she could bear it well; and the captain sang out from the quarter-deck — "Another reef out of that fore topsail, and give it to her." Two hands sprang aloft; the frozen reef points and earings were cast adrift, the halyards manned, and the sail gave out her increased canvas to the gale. All hands were kept on deck to watch the effect of the change. It was as much as she could well carry, and with a heavy sea astern, it took two men at the wheel to steer her.

She flung the foam from her bows; the spray breaking aft as far as the gangway. She was going at a prodigious rate. Still, everything held. Preventer braces were reeved and hauled taut; tackles got upon the backstays; and everything done to keep all snug and strong. The captain walked the deck at a rapid stride, looked aloft at the sails, and then to windward; the mate stood in the gangway, rubbing his hands, and talking aloud to the ship—"Hurrah, old bucket! the Boston girls have got hold of the towrope!" and the like; and we were on the forecastle looking to see how the spars stood it, and guessing the rate at which she was going, — when the captain called out — "Mr. Brown, get up the topmast studding sail! What she can't carry she may drag!"

The mate looked a moment; but he would let no one be before him in daring. He sprang forward, — "Hurrah, men! rig out the topmast studding sail boom! Lay aloft, and I'll send the rigging up to you!" We sprang aloft into the top; lowered a girtline down, by which we hauled up the rigging; rove the tacks and halyards; ran out the boom and lashed it fast, and sent down the lower halyards as a preventer. It was a clear starlight night, cold and blowing; but everybody worked with a will. Some, indeed, looked as though they thought the "old man" was mad, but no one said a word.

We had had a new topmast studding sail made with a reef in it,—a thing hardly ever heard of, and which the sailors had ridiculed a good deal, saying that when it was time to reef a studding sail it was time to take it in. But we found a use for it now; for, there being a reef in the topsail, the studding sail could not be set without one in it also. To be sure, a studding sail with reefed topsails was rather a novelty; yet there was some reason in it, for if we carried that away, we should lose only a sail and a boom; but a whole topsail might have carried away the mast and all.

While we were aloft, the sail had been got out, bent to the yard, reefed, and ready for hoisting. Waiting for a good opportunity, the halyards were manned and the yard hoisted fairly up to the block; but when the mate came to shake the cat's-paw out of the downhaul, and we began to boom end the sail, it shook the ship to her center. The boom buckled up and bent like a whipstick, and we looked every moment to see something go; but, being of the short, tough upland spruce, it bent like whalebone, and nothing could break it. The carpenter said it was the best stick he had ever seen.

The strength of all hands soon brought the tack to the boom end, and the sheet was trimmed down, and the preventer and the weather brace hauled taut to take off the strain. Every rope-yarn seemed stretched to the utmost, and every thread of canvas; and with this sail added to her, the ship sprang through the water like a thing possessed. The sail being nearly all forward, it lifted her out of the water, and she seemed actually to jump from sea to sea. From the time her keel was laid, she had never been so driven; and had it been life or death with everyone of us, she could not have borne another stitch of canvas.

Finding that she would bear the sail, the hands were sent below, and our watch remained on deck. Two men at the wheel had as much as they could do to keep her within three points of her course, for she steered as wild as a young colt. The mate walked the deck, looking at the sails, and then over the side to see the foam fly by her,—slapping his hands upon his thighs and talking to the ship—"Hurrah, you jade, you've got the scent! you know where you're going!" And when she leaped over the seas, and almost out of the water, and trembled to her very keel, the spars and masts snapping and creaking,—"There she goes!—There she goes—handsomely!—As long as she cracks, she holds!"—while we stood with the rigging laid down fair for letting go, and ready to take in sail and clear away if anything went.

At four bells we hove the log, and she was going eleven knots fairly; and had it not been for the sea from aft which sent the chip home, and threw her continually off her course, the log would have shown her to have been going somewhat faster. I went to the wheel with a young fellow from the Kennebec, who was a good helmsman: and for two hours we had our hands full. A few minutes showed us that our monkey jackets must come off; and, cold as it was, we stood in our shirt sleeves in a perspiration, and were glad enough to have it eight bells and the wheels relieved. We turned in and slept as well as we could, though the sea made a constant roar under her bows, and washed over the forecastle like a small cataract.

NOTES.—The **Falkland Islands** are a group in the Atlantic just east of Cape Horn.

Bells. On shipboard time is counted in bells, the bell being struck every half hour.

CVII. IMPEACHMENT OF WARREN HASTINGS.

Thomas Babington Macaulay, 1800-1859, was born in the village of Rothley, Leicestershire. On his father's side, he descended from Scotch Highlanders and ministers of the kirk. His education began at home, and was completed at Trinity College, Cambridge. While a student, he gained much reputation as a writer and a debater. In 1826 he was admitted to the bar. In 1825 began his connection with the "Edinburgh Review," which continued twenty years. Some of his most brilliant essays appeared first in its pages. He was first chosen to Parliament in 1830, and was reëlected several times. In 1840 his essays and some other writings were collected and published with the title of "Miscellanies." His "Lays of Ancient Rome" was published in 1842. His "History of England" was published near the close of his life. In 1857 he was given the title of Baron Macaulay. "His style is vigorous, rapid in its movement, and brilliant; and yet, with all its splendor, has a crystalline clearness. Indeed, the fault generally found with his style is, that it is so constantly brilliant that the vision is dazzled and wearied with its excessive brightness." He has sometimes been charged with sacrificing facts to fine sentences.

In his statesmanship, Macaulay was always an earnest defender of liberty. His first speech in Parliament was in support of a bill to remove the civil disabilities of the Jews, and his whole parliamentary career was consistent with this wise and liberal beginning.

THE place in which the impeachment of Warren Hastings was conducted, was worthy of such a trial. It was the great hall of William Rufus; the hall which had resounded with acclamations at the inauguration of thirty kings; the hall which had witnessed the just sentence of Bacon, and the just absolution of Somers; the hall where the eloquence of Strafford had for a moment awed and melted a victorious party inflamed with just resentment; the hall where Charles had confronted the High Court of Justice with the placid courage which half redeemed his fame.

Neither military nor civil pomp was wanting. The avenues were lined with grenadiers. The streets were kept clear by cavalry. The peers, robed in gold and ermine, were marshaled by heralds. The judges, in their vestments of state, attended to give advice on points of law. The long galleries were crowded by such an audience as has rarely excited the fears or the emulation of an orator. There were gathered together, from all parts of a great, free, enlightened, and prosperous realm, grace and female loveliness, wit and learning, the representatives of every science and of every art.

There were seated around the queen, the fair-haired, young daughters of the house of Brunswick. There the embassadors of great kings and commonwealths gazed with admiration on a spectacle which no other country in the world could present. There Siddons, in the prime of her majestic beauty, looked with emotion on a scene surpassing all the imitations of the stage. There Gibbon, the historian of the Roman Empire, thought of the days when Cicero pleaded the cause of Sicily against Verres; and when, before a senate which had still some show of freedom, Tacitus thundered against the oppressor of Africa. There, too, were seen, side by side, the greatest painter and the greatest scholar of the age; for the spectacle had allured Reynolds from his easel and Parr from his study.

The sergeants made proclamation. Hastings advanced to the bar, and bent his knee. The culprit was indeed not unworthy of that great presence. He had ruled an extensive and populous country; had made laws and treaties; had sent forth armies; had set up and pulled down princes; and in his high place he had so borne himself, that all had feared him, that most had loved him, and that hatred itself could deny him no title to glory, except virtue. A person, small and emaciated, yet deriving dignity from a carriage which, while it indicated deference to the court, indicated, also, habitual self-possession and self-respect; a high and intellectual forehead; a brow, pensive, but not gloomy; a mouth of inflexible decision; a face, pale and worn, but serene, on which a great and well-balanced mind was legibly written: such was the aspect with which the great proconsul presented himself to his judges.

The charges, and the answers of Hastings, were first read. This ceremony occupied two whole days. On the third, Burke rose. Four sittings of the court were occupied by his opening speech, which was intended to be a general introduction to all the charges. With an exuberance of thought and a splendor of diction, which more than satisfied the highly raised expectations of the audience, he described the character and institutions of the natives of India; recounted the circumstances in which the Asiatic Empire of Britain had originated; and set forth the constitution of the Company and of the English Presidencies.

Having thus attempted to communicate to his hearers an idea of eastern society, as vivid as that which existed in his own mind, he proceeded to arraign the administration of Hastings, as systematically conducted in defiance of morality and public law. The energy and pathos of the great orator extorted expressions of unwonted admiration from all; and, for a moment, seemed to pierce even the resolute heart of the defendant. The ladies in the galleries, unaccustomed to such displays of eloquence, excited by the solemnity of the occasion, and perhaps not unwilling to display their taste and sensibility, were in a state of uncontrollable emotion. Handkerchiefs were pulled out; smelling bottles were handed round; hysterical sobs and screams were heard, and some were even carried out in fits.

At length the orator concluded. Raising his voice, till the old arches of Irish oak resounded—"Therefore," said he, "hath it with all confidence been ordered by the Commons of Great Britain, that I impeach Warren Hastings of high crimes and misdemeanors. I impeach him in the name of the Commons House of Parliament, whose trust he has betrayed. I impeach him in the name of the English nation, whose ancient honor he has sullied. I impeach him in the name of the people of India, whose rights he has trodden under foot, and whose country he has turned into a desert. Lastly, in the name of human nature itself, in the name of both sexes, in the name of every age, in the name of every rank, I impeach the common enemy and oppressor of all."

NOTES.—**Warren Hastings** (b. 1732, d. 1818) was Governor-general of British India. He was impeached for maladministration, but, after a trial which extended from Feb. 13th, 1788, to April 23d, 1795, and occupied one hundred and forty-eight days, he was acquitted by a large majority on each separate count of the impeachment.

William Rufus, or William II. (b. 1056, d. 1100), built Westminster Hall in which the trial was held. **Bacon**; see biographical notice, pages 332 and 333. **Somers**, John (b. 1651. d. 1716) was impeached for maladministration while holding the office of lord chamberlain. **Strafford**, Thomas Wentworth, earl of, (b. 1593, d. 1641,) was impeached for his mismanagement while governor of Ireland. He conducted his own defense with such eloquence that the original impeachment was abandoned, although he was immediately condemned for high treason and executed. **Charles I.** (b. 1600, d. 1649), after a war with Parliament, in which the rights of the people were at issue, was captured, tried, and condemned to death.

The **House of Brunswick** is one of the oldest families of Germany. A branch of this family occupies the British throne. **Siddons,** Sarah (b. 1755, d. 1831), was a famous English actress. **Gibbon,** Edward (b. 1737, d. 1794), was a celebrated English historian. **Cicero**; see note on page 156. **Tacitus** (b. about 55, d. after 117 A. D.) was a Roman orator and historian, who conducted the prosecution of Marius, proconsul of Africa. **Reynolds,** Sir Joshua (b. 1723, d. 1792), an English portrait painter of note. **Parr,** Samuel (b. 1747, d. 1825), was an English author. **Burke,** Edmund; see biographical sketch accompanying the following lesson.

CVIII. DESTRUCTION OF THE CARNATIC.

Edmund Burke, 1730–1797, one of the most able and brilliant of England's essayists, orators, and statesmen, was born in Dublin, and was the son of an able lawyer. He graduated at Trinity College, Dublin, in 1748. As a student, he was distinguished for ability and industry. From 1750 to 1766 he was in London writing for periodicals, publishing books, or serving as private secretary. His work on "The Sublime and Beautiful" appeared in 1756. From 1766 to 1794 he was a member of Parliament, representing at different times different constituencies. On the first day of his appearance in the House of Commons he made a successful speech. "In the three principal questions which excited his interest, and called forth the most splendid displays of his eloquence—the contest with the American Colonies, the impeachment of Warren Hastings, and the French Revolution—we see displayed a philănthropy the most pure, illustrated by a genius the most resplendent." Mr. Burke's foresight, uprightness, integrity, learning, magnanimity, and eloquence made him one of the most conspicuous men of his time; and his writings stand among the noblest contributions to English literature.

WHEN at length Hyder Ali found that he had to do with men who either would sign no convention, or whom no treaty and no signature could bind, and who were the determined enemies of human intercourse itself, he decreed to make the country possessed by these incorrigible and predestinated criminals a memorable example to mankind. He resolved, in the gloomy recesses of a mind capacious of such things, to leave the whole Carnatic an everlasting monument of vengeance, and to put perpetual desolation as a barrier between him and those against whom the faith which holds the moral elements of the world together was no protection.

He became at length so confident of his force, so collected in his might, that he made no secret whatsoever of his dreadful resolution. Having terminated his disputes with every enemy and every rival, who buried their mutual animosities in their common detestation against the creditors of the Nabob of Arcot, he drew from every quarter whatever a savage ferocity could add to his new rudiments in the arts of destruction; and compounding all the materials of fury, havoc, and desolation into one black cloud, he hung for a while on the declivities of the mountains.

Whilst the authors of all these evils were idly and stupidly gazing on this menacing meteor which blackened all their horizon, it suddenly burst, and poured down the whole of its contents upon the plains of the Carnatic.

Then ensued a scene of woe, the like of which no eye had seen, no heart conceived, and which no tongue can adequately tell. All the horrors of war before known or heard of, were mercy to that new havoc. A storm of universal fire blasted every field, consumed every house, destroyed every temple. The miserable inhabitants, flying from their flaming villages, in part were slaughtered; others, without regard to sex, to age, to the respect of rank, or sacredness of function, — fathers torn from children, husbands from wives, enveloped in a whirlwind of cavalry, and, amidst the goading spears of drivers, and the trampling of pursuing horses, — were swept into captivity, in an unknown and hostile land.

Those who were able to evade this tempest, fled to the walled cities; but escaping from fire, sword, and exile, they fell into the jaws of famine.

The alms of the settlement of Madras, in this dreadful exigency, were certainly liberal, and all was done by charity that private charity could do; but it was a people in beggary; it was a nation which stretched out its hands for food.

For months together these creatures of sufferance, whose very excess and luxury in their most plenteous days had fallen short of the allowance of our austerest fasts, silent, patient, resigned, without sedition or disturbance, almost without complaint, perished by a hundred a day in the streets of Madras; every day seventy at least laid their bodies in the streets, or on the glacis of Tanjore, and expired of famine in the granary of India.

I was going to wake your justice toward this unhappy part of our fellow-citizens, by bringing before you some of the circumstances of this plague of hunger. Of all the calamities which beset and waylay the life of man, this comes the nearest to our heart, and is that wherein the proudest of us all feels himself to be nothing more than he is.

But I find myself unable to manage it with decorum. These details are of a species of horror so nauseous and disgusting; they are so degrading to the sufferers and to the hearers; they are so humiliating to human nature itself, that, on better thoughts, I find it more advisable to throw a pall over this hideous object, and to leave it to your general conceptions.

For eighteen months, without intermission, this destruction raged from the gates of Madras to the gates of Tanjore; and so completely did these masters in their art, Hyder Ali, and his more ferocious son, absolve themselves of their impious vow, that when the British armies traversed, as they did, the Carnatic, for hundreds of miles in all directions, through the whole line of their march they did not see one man — not one woman — not one child — not one four-footed beast of any description whatever! One dead, uniform silence reigned over the whole region.

With the inconsiderable exceptions of the narrow vicinage of some few forts, I wish to be understood as speaking literally;—I mean to produce to you more than three witnesses, who will support this assertion in its full extent. That hurricane of war passed through every part of the central provinces of the Carnatic. Six or seven districts to the north and to the south (and these not wholly untouched) escaped the general ravage.

NOTES.—This selection is an extract from Burke's celebrated speech in Parliament, in 1785, on the Nabob of Arcot's debts; it bore upon the maladministration of Hastings.

Arcot, a district in India, had been ceded to the British on condition that they should pay the former ruler's debts. These were found to be enormous, and the creditors proved to be individuals in the East India Company's employ. The creditors, for their private gain, induced the Nabob to attempt the subjugation of other native princes, among whom was **Hyder Ali.** The latter at first made successful resistance, and compelled the Nabob and his allies to sign a treaty. The treaty was not kept, and the destruction above recounted took place.

The **Carnatic** is a province in British India, on the eastern side of the peninsula; it contains about 50,000 square miles. **Madras** is a city, and **Tanjore** a town, in this province.

CIX. THE RAVEN.

Edgar Allan Poe, 1809-1849, was born in Boston, and died in Baltimore. He was left a destitute orphan at an early age, and was adopted by Mr. John Allan, a wealthy citizen of Richmond. He entered the University of Virginia, at Charlottesville, where he excelled in his studies, and was always at the head of his class; but he was compelled to leave on account of irregularities. He was afterwards appointed a cadet at West Point, but failed to graduate there for the same reason. Poe now quarreled with his benefactor and left his house never to return. During the rest of his melancholy career, he obtained a precarious livelihood by different literary enterprises. His ability as a writer gained him positions with various periodicals in Richmond, New York, and Philadelphia, and during this time he wrote some of his finest prose. The appearance of "The Raven" in 1845, however, at once made Poe a

literary lion. He was quite successful for a time, but then fell back into his dissipated habits which finally caused his death. In his personal appearance, Poe was neat and gentlemanly; his face was expressive of intellect and sensibility; and his mental powers in some directions were of a high order. His writings show care, and a great degree of skill in their construction; but their effect is generally morbid.

ONCE upon a midnight dreary,
While I pondered, weak and weary,
Over many a quaint and curious
Volume of forgotten lore—
While I nodded, nearly napping,
Suddenly there came a tapping,
As of some one gently rapping,
Rapping at my chamber door.
"'Tis some visitor," I muttered,
"Tapping at my chamber door—
Only this, and nothing more."

Ah, distinctly I remember,
It was in the bleak December,
And each separate dying ember
Wrought its ghost upon the floor.
Eagerly I wished the morrow;—
Vainly I had sought to borrow
From my books surcease of sorrow—
Sorrow for the lost Lenore—
For the rare and radiant maiden
Whom the angels name Lenore—
Nameless here for evermore.

And the silken, sad, uncertain
Rustling of each purple curtain
Thrilled me,—filled me with fantastic
Terrors, never felt before;
So that now, to still the beating

Of my heart, I stood repeating,
"'Tis some visitor entreating
Entrance at my chamber door—
Some late visitor entreating
Entrance at my chamber door;
This it is, and nothing more."

Presently my soul grew stronger;
Hesitating then no longer,
"Sir," said I, "or Madam, truly
Your forgiveness I implore;
But the fact is I was napping,
And so gently you came rapping,
And so faintly you came tapping,
Tapping at my chamber door,
That I scarce was sure I heard you."—
Here I opened wide the door;—
Darkness there, and nothing more.

Deep into that darkness peering,
Long I stood there, wondering, fearing,
Doubting, dreaming dreams no mortals
Ever dared to dream before;
But the silence was unbroken,
And the stillness gave no token,
And the only word there spoken
Was the whispered word, "Lenore!"
This I whispered, and an echo
Murmured back the word, "Lenore!"—
Merely this, and nothing more.

Back into the chamber turning,
All my soul within me burning,
Soon again I heard a tapping,
Something louder than before.

"Surely," said I, "surely, that is
Something at my window lattice;
Let me see then, what thereat is,
And this mystery explore—
Let my heart be still a moment,
And this mystery explore;—
'Tis the wind, and nothing more."

Open here I flung the shutter,
When, with many a flirt and flutter,
In there stepped a stately Raven
Of the saintly days of yore;
Not the least obeisance made he;
Not a minute stopped or stayed he,
But, with mien of lord or lady,
Perched above my chamber door—
Perched upon a bust of Pallas
Just above my chamber door—
Perched, and sat, and nothing more.

Then this ebony bird beguiling
My sad fancy into smiling,
By the grave and stern decorum
Of the countenance it wore,
"Though thy crest be shorn and shaven,
Thou," I said, "art sure no craven,
Ghastly, grim, and ancient Raven,
Wandering from the nightly shore,
Tell me what thy lordly name is
On the night's Plutonian shore!"
Quoth the Raven, "Nevermore."

Much I marveled this ungainly
Fowl to hear discourse so plainly,
Though its answer little meaning—

Little relevancy bore;
For we can not help agreeing
That no living human being
Ever yet was blest with seeing
Bird above his chamber door—
Bird or beast upon the sculptured
Bust above his chamber door,
With such name as "Nevermore."

But the Raven, sitting lonely
On that placid bust, spoke only
That one word, as if his soul in
That one word he did outpour.
Nothing farther then he uttered,
Not a feather then he fluttered,
Till I scarcely more than muttered,
"Other friends have flown before—
On the morrow he will leave me,
As my Hopes have flown before."
Then the bird said, "Nevermore."

Startled at the stillness broken
By reply so aptly spoken,
'Doubtless," said I, "what it utters
Is its only stock and store,
Caught from some unhappy master
Whom unmerciful Disaster
Followed fast and followed faster
Till his songs one burden bore—
Till the dirges of his Hope tnat
Melancholy burden bore
Of 'Never—nevermore.'"

But the Raven still beguiling
All my sad soul into smiling,
Straight I wheeled a cushioned seat in

Front of bird, and bust, and door;
Then, upon the velvet sinking,
I betook myself to linking
Fancy unto fancy, thinking
What this ominous bird of yore—
What this grim, ungainly, ghastly,
Gaunt, and ominous bird of yore
Meant in croaking "Nevermore."

This I sat engaged in guessing,
But no syllable expressing
To the fowl whose fiery eyes now
Burned into my bosom's core;
This and more I sat divining,
With my head at ease reclining
On the cushion's velvet lining
That the lamplight gloated o'er,
But whose velvet violet lining,
With the lamplight gloating o'er
She shall press, ah, nevermore!

Then, methought, the air grew denser,
Perfumed from an unseen censer
Swung by Seraphim, whose footfalls
Tinkled on the tufted floor.
"Wretch," I cried, "thy God hath lent thee—
By these angels he hath sent thee
Respite—respite and nepenthe
From thy memories of Lenore!
Quaff, oh quaff this kind nepenthe,
And forget this lost Lenore!"
Quoth the Raven, "Nevermore."

"Prophet!" said I, "thing of evil!—
Prophet still, if bird or devil!—
Whether Tempter sent, or whether

Tempest tossed thee here ashore,
Desolate, yet all undaunted,
On this desert land enchanted—
On this home by Horror haunted—
Tell me truly, I implore—
Is there—is there balm in Gilead?
Tell me—tell me, I implore!"
Quoth the Raven, "Nevermore."

"Prophet!" said I, "thing of evil,—
Prophet still, if bird or devil!—
By that heaven that bends above us,
By that God we both adore,
Tell this soul with sorrow laden,
If, within the distant Aidenn,
It shall clasp a sainted maiden
Whom the angels name Lenore—
Clasp a rare and radiant maiden,
Whom the angels name Lenore."
Quoth the Raven, "Nevermore."

"Be that word our sign of parting,
Bird or fiend," I shrieked, upstarting;
"Get thee back into the tempest
And the night's Plutonian shore!
Leave no black plume as a token
Of that lie thy soul hath spoken!
Leave my loneliness unbroken!—
Quit the bust above my door!
Take thy beak from out my heart, and
Take thy form from off my door!"
Quoth the Raven, "Nevermore."

And the Raven, never flitting,
Still is sitting, still is sitting

On the pallid bust of Pallas
Just above my chamber door;
And his eyes have all the seeming
Of a demon's that is dreaming,
And the lamplight o'er him streaming
Throws his shadow on the floor;
And my soul from out that shadow,
That lies floating on the floor,
Shall be lifted—nevermore!

NOTES.—**Pallas,** or Minerva, in ancient mythology, was the goddess of wisdom.

Plutonian, see note on Pluto, page 242.

Gilead is the name of a mountain group of Palestine, celebrated for its balsam or balm made from herbs. It is here used figuratively.

Aidenn is an Anglicized and disguised spelling of the Arabic form of the word Eden: it is here used as a synonym for heaven.

CX. A VIEW OF THE COLOSSEUM.

Orville Dewey, 1794–1882, a well known Unitarian clergyman and author, was born in Sheffield, Massachusetts, graduated with distinction at Williams College in 1814, and afterward studied theology at Andover. For a while he was assistant to Dr. W. E. Channing in Boston, and later, was a pastor in New Bedford, New York City, and Boston. He made two or three voyages to Europe, and published accounts of his travels.

"Discourses on Human Life," "Discourses on the Nature of Religion," "Discourses on Commerce and Business," are among his published works. His writings are both philosophical and practical; and, as a preacher, he was esteemed original, earnest, and impressive.

ON the eighth of November, from the high land, about fourteen miles distant, I first saw Rome; and although there is something very unfavorable to impression in the expectation that you are to be greatly impressed, or that you ought to be, or that such is the fashion; yet Rome is too mighty a name to be withstood by such or any other influences. Let you come upon that hill in what mood you may, the scene will lay hold upon you as with the hand of a giant. I scarcely know how to describe the impression, but it seemed to me as if something strong and stately, like the slow and majestic march of a mighty whirlwind, swept around those eternal towers; the storms of time, that had prostrated the proudest monuments of the world, seemed to have left their vibrations in the still and solemn air; ages of history passed before me; the mighty procession of nations, kings, consuls, emperors, empires, and generations had passed over that sublime theater. The fire, the storm, the earthquake, had gone by; but there was yet left the still, small voice like that at which the prophet "wrapped his face in his mantle."

I went to see the Colosseum by moonlight. It is the monarch, the majesty of all ruins; there is nothing like it. All the associations of the place, too, give it the most impressive character. When you enter within this stupendous circle of ruinous walls and arches, and grand terraces of masonry, rising one above another, you stand upon the arena of the old gladiatorial combats and Christian martyrdom; and as you lift your eyes to the vast amphitheater, you meet, in imagination, the eyes of a hundred thousand Romans, assembled to witness these bloody spectacles. What a multitude and mighty array of human beings; and how little do we know in modern times of great assemblies! One, two, and three, and, at its last enlargement by Constantine, more than three hundred thousand persons could be seated in the Circus Maximus!

But to return to the Colosseum; we went up under the conduct of a guide upon the walls and terraces, or embankments, which supported the ranges of seats. The seats have long since disappeared; and grass overgrows the spots where the pride, and power, and wealth, and beauty of

Rome sat down to its barbarous entertainments. What thronging life was here then! What voices, what greetings, what hurrying footsteps upon the staircases of the eighty arches of entrance! And now, as we picked our way carefully through the decayed passages, or cautiously ascended some moldering flight of steps, or stood by the lonely walls—ourselves silent, and, for a wonder, the guide silent, too—there was no sound here but of the bat, and none came from without but the roll of a distant carriage, or the convent bell from the summit of the neighboring Esquiline.

It is scarcely possible to describe the effect of moonlight upon this ruin. Through a hundred lonely arches and blackened passageways it streamed in, pure, bright, soft, lambent, and yet distinct and clear, as if it came there at once to reveal, and cheer, and pity the mighty desolation. But if the Colosseum is a mournful and desolate spectacle as seen from within—without, and especially on the side which is in best preservation, it is glorious. We passed around it; and, as we looked upward, the moon shining through its arches, from the opposite side, it appeared as if it were the coronet of the heavens, so vast was it—or like a glorious crown upon the brow of night.

I feel that I do not and can not describe this mighty ruin. I can only say that I came away paralyzed, and as passive as a child. A soldier stretched out his hand for "*un dono*," as we passed the guard; and when my companion said I did wrong to give, I told him that I should have given my cloak, if the man had asked it. Would you break any spell that worldly feeling or selfish sorrow may have spread over your mind, go and see the Colosseum by moonlight.

NOTES.—The **Colosseum** (pro. Col-os-sē'um) was commenced by the Roman emperor Vespasian, and was completed by Titus, his son, 79 A. D. Its construction occupied but three years, notwithstanding its size; a great part of its walls are standing to-day.

The **Circus Maximus** was an amphitheater built by Tarquin the Elder about 600 B. C.

Constantine. See note on page 175.

The **Esquiline** is one of the seven hills upon which Rome is built.

Un dono, an Italian phrase meaning *a gift* or *alms.*

CXI. THE BRIDGE.

I stood on the bridge at midnight,
  As the clocks were striking the hour,
And the moon rose o'er the city,
  Behind the dark church tower.

I saw her bright reflection
  In the waters under me,
Like a golden goblet falling
  And sinking into the sea.

And far in the hazy distance
  Of that lovely night in June,
The blaze of the flaming furnace
  Gleamed redder than the moon.

Among the long, black rafters
  The wavering shadows lay,
And the current that came from the ocean
  Seemed to lift and bear them away;

As, sweeping and eddying through them,
  Rose the belated tide,
And, streaming into the moonlight,
  The seaweed floated wide.

And like those waters rushing
  Among the wooden piers,
A flood of thoughts came o'er me
  That filled my eyes with tears

How often, oh, how often,
  In the days that had gone by,
I had stood on that bridge at midnight
  And gazed on that wave and sky!

How often, oh, how often,
  I had wished that the ebbing tide
Would bear me away on its bosom
  O'er the ocean wild and wide.

For my heart was hot and restless,
  And my life was full of care,
And the burden laid upon me
  Seemed greater than I could bear.

But now it has fallen from me,
  It is buried in the sea;
And only the sorrow of others
  Throws its shadow over me.

Yet, whenever I cross the river
  On its bridge with wooden piers,
Like the odor of brine from the ocean
  Comes the thought of other years.

And I think how many thousands
  Of care-encumbered men,
Each bearing his burden of sorrow,
  Have crossed the bridge since then.

I see the long procession
  Still passing to and fro,
The young heart hot and restless,
  And the old, subdued and slow!

And forever and forever,
  As long as the river flows,
As long as the heart has passions,
  As long as life has woes;

The moon and its broken reflection
  And its shadows shall appear
As the symbol of love in heaven,
  And its wavering image here.

—*Longfellow.*

CXII. OBJECTS AND LIMITS OF SCIENCE.

Robert Charles Winthrop, 1809-1894, was a descendant of John Winthrop, the first Governor of the Colony of Massachusetts Bay. He was born in Boston, studied at the public Latin School, graduated at Harvard in 1828, and studied law with Daniel Webster. Possessing an ample fortune, he made little effort to practice his profession. In 1834 he was elected to the Legislature of his native state, and was reëlected five times; three years he was Speaker of the House of Representatives. In 1840 he was chosen to Congress, and sat as Representative for ten years. In 1847 he was chosen Speaker of the House. He also served a short time in the Senate. His published writings are chiefly in the form of addresses and speeches; they are easy, finished, and scholarly. As a speaker, Mr. Winthrop was ready, full-voiced, and self-possessed.

THERE are fields enough for the wildest and most extravagant theorizings, within man's own appropriate domain, without overleaping the barriers which separate things human and divine. Indeed, I have often thought that modern science had afforded a most opportune and providential safety valve for the intellectual curiosity and ambition of man, at a moment when the progress of education, invention, and liberty had roused and stimulated him to a pitch of such unprecedented eagerness and ardor. Astronomy, Chemistry, and, more than all, Geology, with their incidental branches of study, have opened an inexhaustible field for investigation and speculation. Here, by the aid of modern instruments and modern modes of analysis, the most ardent and earnest spirits may find ample room and verge enough for their insatiate activity and audacious enterprise, and may pursue their course not only without the slightest danger of doing mischief to others, but with the certainty of promoting the great end of scientific truth.

Let them lift their vast reflectors or refractors to the skies, and detect new planets in their hiding places. Let them waylay the fugitive comets in their flight, and compel them to disclose the precise period of their orbits, and to give bonds for their punctual return. Let them drag out reluctant satellites from "their habitual concealments." Let them resolve the unresolvable nebulæ of Orion or Andromeda. They need not fear. The sky will not fall, nor a single star be shaken from its sphere.

Let them perfect and elaborate their marvelous processes of making the light and the lightning their ministers, for putting "a pencil of rays" into the hand of art, and providing tongues of fire for the communication of intelligence. Let them foretell the path of the whirlwind, and calculate the orbit of the storm. Let them hang out their gigantic pendulums, and make the earth do the work of describing and measuring her own motions. Let them annihilate human pain, and literally "charm ache with air, and agony with ether." The blessing of God will attend all their toils, and the gratitude of man will await all their triumphs.

Let them dig down into the bowels of the earth. Let them rive asunder the massive rocks, and unfold the history of creation as it lies written on the pages of their piled up strata. Let them gather up the fossil fragments of a lost Fauna, reproducing the ancient forms which inhabited the land or the seas, bringing them together, bone to his bone, till Leviathan and Behemoth stand before us in bodily presence and in their full proportions, and we almost tremble lest these dry bones should live again! Let them put nature to the rack, and torture her, in all her forms, to the betrayal of her inmost secrets and confidences. They need not forbear. The foundations of the round world have been laid so strong that they can not be moved.

But let them not think by searching to find out God. Let them not dream of understanding the Almighty to perfection. Let them not dare to apply their tests and solvents, their modes of analysis or their terms of definition, to the secrets of the spiritual kingdom. Let them spare the foundations of faith. Let them be satisfied with what is revealed of the mysteries of the Divine Nature. Let them not break through the bounds to gaze after the Invisible.

NOTES.—**Orion** and **Andromeda** are the names of two constellations.

The **Leviathan** is described in Job, chap. xli, and the **Behemoth** in Job, chap. xl. It is not known exactly what beasts are meant by these descriptions.

CXIII. THE DOWNFALL OF POLAND.

O SACRED Truth! thy triumph ceased a while,
And Hope, thy sister, ceased with thee to smile,
When leagued Oppression poured to northern wars
Her whiskered pandours and her fierce hussars,
Waved her dread standard to the breeze of morn,
Pealed her loud drum, and twanged her trumpet horn;

Tumultuous horror brooded o'er her van,
Presaging wrath to Poland—and to man!

Warsaw's last champion, from her height surveyed,
Wide o'er the fields a waste of ruin laid;
"O Heaven!" he cried, "my bleeding country save!
Is there no hand on high to shield the brave?
Yet, though destruction sweep those lovely plains,
Rise, fellow-men! our country yet remains!
By that dread name, we wave the sword on high,
And swear for her to live—with her to die!"

He said, and on the rampart heights arrayed
His trusty warriors, few, but undismayed;
Firm-paced and slow, a horrid front they form,
Still as the breeze, but dreadful as the storm;
Low murmuring sounds along their banners fly,
Revenge or death—the watchword and reply;
Then pealed the notes, omnipotent to charm,
And the loud tocsin tolled their last alarm.

In vain, alas! in vain, ye gallant few!
From rank to rank, your volleyed thunder flew!
Oh, bloodiest picture in the book of time,
Sarmatia fell, unwept, without a crime;
Found not a generous friend, a pitying foe,
Strength in her arms, nor mercy in her woe!
Dropped from her nerveless grasp the shattered spear,
Closed her bright eye, and curbed her high career;
Hope, for a season, bade the world farewell,
And Freedom shrieked as Kosciusko fell!

—*Thomas Campbell.*

NOTES.—**Kosciusko** (b. 1746, d. 1817), a celebrated Polish patriot, who had served in the American Revolution, was besieged at Warsaw, in 1794, by a large force of Russians, Prussians, and Austrians. After the siege was raised, he marched against a force of Russians much larger than his own, and was defeated. He was himself severely wounded and captured.

Sarmatia is the ancient name for a region of Europe which embraced Poland, but was of greater extent.

CXIV. LABOR.

Horace Greeley, 1811-1872, perhaps the most famous editor of America, was born in Amherst, New Hampshire, of poor parents. His boyhood was passed in farm labor, in attending the common school. and in reading every book on which he could lay his hands. His reading was mostly done by the light of pine knots. At fifteen he entered a printing office in Vermont, became the best workman in the office, and continued to improve every opportunity for study. At the age of twenty he appeared in New York City, poorly clothed, and almost destitute of money. He worked at his trade for a year or two, and then set up printing for himself. For several years he was not successful, but struggled on, performing an immense amount of work as an editor. In 1841 he established the "New York Tribune," which soon became one of the most successful and influential papers in the country. In 1848 he was elected to Congress, but remained but a short time. In 1872 he was a candidate for the Presidency, was defeated, and died a few days afterward. Mr. Greeley is a rare example of what may be accomplished by honesty and unflinching industry. Besides the vast amount which he wrote for the newspapers, he published several books; the best known of which is "The American Conflict."

EVERY child should be trained to dexterity in some useful branch of productive industry, not in order that he shall certainly follow that pursuit, but that he may at all events be able to do so in case he shall fail in the more intellectual or artificial calling which he may prefer to it. Let him seek to be a doctor, lawyer, preacher, poet, if he will; but let him not stake his all on success in that pursuit, but have a second line to fall back upon if driven from his first. Let him be so reared and trained that he may enter, if he will, upon some intellectual calling in the sustaining consciousness that he need not debase himself, nor do violence to his convictions, in order to achieve success therein, since he can live and thrive in another (if you choose, humbler) vocation, if driven from that of his choice. This buttress to integrity, this assurance of self-respect, is to be found in a universal training to efficiency in Productive Labor.

The world is full of misdirection and waste; but all the calamities and losses endured by mankind through frost, drought, blight, hail, fires, earthquakes, inundations, are as nothing to those habitually suffered by them through human idleness and inefficiency, mainly caused (or excused) by lack of industrial training. It is quite within the truth to estimate that one tenth of our people, in the average, are habitually idle because (as they say) they can find no employment. They look for work where it can not be had. They seem to be, or they are, unable to do such as abundantly confronts and solicits them. Suppose these to average but one million able-bodied persons, and that their work is worth but one dollar each per day; our loss by involuntary idleness can not be less than $300,000,000 per annum. I judge that it is actually $500,000,000. Many who stand waiting to be hired could earn from two to five dollars per day had they been properly trained to work. "There is plenty of room higher up," said Daniel Webster, in response to an inquiry as to the prospects of a young man just entering upon the practice of law; and there is never a dearth of employment for men or women of signal capacity or skill. In this city, ten thousand women are always doing needlework for less than fifty cents per day, finding themselves; yet twice their number of capable, skillful seamstresses could find steady employment and good living in wealthy families at not less than one dollar per day over and above board and lodging. He who is a good blacksmith, a fair millwright, a tolerable wagon maker, and can chop timber, make fence, and manage a small farm if required, is always sure of work and fair recompense; while he or she who can keep books or teach music fairly, but knows how to do nothing else, is in constant danger of falling into involuntary idleness and consequent beggary. It is a broad, general truth, that no boy was ever yet inured to daily, systematic, productive labor in field or shop throughout the latter half of his minority, who did not prove a useful man, and was not able to find work whenever he wished it.

Yet to the ample and constant employment of a whole community one prerequisite is indispensable, — that a variety of pursuits shall have been created or naturalized therein. A people who have but a single source of profit are uniformly poor, not because that vocation is necessarily ill-chosen, but because no single calling can employ and reward the varied capacities of male and female, old and young, robust and feeble. Thus a lumbering or fishing region with us is apt to have a large proportion of needy inhabitants; and the same is true of a region exclusively devoted to cotton growing or gold mining. A diversity of pursuits is indispensable to general activity and enduring prosperity.

Sixty or seventy years ago, what was then the District, and is now the State, of Maine, was a proverb in New England for the poverty of its people, mainly because they were so largely engaged in timber cutting. The great grain-growing, wheat-exporting districts of the Russian empire have a poor and rude people for a like reason. Thus the industry of Massachusetts is immensely more productive per head than that of North Carolina, or even that of Indiana, as it will cease to be whenever manufactures shall have been diffused over our whole country, as they must and will be. In Massachusetts half the women and nearly half the children add by their daily labor to the aggregate of realized wealth; in North Carolina and in Indiana little wealth is produced save by the labor of men, including boys of fifteen or upward. When this disparity shall have ceased, its consequence will also disappear.

CXV. THE LAST DAYS OF HERCULANEUM.

Edwin Atherstone, 1788-1872, was born at Nottingham, England, and became known to the literary world chiefly through two poems, "The Last Days of Herculaneum" and "The Fall of Nineveh." Both poems are written in blank verse, and are remarkable for their splendor of diction and their great descriptive power. Atherstone is compared to Thomson, whom he resembles somewhat in style.

THERE was a man,
A Roman soldier, for some daring deed
That trespassed on the laws, in dungeon low
Chained down. His was a noble spirit, rough,
But generous, and brave, and kind.
He had a son; it was a rosy boy,
A little faithful copy of his sire,
In face and gesture. From infancy, the child
Had been his father's solace and his care.

Every sport
The father shared and heightened. But at length,
The rigorous law had grasped him, and condemned
To fetters and to darkness.

The captive's lot,
He felt in all its bitterness: the walls
Of his deep dungeon answered many a sigh
And heart-heaved groan. His tale was known, and touched
His jailer with compassion; and the boy,
Thenceforth a frequent visitor, beguiled
His father's lingering hours, and brought a balm
With his loved presence, that in every wound
Dropped healing. But, in this terrific hour,
He was a poisoned arrow in the breast
Where he had been a cure.

With earliest morn
Of that first day of darkness and amaze,
He came. The iron door was closed—for them
Never to open more! The day, the night
Dragged slowly by; nor did they know the fate
Impending o'er the city. Well they heard
The pent-up thunders in the earth beneath,
And felt its giddy rocking; and the air
Grew hot at length, and thick; but in his straw
The boy was sleeping: and the father hoped
The earthquake might pass by: nor would he wake
From his sound rest the unfearing child, nor tell
The dangers of their state.

On his low couch
The fettered soldier sank, and, with deep awe,
Listened the fearful sounds: with upturned eye,
To the great gods he breathed a prayer; then, strove
To calm himself, and lose in sleep awhile
His useless terrors. But he could not sleep:
His body burned with feverish heat; his chains
Clanked loud, although he moved not; deep in earth
Groaned unimaginable thunders; sounds,
Fearful and ominous, arose and died,
Like the sad moanings of November's wind,
In the blank midnight. Deepest horror chilled
His blood that burned before; cold, clammy sweats
Came o'er him; then anon, a fiery thrill
Shot through his veins. Now, on his couch he shrunk
And shivered as in fear; now, upright leaped,
As though he heard the battle trumpet sound,
And longed to cope with death.

He slept, at last,
A troubled, dreamy sleep. Well had he slept

Never to waken more! His hours are few,
But terrible his agony

Soon the storm
Burst forth; the lightnings glanced; the air
Shook with the thunders. They awoke; they sprung
Amazed upon their feet. The dungeon glowed
A moment as in sunshine—and was dark:
Again, a flood of white flame fills the cell,
Dying away upon the dazzled eye
In darkening, quivering tints, as stunning sound
Dies throbbing, ringing in the ear.

With intensest awe,
The soldier's frame was filled; and many a thought
Of strange foreboding hurried through his mind,
As underneath he felt the fevered earth
Jarring and lifting; and the massive walls,
Heard harshly grate and strain: yet knew he not,
While evils undefined and yet to come
Glanced through his thoughts, what deep and cureless wound
Fate had already given.—Where, man of woe!
Where, wretched father! is thy boy? Thou call'st
His name in vain:—he can not answer thee.

Loudly the father called upon his child:
No voice replied. Trembling and anxiously
He searched their couch of straw; with headlong haste
Trod round his stinted limits, and, low bent,
Groped darkling on the earth:—no child was there.
Again he called: again, at farthest stretch
Of his accursèd fetters, till the blood
Seemed bursting from his ears, and from his eyes
Fire flashed, he strained with arm extended far,
And fingers widely spread, greedy to touch

Though but his idol's garment. Useless toil!
Yet still renewed: still round and round he goes,
And strains, and snatches, and with dreadful cries
Calls on his boy.

Mad frenzy fires him now.
He plants against the wall his feet; his chain
Grasps; tugs with giant strength to force away
The deep-driven staple; yells and shrieks with rage:
And, like a desert lion in the snare,
Raging to break his toils,—to and fro bounds.
But see! the ground is opening;—a blue light
Mounts, gently waving,—noiseless;—thin and cold
It seems, and like a rainbow tint, not flame;
But by its luster, on the earth outstretched,
Behold the lifeless child! his dress is singed,
And, o'er his face serene, a darkened line
Points out the lightning's track.

The father saw,
And all his fury fled:—a dead calm fell
That instant on him:—speechless—fixed—he stood,
And with a look that never wandered, gazed
Intensely on the corse. Those laughing eyes
Were not yet closed,—and round those ruby lips
The wonted smile returned.

Silent and pale
The father stands:—no tear is in his eye:—
The thunders bellow;—but he hears them not:—
The ground lifts like a sea;—he knows it not:—
The strong walls grind and gape:—the vaulted roof
Takes shape like bubble tossing in the wind;
See! he looks up and smiles; for death to him
Is happiness. Yet could one last embrace
Be given, 'twere still a sweeter thing to die.

It will be given. Look! how the rolling ground,
At every swell, nearer and still more near
Moves toward the father's outstretched arm his boy.
Once he has touched his garment: — how his eye
Lightens with love, and hope, and anxious fears!
Ha, see! he has him now! — he clasps him round;
Kisses his face; puts back the curling locks,
That shaded his fine brow; looks in his eyes;
Grasps in his own those little dimpled hands;
Then folds him to his breast, as he was wont
To lie when sleeping; and resigned, awaits
Undreaded death.

And death came soon and swift
And pangless. The huge pile sank down at once
Into the opening earth. Walls — arches — roof —
And deep foundation stones — all — mingling — fell!

NOTES.—**Herculaneum** and **Pompeii** were cities of Italy, which were destroyed by an eruption of Vesuvius in the year 79 A. D., being entirely buried under ashes and lava. During the last century they have been dug out to a considerable extent, and many of the streets, buildings, and utensils have been found in a state of perfect preservation.

CXVI. HOW MEN REASON.

MY friend, the Professor, whom I have mentioned to you once or twice, told me yesterday that somebody had been abusing him in some of the journals of his calling. I told him that I did n't doubt he deserved it; that I hoped he did deserve a little abuse occasionally, and would for a number of years to come; that nobody could do anything to make his neighbors wiser or better without being liable to abuse for it; especially that people hated to have their little mistakes made fun of, and perhaps he had been doing something of the kind. The Professor smiled.

Now, said I, hear what I am going to say. It will not take many years to bring you to the period of life when men, at least the majority of writing and talking men, do nothing but praise. Men, like peaches and pears, grow sweet a little while before they begin to decay. I don't know what it is,—whether a spontaneous change, mental or bodily, or whether it is through experience of the thanklessness of critical honesty,—but it is a fact, that most writers, except sour and unsuccessful ones, get tired of finding fault at about the time when they are beginning to grow old.

As a general thing, I would not give a great deal for the fair words of a critic, if he is himself an author, over fifty years of age. At thirty, we are all trying to cut our names in big letters upon the walls of this tenement of life; twenty years later, we have carved it, or shut up our jackknives. Then we are ready to help others, and care less to hinder any, because nobody's elbows are in our way. So I am glad you have a little life left; you will be saccharine enough in a few years.

Some of the softening effects of advancing age have struck me very much in what I have heard or seen here and elsewhere. I just now spoke of the sweetening process that authors undergo. Do you know that in the gradual passage from maturity to helplessness the harshest characters sometimes have a period in which they are gentle and placid as young children? I have heard it said, but I can not be sponsor for its truth, that the famous chieftain, Lochiel, was rocked in a cradle like a baby, in his old age. An old man, whose studies had been of the severest scholastic kind, used to love to hear little nursery stories read over and over to him. One who saw the Duke of Wellington in his last years describes him as very gentle in his aspect and demeanor. I remember a person of singularly stern and lofty bearing who became remarkably gracious and easy in all his ways in the later period of his life.

And that leads me to say that men often remind me of pears in their way of coming to maturity. Some are ripe at twenty, like human Jargonelles, and must be made the most of, for their day is soon over. Some come into their perfect condition late, like the autumn kinds, and they last better than the summer fruit. And some, that, like the Winter Nelis, have been hard and uninviting until all the rest have had their season, get their glow and perfume long after the frost and snow have done their worst with the orchards. Beware of rash criticisms; the rough and stringent fruit you condemn may be an autumn or a winter pear, and that which you picked up beneath the same bough in August may have been only its worm-eaten windfalls. Milton was a Saint Germain with a graft of the roseate Early Catherine. Rich, juicy, lively, fragrant, russet-skinned old Chaucer was an Easter Beurré; the buds of a new summer were swelling when he ripened.

—*Holmes.*

NOTES.—The above selection is from the "Autocrat of the Breakfast Table."

Lochiel. See note on page 214.

The **Duke of Wellington** (b. 1769, d. 1852) was the most celebrated of English generals. He won great renown in India and in the "Peninsular War," and commanded the allied forces when Napoleon was defeated at Waterloo.

Easter Beurré, Saint Germain, Winter Nelis, Early Catherine and **Jargonelles** are the names of certain varieties of pears.

Milton. See biographical notice on page 312.

Chaucer, Geoffrey (b. 1328, d. 1400). is often called "The Father of English Poetry." He was the first poet buried in Westminster Abbey. He was a prolific writer, but his "Canterbury Tales" is by far the best known of his works.

CXVII. THUNDERSTORM ON THE ALPS.

CLEAR, placid Leman! thy contrasted lake,
With the wild world I dwell in, is a thing
Which warns me, with its stillness, to forsake
Earth's troubled waters for a purer spring.
This quiet sail is as a noiseless wing
To waft me from distraction; once I loved
Torn ocean's roar, but thy soft murmuring
Sounds sweet, as if a sister's voice reproved,
That I with stern delights should e'er have been so moved.

All heaven and earth are still — though not in sleep,
But breathless, as we grow when feeling most;
And silent, as we stand in thoughts too deep —
All heaven and earth are still: from the high host
Of stars, to the lulled lake and mountain coast,
All is concentered in a life intense,
Where not a beam, nor air, nor leaf is lost,
But hath a part of being, and a sense
Of that which is of all Creator and defense.

The sky is changed! and such a change! O night,
And storm, and darkness, ye are wondrous strong,
Yet lovely in your strength, as is the light
Of a dark eye in woman! Far along,
From peak to peak, the rattling crags among,
Leaps the live thunder! Not from one lone cloud,
But every mountain now hath found a tongue,
And Jura answers, through her misty shroud,
Back to the joyous Alps, who call to her aloud!

And this is in the night. — Most glorious night!
Thou wert not sent for slumber! let me be

A sharer in thy fierce and far delight,—
A portion of the tempest and of thee!
How the lit lake shines,—a phosphoric sea!
And the big rain comes dancing to the earth!
And now again, 't is black,—and now, the glee
Of the loud hills shakes with its mountain mirth,
As if they did rejoice o'er a young earthquake's birth.

Now, where the swift Rhone cleaves his way between
Heights which appear as lovers who have parted
In hate, whose mining depths so intervene,
That they can meet no more, though broken-hearted;
Though in their souls, which thus each other thwarted,
Love was the very root of the fond rage,
Which blighted their life's bloom, and then—departed.
Itself expired, but leaving them an age
Of years, all winters,—war within themselves to wage.

Now, where the quick Rhone thus hath cleft his way,
The mightiest of the storms hath ta'en his stand!
For here, not one, but many make their play,
And fling their thunderbolts from hand to hand,
Flashing and cast around! Of all the band,
The brightest through these parted hills hath forked
His lightnings,—as if he did understand,
That in such gaps as desolation worked,
There, the hot shaft should blast whatever therein lurked.

—*Byron.*

NOTE.—Lake **Leman** (or Lake of Geneva) is in the southwestern part of Switzerland, separating it in part from Savoy. The **Rhone** flows through it, entering by a deep narrow gap, with mountain groups on either hand, eight or nine thousand feet above the water. The scenery about the lake is magnificent, the **Jura** mountains bordering it on the northwest, and the **Alps** lying on the south and east.

CXVIII. ORIGIN OF PROPERTY.

Sir William Blackstone, 1723-1780, was the son of a silk merchant, and was born in London. He studied with great success at Oxford, and was admitted to the bar in 1745. At first he could not obtain business enough in his profession to support himself, and for a time relinquished practice, and lectured at Oxford. He afterwards returned to London, and resumed his practice with great success, still continuing to lecture at Oxford. He was elected to Parliament in 1761; and in 1770 was made a justice of the Court of Common Pleas, which office he held till his death. Blackstone's fame rests upon his "Commentaries on the Laws of England," published about 1769. He was a man of great ability, sound learning, unflagging industry, and moral integrity. His great work is still a common text-book in the study of law.

In the beginning of the world, we are informed by Holy Writ, the all-bountiful Creator gave to man dominion over all the earth, and "over the fish of the sea, and over the fowl of the air, and over every living thing that moveth upon the earth." This is the only true and solid foundation of man's dominion over external things, whatever airy, metaphysical notions may have been started by fanciful writers upon this subject. The earth, therefore, and all things therein, are the general property of all mankind, exclusive of other beings, from the immediate gift of the Creator. And while the earth continued bare of inhabitants, it is reasonable to suppose that all was in common among them, and that everyone took from the public stock, to his own use, such things as his immediate necessities required.

These general notions of property were then sufficient to answer all the purposes of human life; and might, perhaps, still have answered them, had it been possible for mankind to have remained in a state of primeval simplicity, in which "all things were common to him." Not that this communion of goods seems ever to have been applicable, even in the earliest ages, to aught but the substance of the thing; nor could it be extended to the use of it. For, by the law of nature and reason, he who first began to use it, acquired therein a kind of transient property, that lasted so long as he was using it, and no longer. Or, to speak with greater precision, the right of possession continued for the same time, only, that the act of possession lasted.

Thus, the ground was in common, and no part of it was the permanent property of any man in particular; yet, whoever was in the occupation of any determined spot of it, for rest, for shade, or the like, acquired for the time a sort of ownership, from which it would have been unjust and contrary to the law of nature to have driven him by force; but, the instant that he quitted the use or occupation of it, another might seize it without injustice. Thus, also, a vine or other tree might be said to be in common, as all men were equally entitled to its produce; and yet, any private individual might gain the sole property of the fruit which he had gathered for his own repast: a doctrine well illustrated by Cicero, who compares the world to a great theater, which is common to the public, and yet the place which any man has taken is, for the time, his own.

But when mankind increased in number, craft, and ambition, it became necessary to entertain conceptions of a more permanent dominion; and to appropriate to individuals not the immediate use only, but the very substance of the thing to be used. Otherwise, innumerable tumults must have arisen, and the good order of the world been continually broken and disturbed, while a variety of persons were striving who should get the first occupation of the same thing, or disputing which of them had actually gained it. As human life also grew more and more refined, abundance of conveniences were devised to render it more easy, commodious, and agreeable; as habitations for shelter and safety, and raiment for warmth and decency. But no man would be at the trouble to provide either, so long as he had only a usufructuary property in them, which was to cease the instant that he quitted possession; if, as soon as he walked out of his tent or pulled off his garment, the next stranger who came by would have a right to inhabit the one and to wear the other.

In the case of habitations, in particular, it was natural to observe that even the brute creation, to whom everything else was in common, maintained a kind of permanent property in their dwellings, especially for the protection of their young; that the birds of the air had nests, and the beasts of the fields had caverns, the invasion of which they esteemed a very flagrant injustice, and would sacrifice their lives to preserve them. Hence a property was soon established in every man's house and homestead; which seem to have been originally mere temporary huts or movable cabins, suited to the design of Providence for more speedily peopling the earth, and suited to the wandering life of their owners, before any extensive property in the soil or ground was established.

There can be no doubt but that movables of every kind became sooner appropriated than the permanent, substantial soil; partly because they were more susceptible of a long occupancy, which might be continued for months together without any sensible interruption, and at length, by usage, ripen into an established right; but, principally, because few of them could be fit for use till improved and meliorated by the bodily labor of the occupant; which bodily labor, bestowed upon any subject which before lay in common to all men, is universally allowed to give the fairest and most reasonable title to an exclusive property therein.

The article of food was a more immediate call, and therefore a more early consideration. Such as were not contented with the spontaneous product of the earth, sought for a more solid refreshment in the flesh of beasts, which they obtained by hunting. But the frequent disappointments incident to that method of provision, induced them to gather together such animals as were of a more tame and sequacious nature and to establish a permanent property in their flocks and herds, in order to sustain themselves in a less precarious manner, partly by the milk of the dams, and partly by the flesh of the young.

The support of these their cattle made the article of water also a very important point. And, therefore, the book of Genesis, (the most venerable monument of antiquity, considered merely with a view to history,) will furnish us with frequent instances of violent contentions concerning wells; the exclusive property of which appears to have been established in the first digger or occupant, even in places where the ground and herbage remained yet in common. Thus, we find Abraham, who was but a sojourner, asserting his right to a well in the country of Abimelech, and exacting an oath for his security "because he had digged that well." And Isaac, about ninety years afterwards, reclaimed this his father's property; and, after much contention with the Philistines, was suffered to enjoy it in peace.

All this while, the soil and pasture of the earth remained still in common as before, and open to every occupant; except, perhaps, in the neighborhood of towns, where the necessity of a sole and exclusive property in lands, (for the sake of agriculture,) was earlier felt, and therefore more readily complied with. Otherwise, when the multitude of men and cattle had consumed every convenience on one spot of ground, it was deemed a natural right to seize upon and occupy such other lands as would more easily supply their necessities.

We have a striking example of this in the history of Abraham and his nephew Lot. When their joint substance became so great that pasture and other conveniences grew scarce, the natural consequence was that a strife arose between their servants; so that it was no longer practicable to dwell together. This contention, Abraham thus endeavored to compose: "Let there be no strife, I pray thee, between me and thee. Is not the whole land before thee? Separate thyself, I pray thee, from me. If thou wilt take the left hand, then I will go to the right; or if thou depart to the right hand, then I will go to the left." This plainly implies an acknowledged right in either to occupy whatever ground he pleased that was not preoccupied by other tribes. "And Lot lifted up his eyes, and beheld all the plain of Jordan, that it was well watered everywhere, even as the garden of the Lord. Then Lot chose him all the plain of Jordan, and journeyed east; and Abraham dwelt in the land of Canaan."

As the world by degrees grew more populous, it daily became more difficult to find out new spots to inhabit, without encroaching upon former occupants; and, by constantly occupying the same individual spot, the fruits of the earth were consumed, and its spontaneous products destroyed, without any provision for future supply or succession. It, therefore, became necessary to pursue some regular method of providing a constant subsistence; and this necessity produced, or at least promoted and encouraged the art of agriculture. And the art of agriculture, by a regular connection and consequence, introduced and established the idea of a more permanent property in the soil than had hitherto been received and adopted.

It was clear that the earth would not produce her fruits in sufficient quantities without the assistance of tillage; but who would be at the pains of tilling it, if another might watch an opportunity to seize upon and enjoy the product of his industry, art and labor? Had not, therefore, a separate property in lands, as well as movables, been vested in some individuals, the world must have continued a forest, and men have been mere animals of prey. Whereas, now, (so graciously has Providence interwoven our duty and our happiness together,) the result of this very necessity has been the ennobling of the human species, by giving it opportunities of improving its rational, as well as of exerting its natural faculties.

Necessity begat property; and, in order to insure that

property, recourse was had to civil society, which brought along with it a long train of inseparable concomitants: states, government, laws, punishments, and the public exercise of religious duties. Thus connected together, it was found that a part only of society was sufficient to provide, by their manual labor, for the necessary subsistence of all; and leisure was given to others to cultivate the human mind, to invent useful arts, and to lay the foundations of science.

NOTE.—**Cicero.** See note on page 156.

CXIX. BATTLE OF WATERLOO.

THERE was a sound of revelry by night,
And Belgium's capital had gathered then
Her Beauty and her Chivalry, and bright
The lamps shone o'er fair women and brave men.
A thousand hearts beat happily; and when
Music arose with its voluptuous swell,
Soft eyes looked love to eyes which spake again,
And all went merry as a marriage bell;
But hush! hark!—a deep sound strikes like a rising knell!

Did ye not hear it?—No; 't was but the wind,
Or the car rattling o'er the stony street;
On with the dance! let joy be unconfined;
No sleep till morn, when Youth and Pleasure meet
To chase the glowing Hours with flying feet—
But, hark!—that heavy sound breaks in once more,
As if the clouds its echo would repeat,
And nearer, clearer, deadlier than before!
Arm! arm! it is—it is the cannon's opening roar!

Ah! then and there was hurrying to and fro,
And gathering tears, and tremblings of distress,
And cheeks all pale, which, but an hour ago
Blushed at the praise of their own loveliness;
And there were sudden partings, such as press
The life from out young hearts, and choking sighs
Which ne'er might be repeated: who could guess
If ever more should meet those mutual eyes,
Since upon night so sweet such awful morn could rise.

And there was mounting in hot haste: the steed,
The mustering squadron, and the clattering car
Went pouring forward with impetuous speed,
And swiftly forming in the ranks of war;
And the deep thunder, peal on peal afar;
And near, the beat of the alarming drum
Roused up the soldier ere the morning star;
While thronged the citizens with terror dumb,
Or whispering with white lips—"The foe! They come!
They come!"

And Ardennes waves above them her green leaves,
Dewy with nature's tear-drops, as they pass,
Grieving, if aught inanimate e'er grieves,
Over the unreturning brave!—alas!
Ere evening to be trodden like the grass,
Which, now, beneath them, but above, shall grow,
In its next verdure, when this fiery mass
Of living valor, rolling on the foe,
And burning with high hope, shall molder, cold and low

Last noon beheld them full of lusty life,
Last eve in beauty's circle proudly gay,
The midnight brought the signal sound of strife,
The morn, the marshaling in arms,—the day,

Battle's magnificently stern array!
The thunderclouds close o'er it, which when rent,
The earth is covered thick with other clay,
Which her own clay shall cover, heaped and pent,
Rider and horse,—friend, foe,—in one red burial blent.

—*Byron.*

NOTES.—The **Battle of Waterloo** was fought on June 18th, 1815, between the French army on one side, commanded by Napoleon Bonaparte, and the English army and allies on the other side, commanded by the Duke of Wellington. At the commencement of the battle, some of the officers were at a ball at Brussels, a short distance from Waterloo, and being notified of the approaching contest by the cannonade, left the ballroom for the field of battle.

The wood of Soignies lay between the field of Waterloo and Brussels. It is supposed to be a remnant of the forest of **Ardennes.**

CXX. "WITH BRAINS, SIR."

John Brown, 1810-1882, was born in Lanarkshire, Scotland, and graduated at the University of Edinburgh. His father was John Brown, an eminent clergyman and the author of several books. Dr. Brown's literary reputation rests largely upon a series of papers contributed to the "North British Review." "Rab and his Friends," a collection of papers published in book form, is the most widely known of all his writings.

"PRAY, Mr. Opie, may I ask you what you mix your colors with?" said a brisk *dilettante* student to the great painter. "With brains, sir," was the gruff reply—and the right one. It did not give much of information; it did not expound the principles and rules of the art; but, if the inquirer had the commodity referred to, it would awaken him; it would set him agoing, athinking, and a-painting to good purpose. If he had not the wherewithal, as was likely enough, the less he had to do with colors and their mixture the better.

Many other artists, when asked such a question, would have either set about detailing the mechanical composition of such and such colors, in such and such proportions, rubbed up so and so; or perhaps they would (and so much the better, but not the best) have shown him how they laid them on; but even this would leave him at the critical point. Opie preferred going to the quick and the heart of the matter: "With brains, sir."

Sir Joshua Reynolds was taken by a friend to see a picture. He was anxious to admire it, and he looked it over with a keen and careful but favorable eye. "Capital composition; correct drawing; the color, tone, *chiaroscuro* excellent; but—but—it wants—hang it, it wants—that!" snapping his fingers; and, wanting "that," though it had everything else, it was worth nothing.

Again, Etty was appointed teacher of the students of the Royal Academy, having been preceded by a clever, talkative, scientific expounder of æsthetics, who delighted to tell the young men *how* everything was done, how to copy this, and how to express that. A student came up to the new master, "How should I do this, sir?" "Suppose you try." Another, "What does this mean, Mr. Etty?" "Suppose you look." "But I have looked." "Suppose you look again."

And they did try, and they did look, and looked again; and they saw and achieved what they never could have done had the how or the what (supposing this possible, which it is not, in full and highest meaning) been told them, or done for them; in the one case, sight and action were immediate, exact, intense, and secure; in the other, mediate, feeble, and lost as soon as gained.

NOTES.—**Opie,** John (b. 1761, d. 1807), was born in Wales, and was known as the "Cornish wonder." He became celebrated as a portrait painter, but afterwards devoted himself to historical subjects. He was professor of painting at the Royal Academy.

Reynolds. See note on page 379.

Etty, William (b. 1787, d. 1849), is considered one of the principal artists of the modern English school. His pictures are mainly historical.

The **Royal Academy** of Arts, in London, was founded in 1768. It is under the direction of forty artists of the first rank in their several professions, who have the title of "Royal Academicians." The admission to the Academy is free to all properly qualified students.

CXXI. THE NEW ENGLAND PASTOR.

Timothy Dwight, 1752–1817, was born at Northampton, Massachusetts. His mother was a daughter of the celebrated Jonathan Edwards. It is said that she taught her son the alphabet in one lesson, that he could read the Bible at four years of age, and that he studied Latin by himself at six. He graduated at Yale in 1769, returned as tutor in 1771, and continued six years. He was chaplain in a brigade under General Putnam for a time. In 1778 his father died, and for five years he supported his mother and a family of twelve children by farming, teaching and preaching. From 1783 to 1795 he was pastor at Greenfield, Connecticut. He was then chosen President of Yale College, and remained in office till he died. Dr. Dwight was a man of fine bodily presence, of extended learning, and untiring industry. His presidency of the college was highly successful. His patriotism was no less ardent and true than his piety. In his younger days he wrote considerably in verse. His poetry is not all of a very high order, but some pieces possess merit.

THE place, with east and western sides,
A wide and verdant street divides:
And here the houses faced the day,
And there the lawns in beauty lay.
There, turret-crowned, and central, stood
A neat and solemn house of God.
Across the way, beneath the shade
Two elms with sober silence spread,
The preacher lived. O'er all the place
His mansion cast a Sunday grace;

Dumb stillness sate the fields around;
His garden seemed a hallowed ground;
Swains ceased to laugh aloud, when near,
And schoolboys never sported there.

In the same mild and temperate zone,
Twice twenty years, his course had run,
His locks of flowing silver spread
A crown of glory o'er his head;
His face, the image of his mind,
With grave and furrowed wisdom shined;
Not cold; but glowing still, and bright;
Yet glowing with October light:
As evening blends, with beauteous ray,
Approaching night with shining day.

His Cure his thoughts engrossed alone:
For them his painful course was run:
To bless, to save, his only care;
To chill the guilty soul with fear;
To point the pathway to the skies,
And teach, and urge, and aid, to rise;
Where strait, and difficult to keep,
It climbs, and climbs, o'er Virtue's steep.

CXXII. DEATH OF ABSALOM.

DAVID numbered the people that were with him, and set captains of thousands and captains of hundreds over them. And David sent forth a third part of the people under the hand of Joab, and a third part under the hand of Abishai, the son of Zeruiah, Joab's brother, and a third part under the hand of Ittai, the Gittite.

And the king said unto the people, I will surely go forth with you myself also. But the people answered, thou shalt not go forth; for if we flee away, they will not care for us; neither if half of us die, will they care for us; but now thou art worth ten thousand of us; therefore now it is better that thou succor us out of the city. And the king said unto them, What seemeth you best, I will do.

And the king stood by the gate side, and all the people came out by hundreds and by thousands. And the king commanded Joab, and Abishai, and Ittai, saying, Deal gently for my sake with the young man, even with Absalom. And all the people heard when the king gave all the captains charge concerning Absalom.

So the people went out into the field against Israel; and the battle was in the wood of Ephraim; where the people of Israel were slain before the servants of David, and there was there a great slaughter that day of twenty thousand men. For the battle was there scattered over the face of all the country: and the wood devoured more people that day than the sword devoured.

And Absalom met the servants of David. And Absalom rode upon a mule, and the mule went under the thick boughs of a great oak, and his head caught hold of the oak, and he was taken up between the heaven and the earth; and the mule that was under him went away.

And a certain man saw it, and told Joab, and said, Behold, I saw Absalom hanged in an oak. And Joab said unto the man that told him, And, behold, thou sawest him, and why didst thou not smite him there to the ground? and I would have given thee ten shekels of silver and a girdle. And the man said unto Joab, Though I should receive a thousand shekels of silver in my hand, yet would I not put forth my hand against the king's son; for, in our hearing, the king charged thee, and Abishai, and Ittai, saying, Beware that none touch the young man Absalom. Otherwise, I should have wrought falsehood against mine own life; for there is no matter hid from the king, and thou thyself wouldst have set thyself against me.

Then said Joab, I may not tarry thus with thee. And he took three darts in his hand, and thrust them through the heart of Absalom, while he was yet alive in the midst of the oak. And ten young men that bare Joab's armor, compassed about and smote Absalom, and slew him. And Joab blew the trumpet, and the people returned from pursuing after Israel; for Joab held back the people.

And they took Absalom, and cast him into a great pit in the wood, and laid a very great heap of stones upon him; and all Israel fled, everyone to his tent. Now Absalom, in his lifetime, had taken and reared up for himself a pillar, which is in the king's dale; for he said, I have no son to keep my name in remembrance; and he called the pillar after his own name; and it is called unto this day, Absalom's Place.

Then said Ahimaaz, the son of Zadok, Let me now run, and bear the king tidings, how that the Lord hath avenged him of his enemies. And Joab said unto him, Thou shalt not bear tidings this day, but thou shalt bear tidings another day: but this day thou shalt bear no tidings, because the king's son is dead. Then said Joab to Cushi, Go, tell the king what thou hast seen. And Cushi bowed himself unto Joab, and ran.

Then said Ahimaaz the son of Zadok yet again to Joab, But howsoever, let me, I pray thee, also run after Cushi. And Joab said, Wherefore wilt thou run, my son, seeing that thou hast no tidings ready? But howsoever, said he, let me run. And he said unto him, Run. Then Ahimaaz ran by the way of the plain, and overran Cushi.

And David sat between the two gates; and the watchman went up to the roof over the gate unto the wall, and lifted up his eyes, and looked, and behold, a man running alone. And the watchman cried, and told the king. And the king said, If he be alone, there is tidings in his mouth. And he came apace, and drew near.

And the watchman saw another man running, and the watchman called unto the porter, and said, Behold, another man running alone. And the king said, He also bringeth tidings. And the watchman said, Methinketh the running of the foremost is like the running of Ahimaaz the son of Zadok. And the king said, He is a good man, and cometh with good tidings.

And Ahimaaz called, and said unto the king, All is well. And he fell down to the earth upon his face before the king, and said, Blessed be the Lord thy God, which hath delivered up the men that lifted up their hand against my lord the king. And the king said, Is the young man Absalom safe? And Ahimaaz answered, When Joab sent the king's servant, and me thy servant, I saw a great tumult, but I knew not what it was. And the king said unto him, Turn aside and stand here. And he turned aside, and stood still.

And behold, Cushi came; and Cushi said, Tidings my lord the king; for the Lord hath avenged thee this day of all them that rose up against thee. And the king said unto Cushi, Is the young man Absalom safe? And Cushi answered, The enemies of my lord the king, and all that rise against thee to do thee hurt, be as that young man is.

And the king was much moved, and went up to the chamber over the gate, and wept; and as he went, thus he said, O my son Absalom! my son, my son Absalom! would God I had died for thee, O Absalom, my son, my son!

—*II Samuel, Chap. xviii.*

CXXIII. ABRAHAM DAVENPORT.

'TWAS on a May day of the far old year
Seventeen hundred eighty, that there fell
Over the bloom and sweet life of the Spring,
Over the fresh earth and the heaven of noon,
A horror of great darkness, like the night
In day of which the Norland *sagas* tell,—
The Twilight of the Gods.

The low-hung sky
Was black with ominous clouds, save where its rim
Was fringed with a dull glow, like that which climbs
The crater's sides from the red hell below.
Birds ceased to sing, and all the barnyard fowls
Roosted; the cattle at the pasture bars
Lowed, and looked homeward; bats on leathern wings
Flitted abroad; the sounds of labor died;
Men prayed, and women wept; all ears grew sharp
To hear the doom blast of the trumpet shatter
The black sky, that the dreadful face of Christ
Might look from the rent clouds, not as he looked
A loving guest at Bethany, but stern
As Justice and inexorable Law.

Meanwhile in the old Statehouse, dim as ghosts,
Sat the lawgivers of Connecticut,
Trembling beneath their legislative robes.
"It is the Lord's Great Day! Let us adjourn,"
Some said; and then, as if with one accord,
All eyes were turned to Abraham Davenport.

He rose, slow-cleaving with his steady voice
The intolerable hush. "This well may be
The Day of Judgment which the world awaits;

But be it so or not, I only know
My present duty, and my Lord's command
To occupy till he come. So at the post
Where he hath set me in his providence,
I choose, for one, to meet him face to face,—
No faithless servant frightened from my task,
But ready when the Lord of the harvest calls;
And therefore, with all reverence, I would say,
Let God do his work, we will see to ours.
Bring in the candles." And they brought them in.

Then by the flaring lights the Speaker read,
Albeit with husky voice and shaking hands,
An act to amend an act to regulate
The shad and alewive fisheries. Whereupon,
Wisely and well spake Abraham Davenport,
Straight to the question, with no figures of speech
Save the ten Arab signs, yet not without
The shrewd, dry humor natural to the man:
His awe-struck colleagues listening all the while,
Between the pauses of his argument,
To hear the thunder of the wrath of God
Break from the hollow trumpet of the cloud.

And there he stands in memory to this day,
Erect, self-poised, a rugged face, half seen
Against the background of unnatural dark,
A witness to the ages as they pass,
That simple duty hath no place for fear.

—*Whittier.*

NOTE.—The "Dark Day," as it is known, occurred May 19th, 1780, and extended over all New England. The darkness came on about ten o'clock in the morning, and lasted with varying degrees of intensity until midnight of the next day. The cause of the phenomenon is unknown.

CXXIV. THE FALLS OF THE YOSEMITE.

Thomas Starr King, 1824-1863, was born in New York City. His father was a Universalist minister; and, in 1834, he settled in Charlestown, Massachusetts. The son was preparing to enter Harvard University, when the death of his father devolved upon him the support of his mother, and his collegiate course had to be given up. He spent several years as clerk and teacher, improving meanwhile all possible opportunities for study. In 1846 he was settled over the church to which his father had preached in Charlestown. Two years later, he was called to the Hollis Street Unitarian Church in Boston. Here his eloquence and active public spirit soon made him well known. He also gained much reputation as a public lecturer. In 1860 he left the East to take charge of the Unitarian church in San Francisco. During the remaining years of his life, he exercised much influence in the public affairs of California. He died suddenly, of diphtheria, in the midst of his brilliant career.

Mr. King was a great lover of nature. His "White Hills," describing the mountain scenery of New Hampshire, is the most complete book ever written concerning that interesting region.

The Yosemite valley, in California, is a pass about ten miles long. At its eastern extremity it leads into three narrower passes, each of which extends several miles, winding by the wildest paths into the heart of the Sierra Nevada chain of mountains. For seven miles of the main valley, which varies in width from three quarters of a mile to a mile and a half, the walls on either side are from two thousand to nearly five thousand feet above the road, and are nearly perpendicular. From these walls, rocky splinters a thousand feet in height start up, and every winter drop a few hundred tons of granite, to adorn the base of the rampart with picturesque ruin.

The valley is of such irregular width, and bends so much and often so abruptly, that there is a great variety and frequent surprise in the forms and combinations of the overhanging rocks as one rides along the bank of the stream. The patches of luxuriant meadow, with their dazzling green, and the grouping of the superb firs, two hundred feet high, that skirt them, and that shoot above the stout and graceful oaks and sycamores through which the horse path winds, are delightful rests of sweetness and beauty amid the threatening awfulness.

The Merced, which flows through the same pass, is a noble stream, a hundred feet wide and ten feet deep. It is formed chiefly of the streams that leap and rush through the narrower passes, and it is swollen, also, by the bounty of the marvelous waterfalls that pour down from the ramparts of the wider valley. The sublime poetry of Habakkuk is needed to describe the impression, and, perhaps, the geology, of these mighty fissures: "Thou didst cleave the earth with rivers."

At the foot of the breakneck declivity of nearly three thousand feet by which we reach the banks of the Merced, we are six miles from the hotel, and every rod of the ride awakens wonder, awe, and a solemn joy. As we approach the hotel, and turn toward the opposite bank of the river, what is that

> "Which ever sounds and shines,
> A pillar of white light upon the wall
> Of purple cliffs aloof descried"?

That, reader, is the highest waterfall in the world—the Yosemite cataract, nearly twenty-five hundred feet in its plunge, dashing from a break or depression in a cliff thirty-two hundred feet sheer.

A writer who visited this valley in September, calls the cataract a mere tape line of water dropped from the sky. Perhaps it is so, toward the close of the dry season; but as we saw it, the blended majesty and beauty of it, apart from the general sublimities of Yosemite gorge, would repay a journey of a thousand miles. There was no deficiency of water. It was a powerful stream, thirty-five feet broad, fresh from the Nevada, that made the plunge from the brow of the awful precipice.

At the first leap it clears fourteen hundred and ninety-seven feet; then it tumbles down a series of steep stairways four hundred and two feet, and then makes a jump to the meadows five hundred and eighteen feet more. But it is the upper and highest cataract that is most wonderful to the eye, as well as most musical. The cliff is so sheer that there is no break in the body of the water during the whole of its descent of more than a quarter of a mile. It pours in a curve from the summit, fifteen hundred feet, to the basin that hoards it but a moment for the cascades that follow.

And what endless complexities and opulence of beauty in the forms and motions of the cataract! It is comparatively narrow at the top of the precipice, although, as we said, the tide that pours over is thirty-five feet broad. But it widens as it descends, and curves a little on one side as it widens, so that it shapes itself, before it reaches its first bowl of granite, into the figure of a comet. More beautiful than the comet, however, we can see the substance of this watery loveliness ever renew itself and ever pour itself away.

"It mounts in spray the skies, and thence again
Returns in an unceasing shower, which round
With its unemptied cloud of gentle rain,
Is an eternal April to the ground,
Making it all one emerald;—how profound
The gulf! and how the giant element
From rock to rock leaps with delirious bound,
Crushing the cliffs."

The cataract seems to shoot out a thousand serpentine heads or knots of water, which wriggle down deliberately through the air and expend themselves in mist before half the descent is over. Then a new set burst from the body and sides of the fall, with the same fortune on the remaining distance; and thus the most charming fretwork of watery nodules, each trailing its vapory train for a hundred feet or more, is woven all over the cascade, which swings, now and then, thirty feet each way, on the mountain side, as if it were a pendulum of watery lace. Once in a while, too, the wind manages to get back of the fall, between it and the cliff, and then it will whirl it round and round for two or three hundred feet, as if to try the experiment of twisting it to wring it dry.

Of course I visited the foot of the lowest fall of the Yosemite, and looked up through the spray, five hundred feet, to its crown. And I tried to climb to the base of the first or highest cataract, but lost my way among the steep, sharp rocks, for there is only one line by which the cliff can be scaled. But no nearer view that I found or heard described, is comparable with the picture, from the hotel, of the comet curve of the upper cataract, fifteen hundred feet high, and the two falls immediately beneath it, in which the same water leaps to the level of the quiet Merced.

CXXV. A PSALM OF LIFE.

TELL me not, in mournful numbers,
  Life is but an empty dream!
For the soul is dead that slumbers,
  And things are not what they seem.

Life is real! Life is earnest!
  And the grave is not its goal;
Dust thou art, to dust returnest,
  Was not spoken of the soul.

Not enjoyment, and not sorrow,
  Is our destined end or way;
But to act, that each to-morrow
  Find us farther than to-day.

Art is long, and Time is fleeting,
And our hearts, though stout and brave,
Still, like muffled drums, are beating
Funeral marches to the grave.

In the world's broad field of battle,
In the bivouac of Life,
Be not like dumb, driven cattle!
Be a hero in the strife!

Trust no Future, howe'er pleasant!
Let the dead Past bury its dead!
Act—act in the living Present!
Heart within, and God o'erhead.

Lives of great men all remind us
We can make our lives sublime,
And, departing, leave behind us
Footprints on the sands of time;—

Footprints, that perhaps another,
Sailing o'er life's solemn main,
A forlorn and shipwrecked brother,
Seeing, shall take heart again.

Let us, then, be up and doing,
With a heart for any fate;
Still achieving, still pursuing,
Learn to labor and to wait.

—*Longfellow.*

CXXVI. FRANKLIN'S ENTRY INTO PHILADELPHIA.

Benjamin Franklin, 1706–1790, was born in Boston. He received little schooling, but being apprenticed to his brother, a printer, he acquired a taste for reading and study. In 1723, he went to Philadelphia, where he followed his chosen calling, and in time became the publisher of the "Pennsylvania Gazette" and the celebrated "Poor Richard's Almanac."

As a philosopher Franklin was rendered famous by his discovery of the identity of lightning with electricity. His career in public affairs may be briefly summarized as follows: In 1736 he was made Clerk of the Provincial Assembly; in 1737, deputy postmaster at Philadelphia; and in 1753, Postmaster general for British America. He was twice in England as the agent of certain colonies. After signing the Declaration of Independence, he was sent as Minister Plenipotentiary to France in 1776. On his return, in 1785, he was made "President of the Commonwealth of Pennsylvania," holding the office three years. He was also one of the framers of the Constitution of the United States.

As a writer Franklin commenced his career when only twelve years old by composing two ballads, which, however, he condemned as "wretched stuff." Franklin's letters and papers on electricity, afterwards enlarged by essays on various philosophical subjects, have been translated into Latin, French, Italian, and German. The most noted of his works, and the one from which the following extract is taken, is his "Autobiography." This book is "one of the half dozen most widely popular books ever printed," and has been published in nearly every written language. Franklin founded the American Philosophical Society, and established an institution which has since grown into the University of Pennsylvania. His life is a noble example of the results of industry and perseverance, and his death was the occasion of public mourning.

WALKING in the evening by the side of the river, a boat came by, which I found was going towards Philadelphia, with several people in her. They took me in, and, as there was no wind, we rowed all the way; and about midnight, not having yet seen the city, some of the company were confident we must have passed it, and would row no farther; the others knew not where we were; so we put toward the shore, got into a creek, landed near an old fence, with the rails of which we made a fire, the night being cold, in October, and there we remained till daylight.

Then one of the company knew the place to be Cooper's Creek, a little above Philadelphia, which we saw as soon as we got out of the creek, and arrived there about eight or nine o'clock on the Sunday morning, and landed at the Market Street wharf.

I have been the more particular in this description of my journey, and shall be so of my first entry into that city, that you may in your mind compare such unlikely beginnings with the figure I have since made there.

I was in my working dress, my best clothes being to come round by sea. I was dirty from my journey; my pockets were stuffed out with shirts and stockings, and I knew no soul nor where to look for lodging. I was fatigued with traveling, rowing, and want of rest; I was very hungry; and my whole stock of cash consisted of a Dutch dollar, and about a shilling in copper. The latter I gave the people of the boat for my passage, who at first refused it on account of my rowing; but I insisted on their taking it,—a man being sometimes more generous when he has but a little money than when he has plenty, perhaps through fear of being thought to have but little.

Then I walked up the street gazing about, till, near the market house, I met a boy with bread. I had made many a meal on bread, and, inquiring where he got it, I went immediately to the baker's he directed me to, in Second Street, and asked for biscuit, intending such as we had in Boston: but they, it seems, were not made in Philadelphia. Then I asked for a threepenny loaf, and was told they had none such. So not considering or knowing the difference of money, and the greater cheapness nor the names of his bread, I bade him give threepenny worth of any sort. He gave me, accordingly, three great puffy rolls. I was surprised at the quantity, but took it, and, having no room in my pockets, walked off with a roll under each arm, and eating the other.

Thus I went up Market Street as far as Fourth Street, passing by the door of Mr. Read, my future wife's father: when she, standing at the door, saw me, and thought I made, as I certainly did, a most awkward, ridiculous appearance. Then I turned and went down Chestnut Street and part of Walnut Street, eating my roll all the way, and, coming round, found myself again at Market Street wharf, near the boat I came in, to which I went for a draught of the river water; and, being filled with one of my rolls, gave the other two to a woman and her child that came down the river in the boat with us, and were waiting to go farther.

Thus refreshed, I walked again up the street, which by this time had many clean-dressed people in it, who were all walking the same way. I joined them, and thereby was led into the great meetinghouse of the Quakers, near the market. I sat down among them, and, after looking round awhile and hearing nothing said, being very drowsy through labor and want of rest the preceding night, I fell fast asleep, and continued so till the meeting broke up, when one was kind enough to rouse me. This was, therefore, the first house I was in, or slept in, in Philadelphia.

Walking down again toward the river, and looking in the faces of people, I met a young Quaker man, whose countenance I liked, and, accosting him, requested he would tell me where a stranger could get lodging. We were then near the sign of the Three Mariners. "Here," says he, "is one place that entertains strangers, but it is not a reputable house; if thee wilt walk with me, I'll show thee a better." He brought me to the Crooked Billet, in Water Street. Here I got a dinner; and, while I was eating it, several sly questions were asked me, as it seemed to be suspected from my youth and appearance that I might be some runaway. After dinner my sleepiness returned, and, being shown to a bed, I lay down without undressing, and slept till six in the evening; was called to supper, went to bed again very early, and slept soundly till next morning.

NOTE.—The river referred to is the Delaware. Franklin was on his way from Boston to Philadelphia, and had just walked from Amboy to Burlington, New Jersey, a distance of fifty miles.

CXXVII. LINES TO A WATERFOWL.

WHITHER 'midst falling dew,
While glow the heavens with the last steps of day,
Far, through their rosy depths, dost thou pursue
Thy solitary way?

Vainly the fowler's eye
Might mark thy distant flight to do thee wrong,
As, darkly painted on the crimson sky,
Thy figure floats along.

Seek'st thou the plashy brink
Of weedy lake, or marge of river wide,
Or where the rocky billows rise and sink
On the chafed ocean side?

There is a Power whose care
Teaches thy way along that pathless coast.
The desert and illimitable air,
Lone wandering, but not lost.

All day, thy wings have fanned,
At that far height, the cold, thin atmosphere,
Yet stoop not, weary, to the welcome land
Though the dark night is near.

And soon that toil shall end,
Soon shalt thou find a summer home, and rest,
And scream among thy fellows; reeds shall bend,
Soon, o'er thy sheltered nest.

Thou'rt gone; the abyss of heaven
Hath swallowed up thy form; yet, on my heart,
Deeply has sunk the lesson thou hast given,
And shall not soon depart.

He, who, from zone to zone,
Guides through the boundless sky thy certain flight,
In the long way that I must tread alone,
Will lead my steps aright.

—*Bryant.*

CXXVIII. GOLDSMITH AND ADDISON.

William Makepeace Thackeray, 1811-1863, was born in Calcutta, and is one of the most popular of English novelists, essayists, and humorists. While a boy, he removed from India to England, where he was educated at the Charterhouse in London, and at Cambridge. When twenty-one years of age, he came into possession of about £20,000. He rapidly dissipated his fortune, however, and was compelled to work for his living, first turning his attention to law and then to art, but finally choosing literature as his profession. He was for many years correspondent, under assumed names, of the "London Times," "The New Monthly Magazine," "Punch," and "Fraser's Magazine." His first novel under his own name, "Vanity Fair," appeared in monthly numbers during 1846-8, and is generally considered his finest production: although "Pendennis," "Henry Esmond," and "The Newcomes" are also much admired. His lectures on "English Humorists of the Eighteenth Century," from which the following selections are taken, were delivered in England first in 1851, and afterwards in America, which he visited in 1852 and again in 1855-6. During the latter visit, he first delivered his course of lectures on "The Four Georges," which were later repeated in England. At the close of 1859, Thackeray became editor of the "Cornhill Magazine," and made it one of the most successful serials ever published.

Thackeray has been charged with cynicism in his writings, but he was noted for his happy temper and genial disposition towards all who came in contact with him.

I. Goldsmith.

To be the most beloved of English writers, what a title that is for a man! A wild youth, wayward, but full of tenderness and affection, quits the country village where his boyhood has been passed in happy musing, in idle shelter, in fond longing to see the great world out of doors, and achieve name and fortune—and after years of dire struggle, and neglect, and poverty, his heart turning back as fondly to his native place as it had longed eagerly for change when sheltered there, he writes a book and a poem, full of the recollections and feelings of home; he paints the friends and scenes of his youth, and peoples Auburn and Wakefield with the remembrances of Lissoy.

Wander he must, but he carries away a home relic with him, and dies with it on his breast. His nature is truant; in repose it longs for change: as on the journey it looks back for friends and quiet. He passes to-day in building an air castle for to-morrow, or in writing yesterday's elegy; and he would fly away this hour, but that a cage, necessity, keeps him. What is the charm of his verse, of his style, and humor? His sweet regrets, his delicate compassion, his soft smile, his tremulous sympathy, the weakness which he owns?

Your love for him is half pity. You come hot and tired from the day's battle, and this sweet minstrel sings to you. Who could harm the kind vagrant harper? Whom did he ever hurt? He carries no weapon, save the harp on which he plays to you, and with which he delights great and humble, young and old, the captains in the tents, or the soldiers round the fire, or the women and children in the villages, at whose porches he stops and sings his simple songs of love and beauty. With that sweet story of "The Vicar of Wakefield" he has found entry into every castle and every hamlet in Europe. Not one of us, however busy or hard, but once or twice in our lives has passed an evening with him, and undergone the charm of his delightful music.

II. Addison.

We love him for his vanities as much as his virtues. What is ridiculous is delightful in him; we are so fond of him because we laugh at him so. And out of that laughter, and out of that sweet weakness, and out of those harmless eccentricities and follies, and out of that touched brain, and out of that honest manhood and simplicity — we get a result of happiness, goodness, tenderness, pity, piety; such as doctors and divines but seldom have the fortune to inspire. And why not? Is the glory of Heaven to be sung only by gentlemen in black coats?

When this man looks from the world, whose weaknesses he describes so benevolently, up to the Heaven which shines over us all, I can hardly fancy a human face lighted up with a more serene rapture; a human intellect thrilling with a purer love and adoration than Joseph Addison's. Listen to him: from your childhood you have known the verses; but who can hear their sacred music without love and awe?

"Soon as the evening shades prevail,
The moon takes up the wondrous tale,
And nightly to the listening earth
Repeats the story of her birth;
And all the stars that round her burn,
And all the planets in their turn,
Confirm the tidings as they roll,
And spread the truth from pole to pole.

"What though, in solemn silence, all
Move round this dark terrestrial ball;
What though no real voice nor sound
Among their radiant orbs be found;
In reason's ear they all rejoice,
And utter forth a glorious voice,
Forever singing, as they shine,
The Hand that made us is divine."

It seems to me those verses shine like the stars. They shine out of a great, deep calm. When he turns to Heaven, a Sabbath comes over that man's mind; and his face lights up from it with a glory of thanks and prayers. His sense of religion stirs through his whole being. In the fields, in the town; looking at the birds in the trees; at the children in the streets; in the morning or in the moonlight; over his books in his own room; in a happy party at a country merrymaking or a town assembly, good will and peace to God's creatures, and love and awe of Him who made them, fill his pure heart and shine from his kind face. If Swift's life was the most wretched, I think Addison's was one of the most enviable. A life prosperous and beautiful—a calm death—an immense fame and affection afterwards for his happy and spotless name.

Notes.—**Goldsmith** (see biographical notice, page 215) founded his descriptions of **Auburn** in the poem of "The Deserted Village," and of **Wakefield**, in "The Vicar of Wakefield," on recollections of his early home at **Lissoy**, Ireland.

Addison. See biographical notice, page 295. The quotation is from a "Letter from Italy to Charles Lord Halifax."

Swift, Jonathan (b. 1667, d. 1745), the celebrated Irish satirist and poet, was a misanthrope. His disposition made his life miserable in the extreme, and he finally became insane.

CXXIX. IMMORTALITY OF THE SOUL.

Scene—Cato, *alone, sitting in a thoughtful posture;—in his hand, Plato's book on the immortality of the soul; a drawn sword on the table by him.*

Cato. It must be so. Plato, thou reasonest well!
Else whence this pleasing hope, this fond desire,
This longing after immortality?
Or whence this secret dread, and inward horror,
Of falling into naught? Why shrinks the soul
Back on herself, and startles at destruction?
'Tis the divinity that stirs within us;
'Tis heaven itself that points out an hereafter,
And intimates eternity to man.

Eternity! thou pleasing, dreadful thought!
Through what variety of untried being,
Through what new scenes and changes must we pass?
The wide, unbounded prospect lies before me:
But shadows, clouds, and darkness rest upon it.
Here will I hold. If there's a Power above us,
(And that there is, all Nature cries aloud
Through all her works) he must delight in virtue;
And that which he delights in must be happy.
But when?—or where?—This world was made for Cæsar.
I'm weary of conjectures—this must end them.
(*Seizes the sword.*)
Thus am I doubly armed: my death and life,
My bane and antidote are both before me.
This in a moment brings me to an end;
But this informs me I shall never die.
The soul, secured in her existence, smiles
At the drawn dagger and defies its point.
The stars shall fade away, the sun himself
Grow dim with age, and Nature sink in years;
But thou shalt flourish in immortal youth,
Unhurt amidst the war of elements,
The wrecks of matter, and the crush of worlds.

—*Addison.*

NOTES.—The above selection is Cato's soliloquy just before committing suicide. It is from the tragedy of "Cato."

Cato, Marcus Porcius, (b. 95, d. 46 B.C.) was a Roman general, statesman, and philosopher. He was exceptionally honest and conscientious, and strongly opposed Cæsar and Pompey in their attempts to seize the state. When Utica, the last African city to resist Cæsar, finally yielded, Cato committed suicide.

Plato (b. 429, d. about 348 B.C.) was a celebrated Greek philosopher. His writings are all in the form of dialogues, and have been preserved in a wonderfully perfect state.

CXXX. CHARACTER OF WASHINGTON.

Jared Sparks, 1789-1866, was born at Willington, Connecticut, and graduated at Harvard in 1815. He was tutor in the University for two years, and in 1819 was ordained pastor of the Unitarian Church in Baltimore. In 1823 he returned to Boston, purchased the "North American Review," and was its sole editor for seven years. From 1839 to 1849 he was Professor in Harvard, and for the next three years was President of the University. Mr. Sparks has written extensively on American history and biography, including the lives of Washington and Franklin. He collected the materials for his biographies with great care, and wrought them up with much skill.

THE person of Washington was commanding, graceful, and fitly proportioned; his stature six feet, his chest broad and full, his limbs long and somewhat slender, but well-shaped and muscular. His features were regular and symmetrical, his eyes of a light blue color, and his whole countenance, in its quiet state, was grave, placid, and benignant. When alone, or not engaged in conversation, he appeared sedate and thoughtful; but when his attention was excited, his eye kindled quickly, and his face beamed with animation and intelligence.

He was not fluent in speech, but what he said was apposite, and listened to with the more interest as being known to come from the heart. He seldom attempted sallies of wit or humor, but no man received more pleasure from an exhibition of them by others; and, although contented in seclusion, he sought his chief happiness in society, and participated with delight in all its rational and innocent amusements. Without austerity on the one hand, or an appearance of condescending familiarity on the other, he was affable, courteous, and cheerful; but it has often been remarked that there was a dignity in his person and manner not easy to be defined, which impressed everyone that saw him for the first time with an instinctive deference and awe. This may have arisen, in part, from a conviction of his superiority, as well as from the effect produced by his external form and deportment.

The character of his mind was unfolded in the public and private acts of his life; and the proofs of his greatness are seen almost as much in the one as the other. The same qualities which raised him to the ascendency he possessed over the will of a nation, as the commander of armies and chief magistrate, caused him to be loved and respected as an individual. Wisdom, judgment, prudence, and firmness were his predominant traits. No man ever saw more clearly the relative importance of things and actions, or divested himself more entirely of the bias of personal interest, partiality, and prejudice, in discriminating between the true and the false, the right and the wrong, in all questions and subjects that were presented to him. He deliberated slowly, but decided surely; and when his decision was once formed he seldom reversed it, and never relaxed from the execution of a measure till it was completed. Courage, physical and moral, was a part of his nature; and, whether in battle, or in the midst of popular excitement, he was fearless of danger, and regardless of consequences to himself.

His ambition was of that noble kind which aims to excel in whatever it undertakes, and to acquire a power over the hearts of men by promoting their happiness and winning their affections. Sensitive to the approbation of others, and solicitous to deserve it, he made no concessions to gain their applause, either by flattering their vanity or yielding to their caprices. Cautious without timidity, bold without rashness, cool in counsel, deliberate but firm in action, clear in foresight, patient under reverses, steady, persevering, and self-possessed, he met and conquered every obstacle that obstructed his path to honor, renown and success. More confident in the uprightness of his intention than in his resources, he sought knowledge and advice from other men. He chose his counselors with unerring sagacity; and his quick perception of the soundness of an opinion, and of the strong points in an argument, enabled him to draw to his aid the best fruits of their talents, and the light of their collected wisdom.

His moral qualities were in perfect harmony with those of his intellect. Duty was the ruling principle of his conduct; and the rare endowments of his understanding were not more constantly tasked to devise the best methods of effecting an object, than they were to guard the sanctity of conscience. No instance can be adduced in which he was actuated by a sinister motive or endeavored to attain an end by unworthy means. Truth, integrity, and justice were deeply rooted in his mind; and nothing could rouse his indignation so soon, or so utterly destroy his confidence, as the discovery of the want of these virtues in anyone whom he had trusted. Weaknesses, follies, indiscretions he could forgive; but subterfuge and dishonesty he never forgot, rarely pardoned.

He was candid and sincere, true to his friends, and faithful to all; neither practicing dissimulation, descending to artifice, nor holding out expectations which he did not intend should be realized. His passions were strong, and sometimes they broke out with vehemence: but he had the power of checking them in an instant. Perhaps self-control was the most remarkable trait of his character. It was, in part, the effect of discipline; yet he seems by nature to have possessed this power in a degree which has been denied to other men.

A Christian in faith and practice, he was habitually devout. His reverence for religion is seen in his example, his public communications, and his private writings. He uniformly ascribed his successes to the beneficent agency of the Supreme Being. Charitable and humane, he was liberal to the poor, and kind to those in distress. As a husband, son, and brother, he was tender and affectionate. Without vanity, ostentation, or pride, he never spoke of himself or his actions unless required by circumstances which concerned the public interests.

As he was free from envy, so he had the good fortune to escape the envy of others by standing on an elevation which none could hope to attain. If he had one passion more strong than another it was love of his country. The purity and ardor of his patriotism were commensurate with the greatness of its object. Love of country in him was invested with the sacred obligation of a duty; and from the faithful discharge of this duty he never swerved for a moment, either in thought or deed, through the whole period of his eventful career.

Such are some of the traits in the character of Washington, which have acquired for him the love and veneration of mankind. If they are not marked with the brilliancy, extravagance, and eccentricity, which, in other men, have excited the astonishment of the world, so neither are they tarnished by the follies, nor disgraced by the crimes of those men. It is the happy combination of rare talents and qualities, the harmonious union of the intellectual and moral powers, rather than the dazzling splendor of any one trait, which constitute the grandeur of his character. If the title of great man ought to be reserved for him who can not be charged with an indiscretion or a vice; who spent his life in establishing the independence, the glory, and durable prosperity of his country; who succeeded in all that he undertook; and whose successes were never won at the expense of honor, justice, integrity, or by the sacrifice of a single principle,—this title will not be denied to Washington.

How sweetly on the ear such echoes sound!
While the mere victors may appall or stun
The servile and the vain, such names will be
A watchword till the Future shall be free.

—*Byron.*

CXXXI. EULOGY ON WASHINGTON.

General Henry Lee, 1756-1818, a member of the celebrated Lee family of Virginia, was born in Westmoreland County in that state, and died on Cumberland Island, Georgia. He graduated at Princeton in his eighteenth year. In 1777 he marched with a regiment of cavalry to join the patriot army, and served with fidelity and success till the close of the war. He was noted for his bravery, skill, and celerity, and received the nickname of "Light-horse Harry." He was a great favorite with both General Greene and General Washington. In 1786 Virginia appointed him one of her delegates to Congress; he also took an active part in favor of the adoption of the constitution in the Virginia Convention of 1788. On the breaking out of the "Whisky Rebellion" in Pennsylvania, in 1794, the President sent General Lee with an army to suppress the disturbance. The insurgents submitted without resistance. In 1799 he was again a member of Congress; and, on the death of Washington, that body appointed him to pronounce a eulogy upon the life and character of the great and good man. The following extract contains the closing part of the oration.

Who is there that has forgotten the vales of Brandywine, the fields of Germantown, or the plains of Monmouth? Everywhere present, wants of every kind obstructing, numerous and valiant armies encountering, himself a host, he assuaged our sufferings, limited our privations, and upheld our tottering Republic. Shall I display to you the spread of the fire of his soul by rehearsing the praises of the hero of Saratoga, and his much-loved compeer of the Carolinas? No; our Washington wears not borrowed glory. To Gates—to Greene, he gave without reserve the applause due to their eminent merit; and long may the chiefs of Saratoga and of Eutaw receive the grateful respect of a grateful people.

Moving in his own orbit, he imparted heat and light to his most distant satellites; and, combining the physical and moral force of all within his sphere, with irresistible weight he took his course, commiserating folly, disdaining vice, dismaying treason, and invigorating despondency; until the auspicious hour arrived, when, united with the intrepid forces of a potent and magnanimous ally, he brought to submission Cornwallis, since the conqueror of India; thus finishing his long career of military glory with a luster corresponding to his great name, and in this his last act of war, affixing the seal of fate to our nation's birth.

First in war, first in peace, and first in the hearts of his countrymen, he was second to none in humble and endearing scenes of private life. Pious, just, humane, temperate, sincere, uniform, dignified, and commanding, his example was edifying to all around him, as were the effects of that example lasting.

To his equals, he was condescending; to his inferiors, kind; and to the dear object of his affections, exemplarily tender. Correct throughout, vice shuddered in his presence, and virtue always felt his fostering hand; the purity of his private character gave effulgence to his public virtues.

His last scene comported with the whole tenor of his life. Although in extreme pain, not a sigh, not a groan, escaped him; and with undisturbed serenity he closed his well-spent life. Such was the man America has lost! Such was the man for whom our nation mourns!

NOTES.—At **Brandywine** Creek, in Pennsylvania, 18,000 British, under Howe, defeated 13,000 Americans under Washington.

Germantown, near Philadelphia, was the scene of an American defeat by the British, the same generals commanding as at Brandywine.

The battle of **Monmouth**, in New Jersey, resulted in victory for the Americans.

The **hero of Saratoga** was General Gates, who there compelled the surrender of General Burgoyne.

At **Eutaw** Springs, General Greene defeated a superior force of British.

Cornwallis, Charles, second earl and first marquis (b. 1738, d. 1805), surrendered his forces to a combined American and French army and French fleet at Yorktown, in 1781, virtually ending the war.

CXXXII. THE SOLITARY REAPER.

William Wordsworth, 1770-1850, the founder of the "Lake School" of poets, was born at Cockermouth, Cumberland, England. From his boyhood he was a great lover and student of nature, and it is to his beautiful descriptions of landscape, largely, that he owes his fame. He was a graduate of Cambridge University, and while there commenced the study of Chaucer, Spenser, Milton, and Shakespeare, as models for his own writings. Two legacies having been bequeathed him, Wordsworth determined to make poetry the aim of his life, and in 1795 located at Racedown with his sister Dorothy, where he commenced the tragedy of "The Borderers." A visit from Coleridge at this period made the two poets friends for life. In 1802 Wordsworth married Miss Mary Hutchinson, and in 1813 he settled at Rydal Mount, on Lake Windermere, where he passed the remainder of his life.

Wordsworth's poetry is remarkable for its extreme simplicity of language. At first his efforts were almost universally ridiculed, and in 1819 his entire income from literary work had not amounted to £140. In 1830 his merit began to be recognized; in 1839 Oxford University conferred upon him the degree of D. C. L.; and in 1843 he was made poet laureate.

"The Excursion" is by far the most beautiful and the most important of Wordsworth's productions. "Salisbury Plain," "The White Doe of Rylstone," "Yarrow Revisited," and many of his sonnets and minor poems are also much admired.

Behold her, single in the field,
Yon solitary Highland lass!
Reaping and singing by herself;
Stop here, or gently pass!
Alone she cuts and binds the grain,
And sings a melancholy strain;
Oh listen! for the vale profound
Is overflowing with the sound.

No nightingale did ever chant
More welcome notes to weary bands
Of travelers in some shady haunt,
Among Arabian sands:
A voice so thrilling ne'er was heard
In springtime from the cuckoo bird,
Breaking the silence of the seas
Among the farthest Hebrides.

Will no one tell me what she sings?—
Perhaps the plaintive numbers flow
For old, unhappy, far-off things,
And battles long ago:
Or is it some more humble lay,
Familiar matter of to-day?
Some natural sorrow, loss, or pain,
That has been, and may be again?

Whate'er the theme, the maiden sang
As if her song could have no ending;
I saw her singing at her work,
And o'er the sickle bending;—
I listened motionless and still;
And, as I mounted up the hill,
The music in my heart I bore,
Long after it was heard no more.

CXXXIII. VALUE OF THE PRESENT.

Ralph Waldo Emerson, 1803-1882, the celebrated essavist and philosopher, was born in Boston. His father was a Unitarian minister, and the son, after graduating at Harvard University, entered the ministry also, and took charge of a Unitarian congregation in Boston. His peculiar ideas on religious topics soon caused him to retire from the ministry, and he then devoted himself to literature. As a lecturer, Emerson attained a wide reputation, both in this country and in England, and he is considered as one of the most independent and original thinkers of the age. His style is brief and pithy, dazzling by its wit, but sometimes paradoxical. He wrote a few poems, but they are not generally admired, being didactic in style, bare, and obscure. Among his best known publications are his volume "Nature," and his lectures, "The Mind and Manners of the Nineteenth Century," "The Superlative in Manners and Literature," "English Character and Manners," and "The Conduct of Life." In 1850 appeared "Representative Men," embracing sketches of Plato, Swedenborg, Montaigne, Shakespeare, Napoleon, and Goethe.

Such are the days,—the earth is the cup, the sky is the cover, of the immense bounty of nature which is offered us for our daily aliment; but what a force of *illusion* begins life with us, and attends us to the end! We are coaxed, flattered, and duped, from morn to eve, from birth to death; and where is the old eye that ever saw through the deception? The Hindoos represent Maia, the illusory energy of Vishnu, as one of his principal attributes. As if, in this gale of warring elements, which life is, it was necessary to bind souls to human life as mariners in a tempest lash themselves to the mast and bulwarks of a ship, and Nature employed certain illusions as her ties and straps,—a rattle, a doll, an apple, for a child; skates, a river, a boat, a horse, a gun, for the growing boy;—and I will not begin to name those of the youth and adult, for they are numberless. Seldom and slowly the mask falls, and the pupil is permitted to see that all is one stuff, cooked and painted under many counterfeit appearances. Hume's doctrine was that the circumstances vary, the amount of happiness does not; that the beggar cracking fleas in the sunshine under a hedge, and the duke rolling by in his chariot, the girl equipped for her first ball, and the orator returning triumphant from the debate, had different means, but the same quantity of pleasant excitement.

This element of illusion lends all its force to hide the values of present time. Who is he that does not always find himself doing something less than his best task? "What are you doing?" "Oh, nothing; I have been doing thus, or I shall do so or so, but now I am only—" Ah! poor dupe, will you never slip out of the web of the master juggler?—never learn that, as soon as the irrecoverable years have woven their blue glory between to-day and us, these passing hours shall glitter and draw us, as the wildest romance and the homes of beauty and poetry? How difficult to deal erect with them! The events they bring, their trade, entertainments, and gossip, their urgent work, all throw dust in the eyes and distract attention. He is a strong man who can look them in the eye, see through this juggle, feel their identity, and keep his own; who can know surely that one will be like another to the end of the world, nor permit love, or death, or politics, or money, war, or pleasure, to draw him from his task.

The world is always equal to itself, and every man in moments of deeper thought is apprised that he is repeating the experiences of the people in the streets of Thebes or Byzantium. An everlasting Now reigns in nature, which hangs the same roses on our bushes which charmed the Roman and the Chaldean in their hanging gardens. "To what end, then," he asks, "should I study languages, and traverse countries, to learn so simple truths?"

History of ancient art, excavated cities, recovery of books and inscriptions,—yes, the works were beautiful, and the history worth knowing; and academies convene to settle the claims of the old schools. What journeys and measurements,—Niebuhr and Müller and Layard,—to identify the plain of Troy and Nimroud town! And your homage to Dante costs you so much sailing; and to ascertain the discoverers of America needs as much voyaging as the discovery cost. Poor child! that flexible clay of which these old brothers molded their admirable symbols was not Persian, nor Memphian, nor Teutonic, nor local at all, but was common lime and silex and water, and sunlight, the heat of the blood, and the heaving of the lungs; it was that clay which thou heldest but now in thy foolish hands, and threwest away to go and seek in vain in sepulchers, mummy pits, and old bookshops of Asia Minor, Egypt, and England. It was the deep to-day which all men scorn; the rich poverty, which men hate; the populous, all-loving solitude, which men quit for the tattle of towns. *He* lurks, *he* hides,—*he* who is success, reality, joy, and power. One of the illusions is that the present hour is not the critical, decisive hour. Write it on your heart that every day is the best day in the year. No man has learned anything rightly, until he knows that every day is Doomsday. 'Tis the old secret of the gods that they come in low disguises. 'Tis the vulgar great who come dizened with gold and jewels. Real kings hide away their crowns in their wardrobes, and affect a plain and poor exterior. In the Norse legend of our ancestors, Odin dwells in a fisher's hut, and patches a boat. In the Hindoo legends, Hari dwells a peasant among peasants. In the Greek legend, Apollo lodges with the shepherds of Admetus; and Jove liked to rusticate among the poor Ethiopians. So, in our history, Jesus is born in a barn, and his twelve peers are fishermen. 'T is the very principle of science that Nature shows herself best in leasts; 't was the maxim of Aristotle and Lucretius; and, in modern times, of Swedenborg and of Hahnemann. The order of changes in the egg determines the age of fossil strata. So it was the rule of our poets, in the legends of fairy lore, that the fairies largest in power were the least in size.

In the Christian graces, humility stands highest of all, in the form of the Madonna; and in life, this is the secret of the wise. We owe to genius always the same debt, of lifting the curtain from the common, and showing us that divinities are sitting disguised in the seeming gang of gypsies and peddlers. In daily life, what distinguishes the master is the using those materials he has, instead of looking about for what are more renowned, or what others have used well. "A general," said Bonaparte, "always has troops enough, if he only knows how to employ those he has, and bivouacs with them." Do not refuse the employment which the hour brings you, for one more ambitious. The highest heaven of wisdom is alike near from every point, and thou must find it, if at all, by methods native to thyself alone.

NOTES.—The Brahmanic religion teaches a Trinity, of which **Vishnu** is the savior of mankind.

Thebes, the ancient capital of Upper Egypt, was at its most flourishing period about 1500 B. C. **Byzantium** was an important Greek city during the second and third centuries B. C.

Niebuhr (b. 1776, d. 1831), **Müller** (b. 1797, d. 1840), and **Layard** (b. 1817, d. 1894), are celebrated archæologists. The first two were Germans, and the last an Englishman.

CXXXIV. HAPPINESS.

Alexander Pope, 1688–1744, was the shining literary light of the so-called Augustan reign of Queen Anne, the poetry of which was distinguished by the highest degree of polish and elegance. Pope was the son of a retired linen draper, who lived in a pleasant country house near the Windsor Forest. He was so badly deformed that his life was "one long disease;" he was remarkably precocious, and had a most intelligent face, with great, flaming, tender eyes. In disposition Pope was the reverse of admirable. He was extremely sensitive, petulant, and supercilious; fierce and even coarse in his attacks on opponents; boastful of his self-acquired wealth and of his intimacy with the nobility. The great redeeming feature of his character was his tender devotion to his aged parents.

As a poet, however, Pope challenges the highest admiration. At the age of sixteen he commenced his "Pastorals," and when only twenty-one published his "Essay on Criticism," pronounced "the finest piece of argumentative and reasoning poetry in the English language." His reputation was now firmly established, and his literary activity ceased only at his death; although, during the latter portion of his life, he was so weak physically that he was unable to dress himself or even to rise from bed without assistance. Pope's great admiration was Dryden, whose style he studied and copied. He lacks the latter's strength, but in elegance and polish he remains unequaled.

Pope's most remarkable work is "The Rape of the Lock;" his greatest, the translation into English verse of Homer's "Iliad" and "Odyssey." His "Epistle of Eloisa to Abelard," "The Dunciad," and the "Essay on Man" are also famous productions. He published an edition of "Shakespeare," which was awaited with great curiosity, and received with equal disappointment. During the three years following its appearance, he united with Swift and Arbuthnot in writing the "Miscellanies," an extensive satire on the abuses of learning and the extravagances of philosophy. His "Epistles," addressed to various distinguished men, and covering a period of four years, were copied after those of Horace; they were marked by great clearness, neatness of diction, and good sense, and by Pope's usual elegance and grace. His "Imitations of Horace" was left unfinished at his death.

The following selection is an extract from the "Essay on Man:"

Oh, sons of earth! attempt ye still to rise,
By mountains piled on mountains, to the skies?
Heaven still with laughter the vain toil surveys,
And buries madmen in the heaps they raise.
Know all the good that individuals find,
Or God and nature meant to mere mankind.
Reason's whole pleasure, all the joys of sense,
Lie in three words,—health, peace, and competence.

But health consists with temperance alone;
And peace, O virtue! peace is all thy own.
The good or bad the gifts of fortune gain;
But these less taste them as they worse obtain.
Say, in pursuit of profit or delight,
Who risk the most, that take wrong means or right?
Of vice or virtue, whether blest or curst,
Which meets contempt, or which compassion first?

Count all th' advantage prosperous vice attains,
'T is but what virtue flies from and disdains:
And grant the bad what happiness they would,
One they must want, which is, to pass for good.
Oh, blind to truth, and God's whole scheme below,
Who fancy bliss to vice, to virtue woe!
Who sees and follows that great scheme the best,
Best knows the blessing, and will most be blest.

But fools the good alone unhappy call,
For ills or accidents that chance to all.
Think we, like some weak prince, the Eternal Cause,
Prone for his favorites to reverse his laws?
Shall burning Ætna, if a sage requires,
Forget to thunder, and recall her fires?
When the loose mountain trembles from on high,
Shall gravitation cease, if you go by?

"But sometimes virtue starves while vice is fed."
What, then? Is the reward of virtue bread?
That, vice may merit, 't is the price of toil;
The knave deserves it when he tills the soil,
The knave deserves it when he tempts the main,
Where folly fights for kings or dives for gain.
Honor and shame from no condition rise;
Act well your part, there all the honor lies.

Worth makes the man, and want of it the fellow;
The rest is all but leather or prunella.
A wit 's a feather, and a chief a rod,
An honest man 's the noblest work of God.
One self-approving hour whole years outweighs
Of stupid starers, and of loud huzzas.

Know then this truth (enough for man to know),
"Virtue alone is happiness below."
The only point where human bliss stands still,
And tastes the good without the fall to ill;
Where only merit constant pay receives,
Is blest in what it takes and what it gives.

CXXXV. MARION.

William Gilmore Simms, 1806-1870, one of the most versatile, prolific, and popular of American authors, was born at Charleston, South Carolina. His family was poor, and his means of education were limited, yet he managed to prepare himself for the bar, to which he was admitted when twenty-one years of age. The law proving uncongenial, he abandoned it, and in 1828 became editor of the "Charleston City Gazette." From this time till his death his literary activity was unceasing, and his writings were so numerous that it is possible only to group them under their various heads. They comprise Biography; History; Historical Romance, both Foreign and Domestic, the latter being further divided into Colonial, Revolutionary, and Border Romances; Pure Romance; The Drama; Poetry; and Criticism; besides miscellaneous books and pamphlets.

In the midst of this remarkable literary activity, Mr. Simms still found time to devote to the affairs of state, being for several years a member of the South Carolina Legislature. He was also a lecturer, and was connected editorially with several magazines. Most of his time was spent at his summer house in Charleston, and at his winter residence, "Woodlands," on a plantation at Midway, S. C.

The following selection is from "The Life and Times of Francis Marion."

Art had done little to increase the comforts or the securities of his fortress. It was one, complete to his hands, from those of nature — such an one as must have delighted the generous English outlaw of Sherwood Forest; insulated by deep ravines and rivers, a dense forest of mighty trees, and interminable undergrowth. The vine and brier guarded his passes. The laurel and the shrub, the vine and sweet-scented jessamine roofed his dwelling, and clambered up between his closed eyelids and the stars. Obstructions scarcely penetrable by any foe, crowded the pathways to his tent; and no footstep not practiced in the secret, and to "the manner born," might pass unchallenged to his midnight rest. The swamp was his moat; his bulwarks were the deep ravines, which, watched by sleepless rifles, were quite as impregnable as the castles on the Rhine. Here, in the possession of his fortress, the partisan slept secure.

His movements were marked by equal promptitude and wariness. He suffered no risks from a neglect of proper precaution. His habits of circumspection and resolve ran together in happy unison. His plans, carefully considered beforehand, were always timed with the happiest reference to the condition and feelings of his men. To prepare that condition, and to train those feelings, were the chief employment of his repose. He knew his game, and how it should be played, before a step was taken or a weapon drawn.

When he himself or any of his parties left the island upon an expedition, they advanced along no beaten paths. They made them as they went. He had the Indian faculty in perfection, of gathering his course from the sun, from the stars, from the bark and the tops of trees, and such other natural guides as the woodman acquires only through long and watchful experience.

Many of the trails thus opened by him, upon these expeditions, are now the ordinary avenues of the country. On starting, he almost invariably struck into the woods, and seeking the heads of the larger water courses, crossed them at their first and small beginnings. He destroyed the bridges where he could. He preferred fords. The former not only facilitated the progress of less fearless enemies, but apprised them of his own approach. If speed was essential, a more direct but not less cautious route was pursued.

He intrusted his schemes to nobody, not even his most confidential officers. He consulted with them respectfully, heard them patiently, weighed their suggestions, and silently approached his conclusions. They knew his determinations only from his actions. He left no track behind him, if it were possible to avoid it. He was often vainly hunted after by his own detachments. He was more apt at finding them than they him. His scouts were taught a peculiar and shrill whistle, which, at night, could be heard at a most astonishing distance. We are reminded of a signal of Roderick Dhu: —

> "He whistled shrill,
> And he was answered from the hill;
> Wild as the scream of the curlew,
> From crag to crag the signal flew."

His expeditions were frequently long, and his men, hurrying forth without due preparation, not unfrequently suffered much privation from want of food. To guard against this danger, it was their habit to watch his cook. If they saw him unusually busied in preparing supplies of the rude, portable food which it was Marion's custom to carry on such occasions, they knew what was before them, and provided themselves accordingly. In no other way could they arrive at their general's intentions. His favorite time for moving was with the setting sun, and then it was known that the march would continue all night.

His men were badly clothed in homespun,— a light wear which afforded little warmth. They slept in the open air, and frequently without a blanket. Their ordinary food consisted of sweet potatoes, garnished, on fortunate occasions, with lean beef. Their swords, unless taken from the enemy, were made out of mill saws, roughly manufactured by a forest blacksmith.

His scouts were out in all directions, and at all hours. They did the double duty of patrol and spies. They hovered about the posts of the enemy, crouching in the thicket, or darting along the plain, picking up prisoners, and information, and spoils together. They cut off stragglers, encountered patrols of the foe, and arrested his supplies on the way to the garrison. Sometimes the single scout, buried in the thick tops of the tree, looked down upon the march of his legions, or hung, perched over the hostile encampment, till it slept; then slipping down, stole through the silent host, carrying off a drowsy sentinel, or a favorite charger, upon which the daring spy flourished conspicuous among his less fortunate companions.

NOTES.—The **outlaw** of Sherwood Forest was Robin Hood. **Roderick Dhu** is a character in Sir Walter Scott's poem, "The Lady of the Lake," from which the quotation is taken.

CXXXVI. A COMMON THOUGHT.

Henry Timrod, 1829-1867, was born at Charleston, South Carolina. He inherited his father's literary taste and ability, and had the advantages of a liberal education. He entered the University of Georgia before he was seventeen years of age, and while there commenced his career as a poet. Poverty and ill health compelled him to leave the university without taking a degree; he then commenced the study of law, and for ten years taught in various private families. At the outbreak of the war, in 1860, he warmly espoused the Southern cause, and wrote many stirring war lyrics. In 1863 he joined the Army of the West, as correspondent of the Charleston "Mercury," and in 1864 he became editor of the "South Carolinian," published first at Columbia and later at Charleston. He also served for a time as assistant secretary to Governor Orr. The advance of Sherman's army reduced him to poverty, and he was compelled to the greatest drudgery in order to earn a bare living. His health soon broke down, and he died of hemorrhage of the lungs. The following little poem seems, almost, to have been written under a presentiment, so accurately does it describe the closing incidents of the poet's life.

The first volume of Timrod's poems appeared in 1860. A later edition, with a memoir of the author, was published in New York in 1873.

SOMEWHERE on this earthly planet
In the dust of flowers that be,
In the dewdrop, in the sunshine,
Sleeps a solemn day for me.

At this wakeful hour of midnight
  I behold it dawn in mist,
And I hear a sound of sobbing
  Through the darkness,—Hist! oh, hist!

In a dim and musky chamber,
  I am breathing life away;
Some one draws a curtain softly,
  And I watch the broadening day.

As it purples in the zenith,
  As it brightens on the lawn,
There's a hush of death about me,
  And a whisper, "He is gone!"

CXXXVII. A DEFINITE AIM IN READING.

Noah Porter, 1811-1892, was born at Farmington, Conn., and graduated at Yale in 1831. He remained in New Haven as a school-teacher, a tutor in college, and a student in the theological department until 1836, when he entered the ministry. In 1846 he was recalled to the college as Clark Professor of Moral Philosophy and Metaphysics; and in 1858 he also assumed the duties of the professorship of Systematic Theology, for a period of seven years. Upon the retirement of President Woolsey in 1871, he was elected to fill the office, which he held until 1886, being the eleventh president of the college.

President Porter's greatest literary work is entitled, "The Human Intellect: With an Introduction upon Psychology and the Human Soul." It is remarkable for the clear thought and sound judgment it displays, as well as for its broad scholarship; and it has been pronounced "the most complete and exhaustive exhibition of the cognitive faculties of the human soul to be found in our language." His other important works are: "The Sciences of Nature *versus* the Science of Man," which is a review of the doctrines of Herbert Spencer; "American Colleges and the American Public;" and the book from which the following selection is taken, namely, "Books and Reading." Besides these he wrote numerous essays, contributions to periodicals, etc. During his professorship he was called upon to act as chief editor in the important work of revising "Webster's Dictionary." The edition of 1864 was the result of his careful oversight, and the subsequent revisions were also under his superintendence.

In reading, we do well to propose to ourselves definite ends and purposes. The more distinctly we are aware of our own wants and desires in reading, the more definite and permanent will be our acquisitions. Hence it is a good rule to ask ourselves frequently, "Why am I reading this book, essay, or poem? or why am I reading it at the present time rather than any other?" It may often be a satisfying answer, that it is convenient; that the book happens to be at hand; or that we read to pass away the time. Such reasons are often very good, but they ought not always to satisfy us. Yet the very habit of proposing these questions, however they may be answered, will involve the calling of ourselves to account for our reading, and the consideration of it in the light of wisdom and duty.

The distinct consciousness of some object at present before us, imparts a manifoldly greater interest to the contents of any volume. It imparts to the reader an appropriate power, a force of affinity, by which he insensibly and unconsciously attracts to himself all that has a near or even a remote relation to the end for which he reads. Anyone is conscious of this who reads a story with the purpose of repeating it to an absent friend; or an essay or a report with the design of using its facts or arguments in a debate; or a poem with the design of reviving its imagery, and reciting its finest passages. Indeed, one never learns to read effectively until he learns to read in such a spirit—not always, indeed, for a definite end, yet always with a mind attent to appropriate and retain and turn to the uses of culture, if not to a more direct application.

The private history of every self-educated man, from Franklin onwards, attests that they all were uniformly not only earnest but *select* in their reading, and that they selected their books with distinct reference to the purposes for which they used them. Indeed, the reason why self-trained men so often surpass men who are trained by others in the effectiveness and success of their reading, is that they know for what they read and study, and have definite aims and wishes in all their dealings with books. The omnivorous and indiscriminate reader, who is at the same time a listless and passive reader, however ardent is his curiosity, can never be a reader of the most effective sort.

Another good rule is suggested by the foregoing. Always have some solid reading in hand; *i. e.*, some work or author which we carry forward from one day to another, or one hour of leisure to the next, with persistence, till we have finished whatever we have undertaken. There are many great and successful readers who do not observe this rule, but it is a good rule notwithstanding.

The writer once called upon one of the most extensive and persevering of modern travelers, at an early hour of the day, to attend him upon a walk to a distant village. It was after breakfast, and though he had but few minutes at command, he was sitting with book in hand — a book of solid history he was perusing day after day. He remarked: "This has been my habit for years in all my wanderings. It is the one habit which gives solidity to my intellectual activities and imparts tone to my life. It is only in this way that I can overcome and counteract the tendency to the dissipation of my powers and the distraction of my attention, as strange persons and strange scenes present themselves from day to day."

To the rule already given — read with a definite aim — we could add the rule — make your aims to be definite by continuously holding them rigidly to a single book at all times, except when relaxation requires you to cease to work, and to live for amusement and play. Always have at least one iron in the fire, and kindle the fire at least once every day.

It is implied in the preceding that we should read upon definite subjects, and with a certain method and proportion in the choice of our books. If we have a single object to accomplish in our reading for the present, that object will of necessity direct the choice of what we read, and we shall arrange our reading with reference to this single end. This will be a nucleus around which our reading will for the moment naturally gather and arrange itself.

If several subjects seem to us equally important and interesting, we should dispose of them in order, and give to each for the time our chief and perhaps our exclusive attention. That this is wise is so obvious as not to require illustration. "One thing at a time," is an accepted condition for all efficient activity, whether it is employed upon things or thoughts, upon men or books. If five or ten separate topics have equal claim upon our interest and attention, we shall do to each the amplest justice, if we make each in its turn the central subject of our reading. There is little danger of weariness or monotony from the workings of such a rule.

Most single topics admit or require a considerable variety of books, each different from the other, and each supplementing the other. Hence it is one of the best of practices in prosecuting a course of reading, to read every author who can cast any light upon the subject which we have in hand. For example, if we are reading the history of the Great Rebellion in England, we should read, if we can, not a single author only, as Clarendon, but a half dozen or a half score, each of whom writes from his own point of view, supplies what another omits, or corrects what he under- or overstates.

But, besides the formal histories of the period, there are the various novels, the scenes and characters of which are placed in those times, such as Scott's Woodstock; there are also diaries, such as those by Evelyn, Pepys, and Burton; and there are memoirs, such as those of Col. Hutchinson; while the last two have been imitated in scores of fictions. There are poems, such as those of Andrew Marvell, Milton, and Dryden. There are also shoals of political tracts and pamphlets, of handbills and caricatures.

We name these various descriptions of works and classes of reading, not because we suppose all of them are accessible to those readers who live at a distance from large public libraries, or because we would advise everyone who may have access to such libraries, to read all these books and classes of books as a matter of course, but because we would illustrate how great is the variety of books and reading matter that are grouped around a single topic, and are embraced within a single period.

Every person must judge for himself how long a time he can bestow upon any single subject, or how many and various are the books in respect to it which it is wise to read; but of this everyone may be assured, that it is far easier, far more agreeable, and far more economical of time and energy, to concentrate the attention upon a single subject at a time than to extend it to half a score, and that six books read in succession or together upon a single topic, are far more interesting and profitable than twice as many which treat of topics remotely related. A lady well known to the writer, of the least possible scholarly pretensions or literary notoriety, spent fifteen months of leisure, snatched by fragments from onerous family cares and brilliant social engagements, in reading the history of Greece as written by a great variety of authors and as illustrated by many accessories of literature and art.

Nor should it be argued that such rules as these, or the habits which they enjoin, are suitable for scholars only, or for people who have much leisure for reading. It should rather be urged that those who can read the fewest books and who have at command the scantiest time, should aim to read with the greatest concentration and method; should occupy all of their divided energy with single centers of interest, and husband the few hours which they can command, in reading whatever converges to a definite, because to a single, impression.

CXXXVIII. ODE TO MT. BLANC.

Samuel Taylor Coleridge, 1772-1834, was born in Devonshire, England, and was educated at Christ's Hospital and Cambridge University. Through poverty he was compelled to enlist in the army, but his literary attainments soon brought him into notice, and he was enabled to withdraw from the distasteful life.

Coleridge's fame arises chiefly from his poems, of which the "Rime of the Ancient Mariner," "Genevieve," and "Christabel" may be classed among the best of English poetry. He also wrote a number of dramas, besides numerous essays on religious and political topics. As a conversationalist Coleridge had a remarkable reputation, and among his ardent admirers and friends may be ranked Southey, Wordsworth, Lovell, Lamb, and De Quincey. He and his friends Southey and Lovell married sisters, and talked at one time of founding a community on the banks of the Susquehanna. Although possessing such brilliant natural gifts, Coleridge fell far short of what he might have attained, through a great lack of energy and application, increased by an excessive use of opium.

HAST thou a charm to stay the morning star
In his steep course? So long he seems to pause
On thy bald, awful head, O sovran Blanc!
The Arve and Arveiron at thy base
Rave ceaselessly; but thou, most awful Form,
Risest from forth thy silent sea of pines,
How silently! Around thee and above,
Deep is the air and dark, substantial, black—
An ebon mass: methinks thou piercest it,
As with a wedge! But when I look again,
It is thine own calm home, thy crystal shrine,
Thy habitation from eternity!
O dread and silent Mount! I gazed upon thee
Till thou, still present to the bodily sense,
Didst vanish from my thoughts: entranced in prayer,
I worshiped the Invisible alone.

Yet, like some sweet, beguiling melody,
So sweet we know not we are listening to it,
Thou, the meanwhile, wast blending with my thought—
Yea, with my life and life's own secret joy

Till the dilating soul, enrapt, transfused,
Into the mighty vision passing — there,
As in her natural form, swelled vast to Heaven!

Awake, my soul! not only passive praise
Thou owest! not alone these swelling tears,
Mute thanks and secret ecstasy! Awake,
Voice of sweet song! Awake, my heart, awake!
Green vales and icy cliffs, all join my hymn.

Thou first and chief, sole sovran of the vale!
Oh, struggling with the darkness all the night,
And visited all night by troops of stars,
Or when they climb the sky, or when they sink —
Companion of the morning star at dawn,
Thyself Earth's rosy star, and of the dawn
Coherald — wake, oh wake, and utter praise!
Who sank thy sunless pillars deep in earth?
Who filled thy countenance with rosy light?
Who made thee parent of perpetual streams?

And you, ye five wild torrents fiercely glad!
Who called you forth from night and utter death,
From dark and icy caverns called you forth,
Down those precipitous, black, jagged rocks,
Forever shattered, and the same forever?
Who gave you your invulnerable life,
Your strength, your speed, your fury, and your joy,
Unceasing thunder, and eternal foam?
And who commanded (and the silence came),
Here let the billows stiffen, and have rest?

Ye icefalls! ye that from the mountain's brow
Adown enormous ravines slope amain —
Torrents, methinks, that heard a mighty voice,
And stopped at once amid their maddest plunge!

Motionless torrents! silent cataracts!
Who made you glorious as the gates of Heaven
Beneath the keen full moon? Who bade the sun
Clothe you with rainbows? Who, with living flowers
Of loveliest blue, spread garlands at your feet?—
God!—let the torrents, like a shout of nations,
Answer! and let the ice plains echo, God!
God! sing ye meadow streams with gladsome voice!
Ye pine groves, with your soft and soul-like sounds!
And they, too, have a voice, yon piles of snow,
And in their perilous fall shall thunder, God!

Ye living flowers that skirt the eternal frost!
Ye wild goats sporting round the eagle's nest!
Ye eagles, playmates of the mountain storm!
Ye lightnings, the dread arrows of the clouds!
Ye signs and wonders of the elements!
Utter forth, God, and fill the hills with praise!

Thou, too, hoar Mount! with thy sky-pointing peaks,
Oft from whose feet the avalanche, unheard,
Shoots downward, glittering through the pure serene,
Into the depth of clouds that veil thy breast—
Thou too again, stupendous Mountain! thou
That as I raise my head, awhile bowed low
In adoration, upward from thy base,
Slow traveling, with dim eyes suffused with tears,
Solemnly seemest, like a vapory cloud,
To rise before me.—Rise, oh ever rise!
Rise like a cloud of incense from the Earth!
Thou kingly spirit throned among the hills,
Thou dread embassador from Earth to Heaven,
Great Hierarch! tell thou the silent sky,
And tell the stars, and tell yon rising sun,
Earth, with her thousand voices, praises God.